Advancing Mobile Learning in Contemporary Educational Spaces

Dominic Mentor
Columbia University, USA

A volume in the Advances in Mobile and Distance Learning (AMDL) Book Series

IGI Global
DISSEMINATOR OF KNOWLEDGE

Published in the United States of America by
IGI Global
Information Science Reference (an imprint of IGI Global)
701 E. Chocolate Avenue
Hershey PA, USA 17033
Tel: 717-533-8845
Fax: 717-533-8661
E-mail: cust@igi-global.com
Web site: http://www.igi-global.com

Library of Congress Cataloging-in-Publication Data

Names: Mentor, Dominic, 1968- editor.
Title: Advancing mobile learning in contemporary educational spaces / Dominic Mentor, Editor.
Description: Hershey, PA : Information Science Reference, [2020] | Includes bibliographical references. | Summary: "This book examines the role and integration of mobile learning in contemporary classrooms"-- Provided by publisher.
Identifiers: LCCN 2019013142 | ISBN 9781522593515 (h/c) | ISBN 9781522593522 (s/c) | ISBN 9781522593539 (eISBN)
Subjects: LCSH: Mobile communication systems in education. | Education--Effect of technological innovations on. | Computer-assisted instruction.
Classification: LCC LB1044.84 .A383 2020 | DDC 371.33--dc23
LC record available at https://lccn.loc.gov/2019013142
This book is published in the IGI Global book series Advances in Mobile and Distance Learning (AMDL) (ISSN: 2327-1892; eISSN: 2327-1906)

British Cataloguing in Publication Data
A Cataloguing in Publication record for this book is available from the British Library.

For electronic access to this publication, please contact: eresources@igi-global.com.

Advances in Mobile and Distance Learning (AMDL) Book Series

ISSN:2327-1892
EISSN:2327-1906

Editor-in-Chief: Patricia Ordóñez de Pablos, Universidad de Oviedo, Spain

MISSION

Private and public institutions have made great strides in the fields of mobile and distance learning in recent years, providing greater learning opportunities outside of a traditional classroom setting. While the online learning revolution has allowed for greater learning opportunities, it has also presented numerous challenges for students and educators alike. As research advances, online educational settings can continue to develop and advance the technologies available for learners of all ages.

The **Advances in Mobile and Distance Learning** (AMDL) Book Series publishes research encompassing a variety of topics related to all facets of mobile and distance learning. This series aims to be an essential resource for the timeliest research to help advance the development of new educational technologies and pedagogy for use in online classrooms.

COVERAGE

- Tablets and Education
- Ubiquitous and Pervasive Learning
- Economics of Distance and M-Learning
- Course Design
- E-Books
- Globalization
- Distance Learning
- Ethical Considerations
- Pedagogy and Design Methodology
- Online Collaborative Learning

IGI Global is currently accepting manuscripts for publication within this series. To submit a proposal for a volume in this series, please contact our Acquisition Editors at Acquisitions@igi-global.com or visit: http://www.igi-global.com/publish/.

The Advances in Mobile and Distance Learning (AMDL) Book Series (ISSN 2327-1892) is published by IGI Global, 701 E. Chocolate Avenue, Hershey, PA 17033-1240, USA, www.igi-global.com. This series is composed of titles available for purchase individually; each title is edited to be contextually exclusive from any other title within the series. For pricing and ordering information please visit http://www.igi-global.com/book-series/advances-mobile-distance-learning/37162. Postmaster: Send all address changes to above address.

Titles in this Series

For a list of additional titles in this series, please visit:
https://www.igi-global.com/book-series/advances-mobile-distance-learning/37162

Administrative Leadership in Open and Distance Learning Programs
Koksal Buyuk (Anadolu University, Turkey) Serpil Kocdar (Anadolu University, Turkey) and Aras Bozkurt (Anadolu Universit, Turkey)
Information Science Reference • ©2018 • 378pp • H/C (ISBN: 9781522526452) • US $195.00

Empowering Learners With Mobile Open-Access Learning Initiatives
Michael Mills (University of Central Arkansas, USA) and Donna Wake (University of Central Arkasas, USA)
Information Science Reference • ©2017 • 357pp • H/C (ISBN: 9781522521228) • US $190.00

Handbook of Research on Mobile Devices and Applications in Higher Education Settings
Laura Briz-Ponce (University of Salamanca, Spain) Juan Antonio Juanes-Méndez (University of Salamanca, Spain) and Francisco José García-Peñalvo (University of Salamanca, Spain)
Information Science Reference • ©2016 • 608pp • H/C (ISBN: 9781522502562) • US $300.00

Handbook of Research on Mobile Learning in Contemporary Classrooms
Dominic Mentor (Teachers College, Columbia University, USA)
Information Science Reference • ©2016 • 475pp • H/C (ISBN: 9781522502517) • US $300.00

Mobile and Blended Learning Innovations for Improved Learning Outcomes
David Parsons (The Mind Lab by Unitec, New Zealand)
Information Science Reference • ©2016 • 366pp • H/C (ISBN: 9781522503590) • US $185.00

Integrating Touch-Enabled and Mobile Devices into Contemporary Mathematics Education
Maria Meletiou-Mavrotheris (European University Cyprus, Cyprus) Katerina Mavrou (European University Cyprus, Cyprus) and Efi Paparistodemou (Cyprus Pedagogical Institute, Cyprus)
Information Science Reference • ©2015 • 341pp • H/C (ISBN: 9781466687141) • US $190.00

For an entire list of titles in this series, please visit:
https://www.igi-global.com/book-series/advances-mobile-distance-learning/37162

IGI Global
DISSEMINATOR OF KNOWLEDGE

701 East Chocolate Avenue, Hershey, PA 17033, USA
Tel: 717-533-8845 x100 • Fax: 717-533-8661
E-Mail: cust@igi-global.com • www.igi-global.com

Table of Contents

Detailed Table of Contents

Action learning teams create opportunities to learn from work on real-life challenges. Use of mobile learning can promote higher-order learning while bringing teams together around new ideas where their thinking may be challenged. In this way, action learning teams use mobile devices as tools of convenience to facilitate their learning. Mobile learning for action learning teams extends beyond simple content delivery and provides a platform that can introduce flow states and make space for critical reflection. The integration of mobile learning into action learning teams allows for the cultivation of individual creativity and maximization of group, virtual or face-to-face meetings.. This chapter will discuss the integration of mobile devices and detail multiple mobile learning exercises that action learning teams may use to promote creativity and critical reflection.

Food preparation for preschool and elementary school children is described from the perspective of enhancing cognition and learning through the use of mobile devices. While most children are too young to use a stove or slice with a knife, with proper guidance and simulated experiences through mobile devices, children can actively contribute to food choice, preparation, and knowledge while educators contribute knowledge from various academic areas. The kitchen can be a positive learning environment for the educator, the student, and potentially, people in the student's

This chapter focuses on using mobile assessment (m-assessment) for teaching and learning in formal and informal education settings. M-assessment has been helpful in making traditional classrooms more interactive and support the individual and class cognitive wealth. Unique affordances of m-assessment can also extend its benefits to digital classrooms, informal and formal learning settings, professional development settings, and other places where learning happens. M-assessment has increased accessibility to assessment, making it available anywhere, anytime, and to anyone.

The aim of this chapter is to investigate the scope and change of m-learning literature over a period of thirteen years. The chapter takes an author co-citation approach where patterns are revealed in mlearning publications between the years 2002 and 2015. The author co-citation analysis is a kind of bibliometrics, and Power Flow Network Library (PFNET) analysis method to obtain the intellectual structure of the m-learning field. A PFNET is a psychometric scaling method where a network of associative nodes is generated using specific values. The method is based on graph theory and has been used to explore knowledge connections. The method is referred to as a pathfinder network for representing segments of knowledge in a branched manner. The method can also be used for developing consensus building or recognizing expertise in a specific area. In this study, it was found that the m-learning research focus on design and effectiveness of m-learning environments, developing a pedagogic framework for m-learning, adoption and proliferation of m-learning-assisted language learning, collaborative m-learning, and augmented reality. Assessing the results of factor analysis together with the PFNET graphics, the subfields show the fundamental orientations of the field, and the main themes will be the focus of future research.

This chapter covers the construction of a digital ecosystem grown organically to digitize a workforce development organization. A criteria matrix was developed from a listening tour to plan and adopt blended, electronic, as well as mobile learning (e&mLearning) as part of its practices. The chapter describes the organic approach to diffuse the technological innovation for cultivating supportive teaching and learning communities. The process started with the establishment and implementation of an academic vision and strategy; the building of a blended, electronic, and mobile learning criteria taxonomy; as well as a theoretically informed, technology integrated educational framework. The results and findings of the implementation of an LMS as the foundation of the e&mLearning vision is shared, followed by recommendations from successes and organizational needs.

The evolution of mobile games and the subsequent evolution of designing mobile gaming experiences to support learning is captured in this chapter. The authors will address the affordances and limitations of mobile gaming as well as the learning theories and frameworks that lay the foundation for learning through mobile games. Current perceptions of mobile gaming in the classroom and the implementation of mobile games in schools are discussed. In addition, this chapter explores mobile games for social change, their prevalence, and their role in facilitating learning. The chapter also discusses differing perceptions on the issue of mobile games in and out of the classroom as well as the potential for marketability.

Machitia is an educator-focused mobile app prototype where educators create, collaborate, and share lesson plans. These lesson plans embed the following liberating and transformative theoretical frameworks and pedagogies called in this chapter "circles of liberation": (1) Dis/ability critical race theory (DisCrit), (2) biliteracy, (3) culturally sustaining pedagogy, (4) radical healing, (5) critical pedagogy, (6) proficiency-based learning, (7) queer theory, and (8) decolonizing theory. After introducing those frameworks, a mapping of currently existing educator-focused platforms prelude the review of mobile technology theoretical frameworks Machitia's design incorporates. Then, the discussion turns to how all the circles of liberation and mobile technology theoretical frameworks manifest as features within Machitia. By the end of the chapter, learners and educators will have a sense of the various possibilities of, and the need for, an education-focused liberation platform.

While there are many educational apps for traditionally taught subjects such as mathematics and science, more specialized curriculum has largely been left unexplored in terms of m-learning. Film studies, an academic discipline that deals with the theoretical, historical, and critical underpinnings of film, is one such subject that has very few mobile applications. This chapter explores creating a mobile application to teach basic approaches to film interpretation and in addition, considers a heutagogical approach in design. Benefits of m-learning include increased delivery options for multimedia, context-based learning support, and the prospects of a more fulfilling learning experience. This chapter provides direction for implementation and evaluation techniques for an introductory film studies module on film noir, which can be integrated into a mobile format to make film theory more relevant and accessible.

How can mobile e-learning be used to support and enhance the goals of working with youth aging out of an orphanage. This chapter describes the use of an ethnographic narrative approach to tell the stories of eight young women who were "aging out" from their orphanage where many of them had spent most of their lives. The chapter examined the way in which a collaborative art community could support the participants as they narrated their lives over a 16-month period through photo-journaling and social media outlets. The focal points of this study were community building via art making and building of personal aesthetic, community engagement, reflection on self-identity, cross-cultural art education, and shared experience via photo-art narratives. The authors also examined the role of collaborative art experiences in helping these young women form social connectedness with their peers. The project was also designed to help sustain the girls digitally and virtually

Preface

Mobile phones and tablets expanded the learning engagement time exponentially, supporting intentional and accidental learning episodes in people's daily lives. Mobile devices has also catered for more opportunities for adult and young learners to be active agents in their learning. Mobile technology allows the everyday learning to be owned, naturally pursued and expanded on, while people's activities using mobile technology can be tracked via history or scores. Additionally, these activities can blend seamlessly and unobtrusively into a user's life. Furthermore, pushing the democratizing of education, users are not just being the receivers of education, but rather the drivers, contributors to, and owners of their education. And this means the material could be called on when needed, tailored to their needs, and they can learn at their own pace. More importantly, they can interact with material and peers in the way they prefer and consume aspects in the level of depth of their choosing. Typically, on a mobile phone, this could be in bite size chunks or larger recreational learning gulps.

Just over 10 years ago, the switch from Palm Pilots and Blackberries to iPhones and Android devices, saw a major shift in mobile technology impacting our socio-cultural world. Technological innovation allowed for a revolutionary transition from personal digital assistants (PDAs), pocket PC's and wireless handheld devices, to advanced cell phones, Palm Pilots and Blackberries, and finally to smartphones. This shift facilitated countless integrated functionalities that assisted users in their personal lives and job responsibilities while at the same time fostering new entrepreneurial and educational endeavors. Mobile phones and tablets (iPads, Kindles, Android tablets) have not only expanded learning engagement time exponentially, supporting both intentional and accidental learning episodes in people's daily lives, but they have also created additional opportunities for individuals to be active agents in their own learning.

The mobile phone as we know it today, equipped with its advanced mini-computer capabilities, has far surpassed the early computers that once sent people into space. We now have powerful, handheld devices with the capacity to simultaneously direct a lost student to his or her class while providing real-time updates from a professor. A smartphone allows people to arrange and coordinate an event while at the same time fostering engagement amongst both participants and outside observers unable to attend in person. Further supporting social connectedness through mobile engagement, to an occasion, a group, a memory, or an event through various interactive and communication affordances creates a sense of affiliation and belonging through virtual means (Mentor, 2018a, 2018b). Additionally, cultivating co-presence and connected presence (Schroeder, 2006) for those in distal or closer proximity, allows people to be virtually present while physically not in attendance. The interaction with these functions implicitly teaches concepts such as navigation and distance estimation, coordination, just in-time recall, or capturing evidence of learning. Simple navigation tools on a mobile device can solve problems in real time and offer students the opportunity to apply new knowledge by merging theoretical considerations with practical learning experiences. Additionally, mobile learning has become a common feature of our daily cultural practices.

Many individuals and organizations have harnessed the power of mobile technologies in their personal, social and work lives, and they have consequently transformed their personal and professional environments from structured and compartmentalized to seamless and overlapping contexts. Many people consider themselves as no longer chained to their desks as they have been liberated by mobile devices that enable them to respond to emails, join conference calls, or access datasets while sitting on the train or taking a walk in the park. The integration of work and personal time has fostered a similar shift in the role of mobile devices in education. Learning is thus not exclusive to typical classrooms because it can occur on the subway while reading a news article, while reviewing datasets on a mobile device, verifying or searching information.

The chapters in this book confirm that, with mobile learning, context is crucial, content is core and collaboration can be a key driver of knowledge sharing and building. With mobile devices, learning occurs during informal and formal environments where interpersonal communication is encouraged. Traditional educational settings often have a mono-directional channel from teacher to learner, which can be reduced by incorporating mobile devices as well as mlearning principles and practices. Or with diverse perspectives be increased, through multi-directional facilitation via mobile engagement. Additionally, various dual-oriented learning processes can be leveraged as the mobile user switches between applications to interact with peers, conduct research, and complete work or recreational tasks. Mobile devices create constant connectedness, stimulating engagement with content, context and people. As mobile technology has become even more ubiquitous and a part of a socio-cultural fabric, the usage of mobile devices offers more educational opportunities and opens a broader world for all mobile device users to explore.

Mobile learning is applicable across fields, including media, orphanages, museums, mass transit, mobile health, well-being, mathematics, language learning, mobile commerce, accessibility for special needs, senior citizens, cooking and nutrition, and assessments in a multitude of fields. The scope of this book goes beyond K – 12, after-school programs, non-profit organizations, workforce development programs, as well as formal and informal organizations; the handbook's audience extends from business corporations to community organizations, including museums and universities. This book on *Advancing Mobile Learning in Contemporary Educational Spaces* presents a wealth of valuable information for trainers, workers and learners in a wide array of contexts, as it offers a variety of examples of how mobile learning could be introduced, leveraged, researched and/or customized. The chapters offer recommendations for improving ease of use, creating customizable learning resources, identifying learning analytics that offer valuable insight and information on learning gaps through data that could inform successes or enhancements of educational programs. The book is also targeted at leaders of K – 12 educational institutions as well as university instructors, department of education, policy makers, administrators, corporate and non-profits, and workforce development professionals as well as medical and mental health practitioners. Thus, expanding the audience of this handbook to any arena where mobile devices are being used informally or formally to aid training, extend learning and support learning and development in a transformational manner that is naturally integrated into daily life.

As we use our mobile devices as learning resources, almost any space, physical and virtual, has the potential to be a contemporary classroom and an educational realm. Various mobile learning implications, curricula ideas, research methods and suggestions are offered in all of the chapters, along with future research recommendations. Integrated mobile learning and embedded examples are offered while ideas and apps with individualized, personalized and customizable options are provided. Nowadays, there are apps that can keep running when there is no internet connection, which overcomes one of the drawbacks of mobile technology. The book also shows how mobile learning can play a complimentary role to formal education while also invoking the concept of learning is everywhere, happening across space and multiscreen lives; it can take place at home, at school, at the workplace, or even during a commute. Learning also happens across time; we learn at different times during the day and different days during the week. The users' data can be synchronized across different devices and the portability of mobile technology will make it easier to record, share, and reflect on in/formal, incidental, accidental or shared learning.

In this handbook, readers will find reports and research examples sequenced as:

- Leveraging mobile learning in action learning teams

- Developing an "app-titude" for cooking: helping to promote positive cooking skills with children
- Mobile assisted language learning from language instructors' perspectives
- mHealth interventions for self-management of chronic disease
- Mobile phones for assessment: anywhere, anytime, by anyone
- Visualizing the evolution of mobile learning research
- Transforming a national workforce development training organization through e-&-m-learning
- Creating a modern day mobile, arcade offering an innovative gamified approach to learning
- An educator focused liberation app prototype informed by critical theories
- Utilizing mobile learning for orphans aging out: young adult development with universal lessons
- M-learning apps for film studies as an approach towards self-determined learning
- Mobile apps with universal design and accessibility for creating inclusive classroom experiences

The themes cover a wide spectrum on how people leverage mobile devices according to their personal and professional needs.

The first chapter offers the value of mobile learning in workforce and talent development which is unpacked through cultivating creativity on action learning teams. Using action reflection, higher-order critical thinking, and cultivating connected communities through e-&m-learning helped developed individuals and their teams. Using m-learning for professional development provides potential for just in time higher-order engagement to enrich action learning teams. Especially as they engage in critical reflection while solving ambiguous, complex challenges through the affordances of mobile devices.

Mobile devices can make almost any contemporary space a classroom, including a kitchen. The chapter "Cooking With 'App-titude'" suggests ideas that help both educators and parents to promote positive cooking and food curiosity. The chapter focuses on how mobile devices offer educational opportunities on making a meal, from identifying the ingredients to sequencing the steps involved. The chapter illustrates the value of mobile interactive learning between parents, and their children drawing connections between the foods we eat and the natural world. For example, illustrating that eggs do not come from the kitchen refrigerator, or revealing the chemistry that allows a cake or bread to rise in an oven. With collaborative activities that engage parents and children with both food preparation and information about nutrition and healthy eating, the kitchen environment has always been a learning space for parents, children, teachers, and chefs, and mobile devices can enrich this learning environment even further.

The chapter on Mobile Assisted Language Learning (MALL) offers a review on research literature for mobile assisted language learning from language teachers' perspectives. The chapter highlights insights and best practices for maximizing MALL successes. It also discusses how mobile devices have transformed the language learning field of practice and showcases the influence and impact of MALL from different teaching attitudes and approaches. The chapter includes information from different geographical regions, and offer research on various technologies, including simple mobile messaging, native mobile applications, and mobile technology that fosters real-world engagement with the target language. Furthermore, numerous perspectives for using mobile technologies for language teaching and learning as well as uniquely focused literature reviews of what we know about how mobile technology for language learning has been used in practice. From this chapter the reader can derive how to best apply these practices in various language learning contexts.

Similarly, the chapter on mobile health (mHealth) is of high value not just for medical practitioners and researchers. The general public who might be experiencing, directly or indirectly, the chronic diseases covered in the literature reviewed will also benefit from this chapter. In the introduction the author offers information about chronic diseases- what they are, their prevalence, the high rates of morbidity and mortality and their burden on healthcare costs. Then she talks about the growth of smartphone use and how many mHealth applications (apps) are targeted to support people with the self-management of chronic diseases. She weighs in on the effectiveness of mobile apps in the self-management of chronic disease touching on blood pressure, diabetes, HIV, depression, cancer, COPD and asthma, as well as FDA regulations and privacy issues. The chapter also addresses what she considers as the most prevalent chronic conditions: high blood pressure, diabetes, pulmonary disorders and depression, as well as how mobile apps are improving self-management opportunities of the disease process. The author's search strategy made use of inclusion and exclusion criteria and to choose those citations with the highest number and derived sense of quality. The author used only randomized controlled trials as they provided the highest level of evidence. She appraised them for quality using standardized tools as part of the systematic reviews to analyze them as high-quality studies. She presents the research processes included and extracted the data results. After which she elaborates on each systematic review in terms of their strengths, weaknesses, and synthesized the results. The author mixed up the format for each disease process since the available evidence and mHealth issues for each disease is so different.

The chapter on using mobile phones for assessment in contemporary classrooms offer a host of examples in a variety of contexts. The chapter explores approaches for promoting the use of mobile devices for formative and summative assessments. The case studies offer relevant examples, including the processes involved in establishing mobile assessments and the possibilities created through the multitude of tools and approaches cited. Various theories, benefits, challenges, as well as practical pedagogy and andragogy for m-assessments are posited in formal and informal education contexts. Moreover, the chapter showcases how m-assessments have contributed to repurposing old assessment practices by making assessment more convenient, offering real time feedback, and streamlining the collection and analysis of learner data. The chapter highlights the value of m-assessments to solicit in-the-moment participant feedback. The case studies offered in both chapters provide examples of how m-assessments can help reflection, capture an outlier or promote a consensus point of view. M-assessment tools can provide shy participants with opportunities to shine in a virtual environment in which they feel comfortable as well as the chance to be anonymous, which can ensure honest feedback. Examples from formative m-assessments as well as those used in workplace training offer details on how mobile apps are repurposed to collect, collate and set organizational priorities or policies in real-time, while at the same time offering transparency and fostering accountability.

The broad scope and change of mobile learning research landscape is covered through a chapter on the visualization of mobile learning research. The author uses the co-citation method to show patterns of the literature on mobile learning from 2002 to 2015. The co-citation analysis method is a kind of bibliometrics to obtain the intellectual structure of the m-learning field during a set period. This research method proved useful in that it revealed research focus on design and the effectiveness of m-learning environments, as well as the importance of developing a pedagogic framework for m-learning. Further results indicated a focus on the adoption and proliferation of m-learning, augmented reality, mobile assisted language learning, and the importance that m-learning practitioners placed on collaboration in m-learning.

A chapter on talent development highlights a multi-city collaborative effort and showcases the implementation of a vision to move a 12-year-old national training organization from paper-based training to blended electronic and mobile learning (e-&-m-learning) engagements. It highlights how an organic approach was used to advance various digital adoptions. The building of a sustainable organizational ecology to transform the workforce development organization to implement e-& m-learning as part of its methods, procedures and practices was met with trepidation and misunderstandings from administrators and operations staff. However, the instructional staff appreciated the criteria-led approach that would save them time, create flexible engagement opportunities with students, material and fellow instructors.

The chapter takes a case study approach, showing the organic change management and social connectedness used to implement the modernization. All of which was in pursuit of cultivating supportive educational engagement and learning communities.

As indicated, the mobile handbook caters to a wide range of audiences and the chapter on mobile games offers similar simulated, meditative and transferable learning opportunities. The authors offer a great exposition on learning with mobile games. The chapter outlines mobile gaming educational theories, the emerging and evolving field of mobile games, and how game design with strategic learning goals can cultivate creative learning. Matters of mobile gaming in K – 12 classroom settings are explored as well as how mobile gaming can be utilized for personal behavioral change like smoking cessation. When reading the Learning with Mobile Games chapter, the reader will also garner the ability for individual or social change through mobile games. The authors show how one could play and contribute to a charity of choice or inform policy. Or how mobile games can also teach simple and complex sequencing and strategy, which can be transferred to school or work processes.

In the chapter based on a mobile app prototype, the author offers an educator focused liberation platform. The mobile app idea promotes saving teachers time and expanding their thinking. The author also advances critical theory in practice which were at the forefront of informing the app's design. As stated earlier, collaboration is crucial, and can be incorporated into m-learning opportunities even with a private mobile device. The prototype app offers space to create, share and collaborate on lesson plans to also benefit teachers' development. Ideas on policy, implementation, and resultant outcomes are discussed as well as theoretical framing and practical methodologies of constructing conducive lesson plans and professional collaboration. The chapter also examines the research on collaboration and technology, using methods of curating, co-construction, and creating transformative learning opportunities, while focusing on elements of on-task learning integrating socio-cultural fiber of our modern-day environments, and experiences. The app prototype and chapter report make mobile collaboration and professional development that much more accessible. The chapter also offers opportunities to create social change, or social activism for democratizing education and civil action.

The chapter on m-learning for promoting film studies can also be used to teach language, mathematics, philosophies, or strategic thinking. The advantages of m-learning in this regard include increased portability, personal engagement time, repetitive engagement, closer inspection as well as a host of context-based learning options. Using the mobile device for film, can offer a more personal interrogation and integration of simulated realism which include cultural scenarios, debates, or covering sensitive topics. With strong, intense and evocative emotions or thinking stimulated, topics can be teased out or stripped for a rich intellectual conversation that could aid or teach emotional intelligence. Like the educator focused liberation platform chapter, the mobile learning in the film study chapter offers opportunities to interrogate identity, empathy, compassion or the valuable trait of hosting multiple perspectives. While movies on a mobile device can be enjoyed with another, it is mostly an individual activity, which host introspective, reflective or just simple entertainment value as one gets caught up in the world of the film.

The chapter on orphans aging out of their orphanage, offers insight into their transition and young adult development through mobile digital story capture. Mobile learning in this context offered access to their lives and brought the real challenges of the world that these young adults faced. With increased mobile and internet usage globally, and increasing access to social media apps and communication tools for these individuals, they could continue feeling supported by the head nun of the orphanage, as well as the principal investigator and one another.

The final chapter in the book promotes universal design and advances how mobile applications may be used to enhance accessibility for individuals with disabilities in all training settings. The chapter also promotes the development of apps that embrace universal design principles. Universal design caters for people with visual, auditory and/or physical disabilities, and not only benefits a particular set of users, but holds many benefits for others. Like wheelchair ramps benefitting people with suitcases, knee injuries or delivery personnel, universal design is beneficial for aging people, who are getting sick and losing functions. An example that can be gleaned from the chapter is the ability of using text to speech tools when driving or while multitasking. The multitude of uses for a diverse set of unintended audiences is also prevalent in the chapter.

In addition, the compiled chapters show that learning can relate to various topics and can have various purposes in different contexts; we can learn for leisure and informal purpose, for self-improvement, and for school or work requirements. And since we, as learners, are always on the move, mobile learning can be central or a supplement to formal or informal learning. The theory of informal or lifelong learning, particularly accidental or incidental learning that occurs through experiences in the world is covered in many of the chapters. Informal and lifelong learning have also been defined with supporting intentional and accidental learning episodes. This means that all the spaces where we engage with our mobile phones are an opportunity for learning. Today people, groups and companies are increasingly becoming location independent and are finding themselves to be strategically disadvantaged if they are confined to a particular place. The consequence of advances in mobile technologies and the changing boundaries have brought the importance of connectedness, data and information to the fore. This is because it is access to information and knowledge through mobile means that helps individuals, groups and companies realize their objectives and help people to make adequate decisions.

The book also indicates that in the workplace and business model of the past, data and information to a large extend were confined to a particular location and it was relatively easy to protect it from falling in the hands of those who should not have it (i.e. maintain confidentiality). Because information was usually processed in a central location, it was also possible to ensure, with a relative degree of certainty, that it's content and form did not change (i.e. maintain integrity) and ensure that it was readily accessible to authorized personnel (i.e. maintain availability). In fact, maintaining confidentiality, integrity and availability were the main tenants for managing security and for establishing responsibility, integrity of people, trustworthiness and ethicality. This still needs to hold true for all mobile learning initiatives.

Due to the different contexts discussed in the handbook, the approaches to mobile learning interventions vary among groups and individuals. As mobile phone use hosts so many variations, any mlearning service will have to be aware of and incorporate this heterogeneity. A variety of approaches is not only possible, but also desirable. Within a mobile learning service then, operating from a general approach, a different angle or direction for unique users should also be anticipated. Therefore, the value of a generalist mobile learning service should not just be singular, absolute, or completely static, but should be operated from a dynamic, contextual and transitory approach.

REFERENCES

Mentor, D. (2018a). mClass Planet of the Apps: The Rise of Mobile Learning. In Computer-Mediated Learning for Workforce Development (pp. 196-215). IGI Global.

Mentor, D. (2018b). TLC for MOOCS: Teaching and Learning Communities for Computer Programming. In Computer-Mediated Learning for Workforce Development (pp. 93-110). IGI Global.

Schroeder, R. (2006). Being there together and the future of connected presence. *Presence (Cambridge, Mass.)*, *15*(4), 438–454. doi:10.1162/pres.15.4.438

Chapter 1
A Case for Leveraging Mobile Learning in Action Learning Teams

Shawn McCann
United State Marine Corps, USA

ABSTRACT

Action learning teams create opportunities to learn from work on real-life challenges. Use of mobile learning can promote higher-order learning while bringing teams together around new ideas where their thinking may be challenged. In this way, action learning teams use mobile devices as tools of convenience to facilitate their learning. Mobile learning for action learning teams extends beyond simple content delivery and provides a platform that can introduce flow states and make space for critical reflection. The integration of mobile learning into action learning teams allows for the cultivation of individual creativity and maximization of group, virtual or face-to-face meetings.. This chapter will discuss the integration of mobile devices and detail multiple mobile learning exercises that action learning teams may use to promote creativity and critical reflection.

INTRODUCTION

In Lewis Carroll's (1871) classic tale, *Through the Looking Glass* the Queen offers Alice jam as an employment benefit, and soon after refers to a rule restricting her access to said jam. "The rule is, jam to-morrow and jam yesterday— but never jam to-day." (Carroll, 2013, Kindle Location 524). Alice is seemingly the only one to see the obvious flaws in the Queen's interpretation of having access to jam. This

DOI: 10.4018/978-1-5225-9351-5.ch001

caricature of company policy, seen through the eyes of Alice, is not to far off from how many employees feel about their own working situation. Unfortunately, it remains much more difficult to see the second and third order of consequences in our everyday norms, customs, or laws than it does to see the error in the Queen's jam policy. However, as Alice sees Wonderland through fresh eyes, there are ways to recognize contradictions within one's own organization, and thinking. With others, as a part of an action learning team, one may go through the looking glass and begin to deconstruct challenges, construct questions, and present solutions that address organizational challenges.

The formation of an action learning team may create the space for learners to come together and participate in a process to uncover their own untested assumptions and create new ways of thinking. In order to uncover the untested assumptions in our own thinking, one may participate in an individual or group critical reflection. Critical reflection serves as a tool to recognize one's own thinking, the assumptions of others, and those of the institution. The team may then endeavor to remedy much of the institutional contradictions (Voronov & Yorks, 2015; Seo & Creed, 2002), which brought about unintended outcomes that possibly originated from unchallenged norms. Mobile learning, when purposefully introduced to an action learning team's process and practice, affords a virtual extension of working space. This provides an additional space to support group development (Tuckman & Jensen, 1977), encourage flow (Csikszentmihalyi, 1997), and provides asynchronous, as well as near *synchronous communication* (Mentor, 2011 dissertation) while working around a problem.

Participating in a group reflection can feel cumbersome, akin to reading the minutes of the previous meeting. The difference in reflection, however, is the space it creates for the individual (first-person space) and the group (second-person space) to make meaning of the shared experience. Adding a critical lens to this process allows the group to go deeper, uncovering institutional contradictions (Seo & Creed, 2002). It may also help the group see the taken-for-granted customs and norms that they may have automatically conformed to while *onboarding* or *adjusting* to the institution (Seo & Creed, 2002). Critical reflection opens space to review and revise the previously invisible institutional contradictions by helping learners to examine assumptions and beliefs that influence interpretations, decisions, and actions.

This chapter considers how action learning teams can leverage higher-order mLearning—high level learning supported by use of handheld mobile devices with smart capabilities and Internet access—to enhance performance. After introducing this argument, the chapter discusses how critically reflective action learning teams' function and contribute to higher-order mLearning.

Action Learning Teams

Action learning allows teams to focus on learning and leadership development while organized around a work-related challenge. The challenge base and learning focus may cultivate higher performance by providing teams with the space to participate in the discourse required for surfacing and testing assumptions as well as institutional norms. Marsick and O'Neil (2007) explain action learning as,

An approach to working with and developing people that uses work on an actual project or problem as the way to learn. Participants work in small groups to take action to solve their problem and learn how to learn from that action. Often a learning coach works with the group in order to help the members learn how to balance their work with the learning from that work. (p. xvii).

The formation of an action learning team often creates the space for learners to engage in critical reflection that uncovers and holds the potential to remedy institutional contradictions. This is especially the case when members take on the challenge to solve ambiguous, complex challenges that demand fresh insight and creative approaches.

The second-person space includes both virtual and face to face interaction amongst members of the team and signifies the *space between* team members, as defined by Torbert and Associates (2004) first-, second-, and third-person perspectives. The first-person perspective is the interpersonal interaction from that person's perspective, bringing their own values and beliefs. A second-person perspective is shared with the group where a commitment exists to share, power, strategy, communication and reflection on group dynamics. A third-person perspective may be viewed as the larger, dispassionate, organizational, or the systems level view of interaction to organize structures and policies. Ideally, the second-person space created by the action learning team would enable unbounded creative thought and privilege the learning over producing evidence of learning.

This chapter makes the case that higher-order mLearning can enhance critical reflection on the go in action learning teams because mobile learning leverages certain inherent elements and affordances of mobile devices. Some of these inherent elements are the call (video and voice), camera (video and still pictures), voice recordings, text capability, and social media access. Examples of these affordances include connecting to data and people outside of working hours and away from the workspace. The critical reflection component allows learners to take up the practice of testing assumptions and collecting relevant data prior to participating in decision-making. This skill set allows learners to operate in complexity, to make sense of how the many parts of an organization interact with them and external forces and develop insight into the resulting challenges.

Lewis Carroll speaks to this need for growth in the face of complexity in the 1871 story *Through the Looking Glass*. Carroll writes, “Now, HERE, you see, it takes all the running YOU can do, to keep in the same place. If you want to get somewhere else, you must run at least twice as fast as that!” (Kindle Locations 234-235). A healthier attempt to keep pace with current conditions of change is to participate in higher-level learning. As an example, double-loop learning creates longer lasting solutions by allowing the learner to uncover the heart of the matter. This type of learning outpaces lower-level, or single-loop learning, which favors incremental change made in relation to undesirable outcomes. Torbert (1999) defines single-loop learning as the change of specific behavioral practices in relation to desired outcomes. Double-loop learning, on the other hand, requires a change in one’s method of deriving strategy and the strategy’s resulting actions as well as how they may define and measure success (Torbert, 1999).

LOWER AND HIGHER-LEVEL WORKPLACE ACTION LEARNING

The label *higher-order* distinguishes learning referenced in this chapter from the knowledge or skill acquisition that remains necessary in the workplace. Fiol and Lyles (1985) inform us that much workplace learning is necessarily lower level, where the outcome of the learning happens within clear boundaries, and its impact remains localized. The higher-level learning may emerge less frequently, but it impacts the whole organization by adjusting the norms, rules and other institutional boundaries that determine how problems are understood, defined and acted upon.

Opportunity for higher-order learning in action learning teams is greatest for those incorporating critical reflection (Marsick & O’Neil, 2007). Critical thinking, specifically Argyris’ (1976) single and double loop learning models, (depicted in **Figure 1**) best highlights the difference between the day-to-day lower level learning and the higher level learning that emerges from ambiguous and ill-defined problems (Fiol & Lyles, 1985).

Lower level learning includes an organization’s processes and procedures. A teacher would submit grading at specific times, using a particular learning management system. The professor would use a particular form to request a sabbatical. Billing sections of the organization would categorize, inform and take action in accordance with the company’s policy. Emergency services like fire or police would enter a home in a well-rehearsed formation to accomplish a specific mission.

Figure 1.

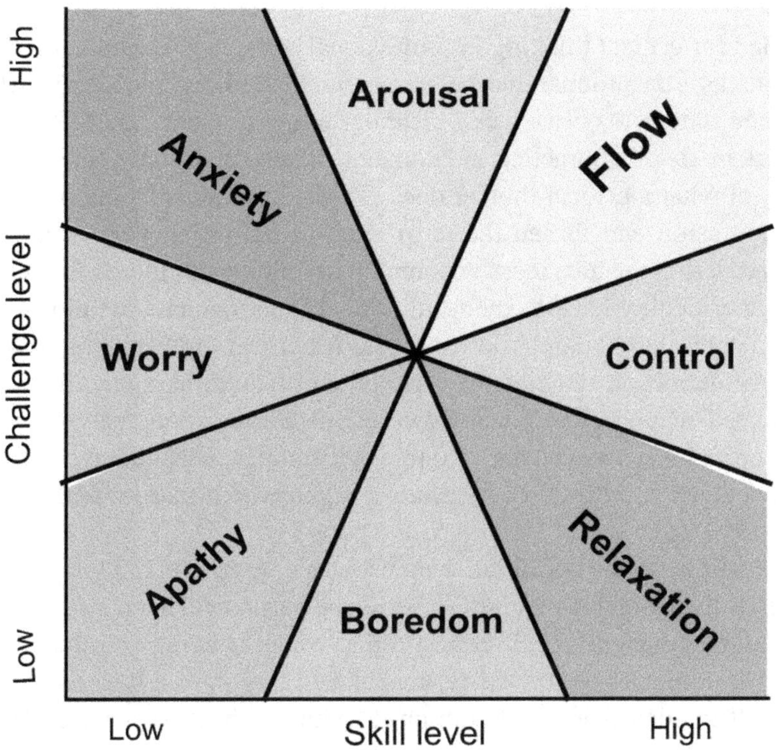

An organization requires higher level learning when policies, processes and procedures fail to cover complex concerns, or contingencies resulting from issues not solved through the application of the standard company sponsored reaction. Teachers need to address troubled students, but no script covers all circumstances. Professors follow formats when applying for grants, but effectively communicating one's work and winning the funding often requires more than a simple form letter. Emergency services enter a home in predetermined formations, but their reaction to instantaneous and life-threatening circumstances should not be constrained only to a book on standing operating procedures. The employee handbook does not cover all of the scenarios that may emerge in our organization. In any case not covered by instruction, one would hope that their direct report employs critical thinking. However, one should not expect critical thinking just to appear or happen without purposeful prior practice.

Double-Loop Learning as Critical Thinking

The demand for critical thinking in professional settings is common on many job descriptions, as is the promise of developing critical thinkers in academia. Although, these written statements often lack a clear definition of the concept (Paul & Elder, 2003). Various fields of practice and domains of academic study constructed their own ideas of what a critical thinker does. These definitions would often diverge, creating confusion and turned the term into a buzz word to mean any type of thinking, rendering the idea inert. A common definition was then created to clarify and bring a criticality back to the term critical thinking. The definition emerged from the Delphi Study commissioned by the American Philosophical Association in 1990. The definition clarified the concept of critical thinking as "the process of purposeful, self-regulatory judgment that results in interpretation, analysis, evaluation, and inference, as well as explanation of the evidential, conceptual, methodological, criteriological, or contextual considerations upon which that judgment is based" (Facione, 1990, p. 2).

This common ground definition demonstrates how critical thinking slows down the action. Critical thinking requires a process and purposeful intention separate from the thinking that created or subscribes to that which is being questioned. Critical thinking therefore may be framed as an application tool, or a method of thinking. The term self-regulatory in this definition also situates the thinker as a tool requiring continuous calibration prior to being capable of rendering judgment. The definition uses a myriad of terms intended to specifically capture and broadly cover a shared process of critical thinking amongst the various fields of practice and multiple domains of academic study. The definitions include the terms: interpretation, analysis, evaluation, inference, and leave space for each practice and domain to determine how their practitioner may action them. Each field and domain also bring their own evidential, conceptual, methodological, criteriological, or contextual considerations to judge and validate the thinking itself.

The pragmatic practice or domain of critical thinking uses a setting of application for their inquiry, as Argyris (1976) does when presenting the double-loop learning model. The double-loop model seeks to generate valid information, while including the participants' feelings during the process (Argyris, 1976). Single-loop learning occurs when one assesses and makes appropriate changes to their actions in response to the undesired effect of previous actions. The second or double-loop refers to the thoughtful examination of the initial intentions, as one's assumptions form their thinking and dictate action. Changing or validating the assumptions alters a learner's inferences and allows them to take more informed actions. Critical reflection requires the thinker to apply a criticality to their assumptions and those that the institution holds in the form of rules, culture, practices and processes. One begins to ask the questions of not just why the rule, cultural artifact, or practice exists, but how it came to being. The precipitous action comes under scrutiny when asking how. Asking the question of when may also hold significant relevance if the policy, practice, or process predated the current automated method of the actioning. In other words, why does the format require two spaces after the punctuation mark?

Critical Reflection, and Critical Thinking

Alice did not find the White Rabbit remarkable until after first sighting the hurried talking animal (Carroll, 1865). Only upon further reflection and observing his use of a pocket watch extracted from his waistcoat pocket, did Alice burn with curiosity. A rabbit is quite unremarkable, and even one saying to itself that it "shall be too late!" did not much stir Alice's inquisitiveness. It was, in the story, the rabbit accessing a chronometer from its jacket, which was completely out of context that brought Alice to her feet and down the rabbit hole. Alice downloaded the information, but the data did not become worthy of reflection until it challenged her normal ways of making meaning. With the rise in availability of data and potential for increased data consumption, the need for a reflective practice becomes more essential to make the impactful information more salient at an increased pace.

Critical reflection is the element of critical thinking that calls for one to identify assumptions and assess their legitimacy with the intent to change the current ways of thinking and acting (Brookfield, 2011). Critical reflection, therefore, assesses the validity of assumptions that form one's meaning perspectives, where one examines their sources and consequences (Mezirow, 1990). The ability to participate in these reflections leads to the appropriate application of critical thinking, presumably free of faulty assumptions, and theoretically results in better decision-making. However, critical reflection often requires the presence of the other to identify blind spots, and surface unconscious assumptions.

Addressing the problem without that criticality may result in a solution produced through the single-loop learning model where solutions emerge without asking significant questions of the problem (Argyris, 1976). Single-loop learning occurs when an instructor makes changes to their lecture in light of poor student performance. A double-loop learning scenario differs in that the instructor, instead of haphazardly adjusting their lecture, dives deeper, questioning their methods, outcomes, and student barriers to gain others' perspectives to challenge their own perceptions of the situation and causes. Double-loop learning would have the instructor share power with the student in a quest to solve the problem, and ensure it remains solved, through the creation of useful information. By integrating one's mobile device, mobile learning can become a part of a more immediate facilitation of, and communication of double-loop learning.

ACTION LEARNING TEAMS GO MOBILE

Action learning, according to Michael Marquardt (2000), involves "effective tools employed by organizations worldwide to develop and build their leaders" (p. 233). In adult workspaces, a sponsor of the team, ideally from the C-Suite, or other types of higher echelon leadership positions specific to the institution, would seek high potential employees from across the organization to create a cross-functional team. Marquardt (2000) asserts action learning assists teams in,

Creating opportunities for participants to find their own answers to problems and to learn from each other's perspectives, successes and mistakes; encouraging a climate where participants will both support and challenge each other; refraining from displaying one's own knowledge and understanding; challenging individual and group assumptions; providing difficult feedback to members; asking questions that assist participants in exploring the reasoning behind their assumptions; acknowledging mistakes publicly and framing them as learning experiences (p. 236).

In this paradigm, an action learning team values the process of learning over an end state, or attempting to solve the problem via single-loop means. Though efficient problem solving remains an institutional priority, an action learning team also intends to further cultivate their members. The organization's talent should continue to dig deeper, developing new thinking, and reaching new levels to approach institutional issues with a double-loop learning framework. The formation of an action learning team encourages members to share their mistakes, assumptions, and challenges over a more common urge to showcase one's strengths, or expertise; and possibly guard one's ego. In an action learning team, the stupid or obvious questions demand attention. When the team labels a question stupid, or obvious without first exploring it, they fail to engage in the critical thinking process, and possibly victim to institutionally sponsored group think.

To avoid group think, teams can first work individually prior to sharing their thinking with the team. The affordance of mobile communication and mobile learning creates opportunities for participants to do on the spot introspection or reflection and communicate it with one another. Team members can post or consume the individual products created and stored to a shared space when it becomes most opportune for them. Not everyone does their best thinking in the office, or event during banker's hours.

While seeking answers or solutions to problems, they can also learn from each other's communicated perspectives, successes and mistakes. Mobile learning encourages a climate where participants will both support and challenge each other; not necessarily forbidding the display of one's own knowledge and understanding but challenging each other's assumptions. On occasion, a communication line or discussion thread can provide reflection in itself. This includes the simple act of texting to test one's own assumptions, or to offer difficult feedback to fellow team members, as well as ask questions that assist participants in exploring the thinking behind their assumptions.

Within mobile learning's virtual space, one could create an environment where—in relative safety—team members can acknowledge mistakes and invite others to share in the learning experiences. The aim of such environments is to shift the focus onto the learning to facilitate growth of the individual and the larger community. Mobile learning provides for first- and second-person space, the platforms and access to content for deeper engagement, reflection, and learning. And all of this can occur when the individual feels prepared to participate, rather than perform during predetermined moments. An introvert can participate more freely when contributing to a virtual space, without having to navigate real-time social norms that accompany group work (Ellis, 2003).

Defining Mobile Learning

There are many definitions of mobile learning, or *mLearning*. The one that resonates with this chapter is MoLeNET's (2011, para. 4), which defines mobile learning as "the exploitation of ubiquitous handheld technologies, together with wireless and mobile phone networks, to facilitate, support, enhance and extend the reach of teaching and learning." One can distinguish mobile devices from other forms of technology by their portability, ability to connect, and ability to reach the learner and the learner's needs (Shippee & Keengwe, 2014). The device, to be considered mobile, should be wearable or fit in the user's palm. If one can store the device in the pocket or pocketbook, it fits under the category of mobile device. Mobile phones, wearables, eReaders, tablets, and small laptops would fit into this category as they would fit onto the person, in a portfolio, or a bag that accompanies the learner throughout their daily routine. The device must also have connectivity to cellular or another wireless network (Shippee & Keengwe, 2014) as to allow the user to interact with the content, a virtual experience, or people. These features create opportunity for the very interactions necessary for the application of critical thinking throughout the conduct of our daily lives.

As a tool, mobile learning opens the door for new modes of learning (Fisher & Baird, 2006). Shippee and Keengwe (2014) aptly describe this possibility by stating how mobile learning "allows learners at every level to reference content faster, yet also to participate, collaborate, and generate content demonstrating a new level of self-reflective meaning making unparalleled in the history of the modern educational experience" (p. 104). Mobile learning provides opportunity for teams to keep pace with the increasing speed of data transmission, and the encroaching complexity on their lives and on the workplace. Creating a workforce able to participate in self-reflective meaning making may enable them to thrive in the complexity. Action learning is one means of creating a staff able to learn from daily difficulties.

Valuing the Learning

In Carroll's (1865) book Alice in Wonderland, the Duchess quips, "If everybody minded their own business, the world would go around a great deal faster than it does" (p. 84). This offending character makes an excellent point, but fails to appreciate complexity and quality of human interactions coming from different realities, and instead fixates only on the speed in which it passes. This is not an uncommon view of the world and how individuals and organizations believe it should operate. Organizational culture tends towards resisting integration and insisting that each function remain stovepiped. An action learning team however, insists on cross function communication. Teams should invite others into a problem-solving scenario that slows the process down, and encourages double-loop learning. Single-loop solutions may prove more expedient, but only in the short term. They give the initial impression of allowing an organization to keep up with the complexity, but the short-term solution may all too quickly require the team to revisit the issue again. An even more disheartening circumstance emerges when another team later down the line must deal with an identical issue stemming from the preceding team's inability to adequately address it the first time. This is where organizations create self-induced complexity in addition to the mounting influence of external forces.

After meeting the members of the Mad Hatter's tea party, Alice commented, "I'll never go there again!... It's the stupidest tea-party I ever was at in all my life!" (Carroll, 1865, p. 111). However, this mad group helped Alice through her journey and development. These friends questioned, stretched, and challenged Alice, and this made her feel uncomfortable. The tea party frustrated Alice by posing questions to her, seemingly unrelated to her challenge. Alice believed that those in attendance were avoiding answering her straightforward questions, intended to solve her problem. Alice expected the quick fix of single-loop feedback, but the veteran partygoers knew well enough to provide a line of inquiry that would result in the challenging of her assumptions.

Alice tells the Caterpillar "one doesn't like changing so often you know" (Carroll, 1865, p. 67). The Caterpillar, a creature often used in metaphors relating to change, replied that it did not know. The Caterpillar understands the difficulty and work that accompanies change, but it also understands the benefits. Without the work and resulting change the caterpillar would never take flight. Critical reflection can prove disorienting for the individual, but beneficial if it results in the more complex loop feedback depicted in **Figure 1**.

Some may perceive the questioning of assumptions and institutional norms as a slowing down of the process and initial productivity. Deliberate actions such as these can create fatigue, especially for those authority figures on the receiving end of the questions. Take the Cheshire Cat as an example, probing Alice with questions to understand her deeper needs.

`Would you tell me, please, which way I ought to go from here?' Alice speaks to Cheshire Cat

`That depends a good deal on where you want to get to,' said the Cat.

`I don't much care where--' said Alice.

`Then it doesn't matter which way you go,' said the Cat.

`--so long as I get somewhere,' Alice added as an explanation.

`Oh, you're sure to do that,' said the Cat, `if you only walk long enough.'

Most anyone trying to solve a significant issue, like Alice lost in a land of strangeness, would want expedient solutions. Though this expedient thinking is favored in many organizations, wins resulting from single-loop learning may be short lived. However, applying a critical lens through mobile learning can frustrate new participants and can trouble leadership. But the space mobile learning provides can allow new participants, as well as leadership to react to, process, and respond rationally. Practitioners of critical thinking can be viewed as rogue, abrasive, or troublemakers. Such thinkers require a relative safe space, like a seat at the Hatter's tea party table, to work through their thinking, and an action learning coach may provide such a space.

An Action Learning Coach at the Tea Party

An action learning coach can facilitate the time and create physical and virtual spaces for the team to further develop their interpersonal relationships and cultivate a more complete understanding of the principles being discussed. This allows the team to dive into double-loop learning, instead of cutting straight to the chase where a solution emerges without fully grasping the underlying issues. Action learning coaches act as facilitators for the group, pausing the action to encourage reflection, provide tools, and introduce methods (Marsick & O'Neil, 2007). The coach inspires action and empowers the team members test out their new beliefs, hunches, and solutions (Marsick & O'Neil, 2007).

The action learning coach aids the group in creating norms for the purposes of holding the safe space for learning to occur. Critical thinking itself can feel counterproductive to the task. Such questioning may land on others as an attack if the exercise is not fully understood by the participants. An action learning coach should facilitate the team's development. And when the processes—introduced by the coach—become a part of the team's regular practice, they can then transfer their role and leadership to the action learning team (Marsick & O'Neil, 2007).

Many groups fail to capitalize on the learning available to them, by diving deeper into the collective discourse, due to a traditional and sole focus on task completion. The term *rabbit hole* normally denotes a tangent detrimental to group productivity. Team members with product producing fixations view any effort not spent on creating a group artifact as a disruption. Another view of a *rabbit hole* may be through the eyes of Alice who followed the White Rabbit to increased self-awareness and personal growth. As Alice recognizes early in her journey, "I almost wish I hadn't gone down that rabbit-hole—and yet—and yet—it's rather curious, you know, this sort of life!" (p. 46). The *rabbit holes,* if managed by the coach may lead down the path of deeper learning. Immersing oneself in the work, and allowing an inspired process to lead them, versus a taskmaster, can lead to creativity and innovation.

MOBILE SUPPORT FOR FLOW

Flow is the state of being entirely immersed in an endeavor that leads to optimal human performance and experience and contributes to creativity. "Creativity results from the interaction of a system consisting of three elements: a culture that contains symbolic rules, a person who brings novelty into the symbolic domain, and a field of experts who recognize and validate the innovation" (Csikszentmihalyi, 1997, p. 1).

In flow, a person is fully concentrated on the task at hand...action and awareness merge in a single beam of focused consciousness... it is very clear what needs to be done from one moment to the next; goals are clearly ordered and sequenced. (Csikszentmihalyi, 2014, p. 24).

Flow requires a response of unambiguous and immediate feedback, which feeds the individual through a continuous loop, and the person may lose self-consciousness to a point where their vulnerable ego disappears (Csikszentmihalyi, 2014). In flow, time becomes distorted, as hours may seem like minutes (Csikszentmihalyi, 2014). One enters into activities that induce flow through their intrinsic motivation, without an expectation of external reward (Csikszentmihalyi, 2014). An integral element of flow, as depicted in **Figure 2**, is the balance of one's perceived skill and challenge (Csikszentmihalyi, 2014). High challenge without the appropriate skill to match may causes anxiety, and an abundance of skill with little challenge produces boredom. Insufficient challenge creates apathy.

Mobile devices aid in the activation of flow by creating a virtual second-person space or augment the physical to enable the group to come together and focus around a task. The mobile device allows the team to push and pull relevant information, and provide timely feedback. These mobile learning attributes enable the team to stay in flow within virtual, or physical environments, in synchronous or asynchronous work. The connectedness, a term used to describe the value, quality and access to friends, family, and larger communities (Mentor, 2011), that mobile devices afford the learner allows a team to maintain a coordination that reduces anxiety in the face of high challenge.

These devices also provide the space, time, and place for learners to conduct a reflection, or review relevant data. This work, done outside of the valuable time of scheduled team meetings, allows the parts of the collective to build their perceived skill levels. Building one's skill allows the learner to reengage and potentially meet a perceived challenge, bringing them back into flow at a later opportunity.

Learners within the team may choose to take more risks in experimentation and participation through their mobile device rather than during face-to-face meetings. Learners may also feel a relative safety provided by the screen of the device's interface, and the physical distance from the other learners. This can be true for introverts as well, participating more in the absence of the social stimuli introverts

Figure 2.

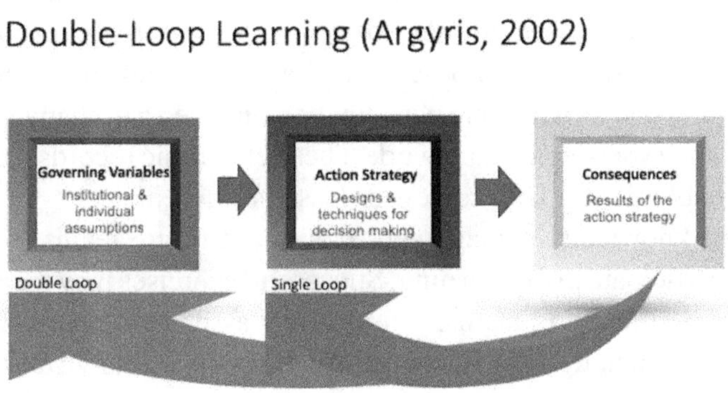

often find exhausting. A virtual space, created through mobile learning, gives the introverts on an action learning team the time and space to participate in critical thinking at their own time and place.

CRITICAL THINKING IN ACTION LEARNING TEAMS

No guarantee exists that all action learning teams will emerge as high performing. What this chapter discusses are ways to engender, ways to create, and ways to make space for teams to participate in critical thinking. Teams tend to focus on the problem or project and neglect the opportunities for learning and leadership development. A reflective journal—via a digital mobile device—is a technique used to draw attention to the learning, work through one's assumptions using a multimodal means to capture thoughts and verbalizations. This modality provides the user an opportunity at a later time to file, sort, as well as transpose, transcribe, and search digitally for words, terms, or phrases.

As an example, a team member might feel that their leadership is causing the team to resent them. Such an assumption might see them withdraw from the team. Journaling this idea may enable them to work through the feeling and see that pulling away from the work would not aid the team in accomplishing the learning or meeting the challenge. Recording the thoughts by writing or talking through their assumptions onto their mobile device may allow them to see the value in where their leadership has brought the team. By sharing this reflection with the learning coach, they afford themselves the opportunity to test their assumption in a relatively safe space, and without feeling exposed to the team. Learners can record their ideas for personal purposes, an after-action review, a learning coach review, or to simply share ideas with the team. It is important for learners to uncover their assumptions and question them, in order to grow their ability to recognize when and why to apply the learning.

Carroll (1865) writes, "That's the reason they're called lessons," the Gryphon remarked: "because they lessen from day to day." (p. 45). The recording of reflections, and events ensures that they do not fade as one's memories might. Something as simple as watching one's own group participation on video can change their original notions of how they show up to team work. These electronic records create an easily recalled, searchable file for later review of lessons learned.

This type of knowledge management would aid future teams, and the larger community to participate in the learning. Successful exercises may then be recreated by upcoming action learning teams. These newly assembled teams will also have access to the *why* when delving into institutional norms. Many rules, norms, and customs exist for important reasons, but those reasons existed at a certain time and place. Having a record of those reasons, and the creators' reasoning allows other teams to assess whether or not the rule, norm, or custom continues to remain relevant.

Carroll (1865) writes of his out of equilibrium protagonist, "Alice felt so desperate that she was ready to ask help of any one; so, when the Rabbit came near…" (p. 18). The mobile device allows for multiple methods of connecting with others, so learners need to look no further than their hip pocket when they are ready to ask for help. Whether they seek help, or need to close a knowledge gap, learners can use their devices to pull support from their team, and coach. Mobile learning allows for timely consultations, or an information push, from the action learning coach. Such pulls and pushes are important uses of mobile devices as critical reflection can cause a learner to fall out of equilibrium, or into disorientation (Taylor, 1987). Designing mobile learning interventions into the method allows the learner to access their team or coach for support outside of the team's face-to-face meetings, as one cannot predict the timeliness and impact of reflection.

THE CASE FOR HIGHER-ORDER MLEARNING REVISITED

Current mobile learning research demonstrates the potential for higher-order learning, but also reveals the need for the distinction between mobile learning and higher-order learning when utilizing mobile devices (Mentor, 2016). Fiol and Lyles (1985) distinguish higher-level learning as being situated in ambiguous and ill-defined contexts and addressing institutional rules and social norms resulting in long-term change. This contrasts their view of lower-level learning that focuses on activities or behaviors, which remains an institutional requirement. However, higher-order learning is also needed to achieve institutional learning goals, outcomes and interventions.

Higher-order mLearning occurs in ambiguous, ill-defined challenges while leveraging the inherent elements and affordances of temporality, space, and connectedness that mobile devices provide (Mentor, 2016; 2018). This definition captures the elements of higher-order learning while recognizing the virtue of mobile technology to bring learners together around new ideas where their thinking may be challenged (Fisher & Baird, 2006). It emphasizes that interventions should allow learners to use mobile learning as a tool of convenience and traverse platforms as needed to facilitate the achievement of the workplace intervention (Fisher & Baird, 2006; Mentor, 2016). Such interventions go beyond delivering content, to provide the platform that reduces the negative effects of time lag and makes space for critical thinking. Web and native mobile applications allow the recording of activities for later review and create an archive of an event that does not degrade like memories, turning the device into a part of the individual's, and group's data management system.

Higher-Order mLearning for Creativity

This chapter argues that flow enables learning to happen while the team engages in the work. This stands in stark contrast to the current and typical organizational assertion that real and substantial learning does not emerge from activities that also bring about joy. Introducing enjoyment into the work environment sounds countercultural and nearly unprofessional. However, the elements of enjoyment, creativity, and imagination lead learners to make connections between ideas and application (James & Brookfield, 2014). Creativity leads to the emergence of unexpected patterns and creation of new ideas (James & Brookfield, 2014).

Coming at a project in new ways can lead a team to generate creative ideas, (James & Brookfield, 2014) and allows for deeper learning around the same themes. Brookfield and James (2014) tell us, "Risk is inherently unsettling, and adrenalin runs as we go into retreat or avoidance. But it is that same adrenalin-infused panic that makes the activity, and then the resultant learning sticks." (James & Brookfield, 2014, Kindle Locations 612-615). Risk is also a necessary ingredient to having flow in mobile learning (Ryu & Parsons, 2012). As an example, the team could construct a LEGO® model of their organizational function, or a cutout art pattern to depict a daily routine. This creative individual assignment might take place back at their workstation, or if at home they can then push or upload a video of the process and final product to their team. Taking simple group activity, like a briefing to the team about each member's role responsibility to the organization and shaking it up can create the risk taking needed to kick start a state of flow.

In order to sustain flow, the group must maintain a balance of challenge and skill. When flow is interrupted, or "when we are too conscious of what we are doing, or when reflection is a form of compliance to meet external performance appraisal or accreditation" (James & Brookfield, 2014, Kindle Locations 678-683), learning and creativity diminish. When the challenge begins to wane, the introduction of risk may reset the balance. As skill fails to rise up to the level of challenge, the learning coach may need to intervene. With pre-established mobile lines of communication open, learning coach interventions of risk or skill building can occur more immediately and more frequently.

When teams experience joy, emanating from their flow experience, the activity becomes autotelic (Csikszentmihalyi, 2014). They want to reengage with the activity, for the sake of the activity, and those people who provided the challenge-skill match. The positive affect gained by those who experience the joy from the flow state also benefit from an increase in capacity to contribute more creativity. Ashby and Isen (1999) say that positive affect promotes cognitive activity that can improve problem solving. Firstly, it makes additional cognitive material available. Secondly it defocuses attention providing a breadth of relevant information in a more complex cognitive context. Lastly, it increases cognitive flexibility, allowing the breadth of relevant elements to come together and create connections. Joy may then feed the group's desire to participate in the work, and to generate an increased capacity for creativity.

The state of flow brings about a connection between the activity, the people involved in the activity and joy. Activities become autotelic in this manner, as the learner seeks out more opportunity to engage in pleasure that results from flow. As in other personal or professional relationships feelings of joy associated with an activity and/or people, make them more likely to engage in that activity and/or with those people. This is in contrast to individuals feeling of dread or anxiety that may cause people to practice avoidance. The virtual platform may also provide a sense of relative safety.

Mobile Learning and Movement

Mobile devices can transform into mobile journals for recording and sharing experiences. They can untether learners from their station and allow for learning to occur, be shared, and recorded during a walk in the park, visit to the museum, or run on the trail. Modern mobile devices hold up better to physical activity, and are more likely to accompany us throughout our daily routine, as opposed to a traditional, non-moisture resistant notebook. Smartphone features allow learners to capture notes, or share work of art from the museum while standing in front of the piece being shared. One might call up the team member with the art history degree for a real time discussion, via a mobile video teleconference, about a specific piece.

Organizing field trips with an action learning team takes much coordination, especially if the team is composed of geographically dispersed members. Mobile devices offer impromptu and often more fiscally responsible alternatives. Though some mobile technology can reach upwards of fifteen hundred dollars, this number is still dwarfed by the cost of other institutional methods. A video teleconference system can reach seventy five thousand dollars (that is a price tag of $150,000 for the pair), and a weekend retreat requires members obtain airfare, hotel rooms, and rental cars.

This virtual method of connecting accounts for a multitude of interests, enabling others to vicariously experience the hobbies of others without subjecting themselves to activities to which they may not be amenable. Skydiving, motorcycle riding, mountain climbing, and white water rafting are examples of physical excursions that not everyone may wish, or are able, to participate in. A GoPro, a popular mobile device that can record and broadcast these risk inherent activities, enables the sharing of these personal and powerful experiences with the group for viewing, discussion, team building, and group reflection. This type of *off duty* interest sharing introduces vulnerability into team dynamics creating the risk, as discussed by Ryu and Parsons (2012), needed for increasing the challenge and skill for flow.

Participating in a personal exercise, whether it includes a walk, a trip to the gym, or a living room yoga routine, allows the team member to tend to the physical quality of the human experience and process the mental and emotional learning. For some team members they may wish to write a poem, paint, or flex their musical muscles. In addition to participating in these creative endeavors alone, one may feel inclined to share these expressions of oneself. Such sharing surfaces vulnerability, which can lead to team cohesion and result in a high performing team.

Creative use of visuals can also help change perspectives. Having the learners create abstract works, in the form of pictures, LEGO® models, or photographs, to depict an organizational strategy, as an example, will help them "understand complex ideas, explore new knowledge, and communicate learning." (James & Brookfield, 2014, Kindle Locations 1515-1516). Introducing new ideas in this fashion allows the team to synthesize concepts and create new learning and solutions. Stirring such creativity in the learner through imagination and play, while providing the space for reflection, can promote significant growth (James & Brookfield, 2014). Such growth as perspective change, and problem reframing is not limited to the individual in the irst-person space, but also provides learning for the group in the second-person space.

As an example of sharing while engaging in opportunistic learning, the author shared the picture found in **Figure 3** with fellow members of a research group. This group used social media and mobile devices to participate in critical thinking via a virtual environment. The author first saw *Treadmill Art* at an art show and views this piece as the embodiment of a newly formed team. A smartphone snapshot and upload to Facebook allowed the author to share the learning with the newly formed and geographically dispersed group. As viewed in **Figure 3**, the author believes the brightly colored belts demonstrate the diversity of the members, and the varying directions and attitudes of the machines represent the members' individual goals and perceptions. When viewing this piece at the event, the author noticed that the centermost and highest treadmill was motionless, while the other treadmills ran at various speeds. The author believed this represents the stifled progress of the team when situated in such a tangled formation. And the florescent ropes are the individual barriers, holding back each individual, reminding them of the risk of failure. Finally, the power cords embody the needs each member requires to be a part of the team.

Each of the elements discussed in the example relate to issues that require attention by the team, and also express an individual's interpretation of group dynamics. The individual could move further in and name each treadmill, rope, belt, and power cord, to help them discuss the un-discussable issues in the group. More simply,

Figure 3.

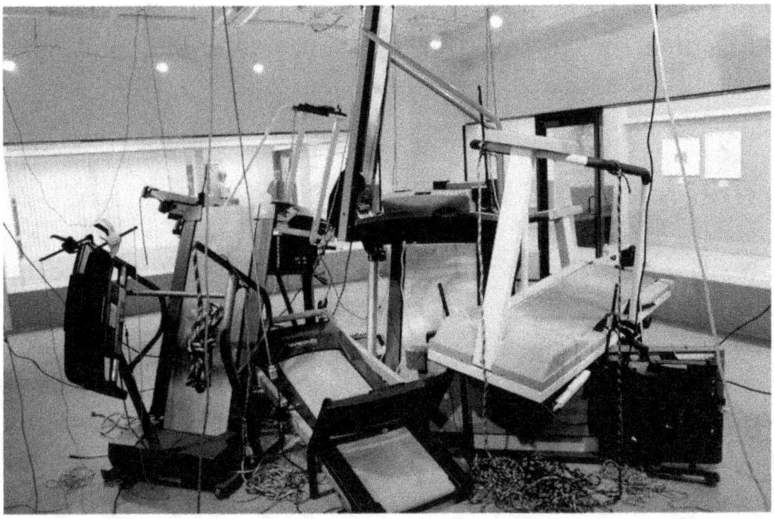

the individual can share the photo, allowing the others to interpret the piece, in an attempt to allow the team to learn more about each other's insights, thinking, and perceptions.

The recording of individual reflections often seems challenging enough, and the prospect of recording them can seem intimidating. However, Brookfield and James (2014) tell us that learners "took enthusiastically to using the Pod to make videos recording their reflections on learning" (James & Brookfield, 2014, Kindle Locations 2593-2597). For purposes of unpacking intrapersonal issues that often accompany critical reflection, the pod provides a space of relative safety to work through such issues. A pod may also provide "a welcome space to vent frustrations and voice panic over all aspects of the learning experience…[and] to clear their heads and identify a suitable course of action or recuperation." (James & Brookfield, 2014, Kindle Locations 2614-2616). The recordings made in the pod may be labeled *for (insert name here) use only*, *for the learning coach only*, or *for the group*.

Designating an inflatable pod, or other self-contained, enclosed space of relative safety for the recording of reflections can move learners to say things that they would not write on paper (James & Brookfield, 2014). Action learning teams open up the idea of work to extend beyond emails, papers, reports, or meetings. Expression does not limit itself to paper and neither should the team. Participating in the work, for action learning teams also means engaging in critical reflection, and attending to group dynamics. Frequent check-ins with the group, initially led by the action learning coach, on the processes and procedures used helps manage a space conducive for cultivating creativity. Accommodating fellow team members while stretching others, via these many options for sharing, reflecting, and processing makes for a learning strategy ripe with creative energy, and an environment set for problem framing and reframing.

WHY ENTER A MOBILE WONDERLAND?

A mobile wonderland is the space that may provide the creative space for strategy development and learning. However, higher-order mLearning, even in a second-person space set for problem framing using mobile devices and critical thinking, is in no way guaranteed. And the act of analyzing assumptions, or asking for acculturation, of an organization is dangerous work. So this begs the question of why one would enter wonderland in the first place. Why would one risk the resistance often aimed at change agents?

Many institutions, whether it is the promise of academia or the mission statement of a nonprofit, or other organization, according to many degree descriptions on .edu websites or job postings on online job boards, wish their talent to be able to think critically and ethically. Even Toyota University, one of over 4,000 organizations with their own internal corporate university, wishes to provide their participants—both internal employees and externally sourced students—with their mindset and procedures of continuous improvement. The resources spent on these endeavors provide the evidence that today's public and private world desires and often requires the creative and innovative thinking that may be facilitated by higher-order mLearning. This type of learning, in conjunction with a leadership-sponsored action learning team, will provide the space for the desired learning and help the organization value the learning, leadership development, and creativity, while experimenting with implementing the resultant innovation.

THROUGH THE MOBILE LOOKING GLASS

This chapter has made the case for ways that mobile learning can leverage reflection and critical reflection in learning groups such as action learning teams as well as in other workplace settings. Higher-order mLearning is not a definite outcome, but mobile devices make it possible to learn, and share learning in creative, unconventional ways that also draw upon and utilize multiple senses and the imagination. Mobile learning introduces opportunities to find joy in and share ideas spontaneously, in real time, that lead to seeing reality through different lenses, which opens the door to reframing challenges and finding innovative solutions. Using the methods discussed in the chapter, learners are afforded the opportunity to *come at* the learning from multiple vantage points. Each activity maintains the common thread of making space for reflection.

Adding flow into the equation turns the work into an autotelic activity. The joy that emanates from the flow states resulting from the work feeds back into the team members' intrinsic motivation. Flow results from the team tightening their bond and desire to continue their work together. This is not to say that an action learning team promotes the feeling of a work environment led by a big purple dinosaur signing children's songs. But what flow does encourage is a shared desire to develop and support the team's skill building in relation to a high challenge. Flow is the element that, if attained, provides the achievement that keeps the organizations talent engaged and coming back to work.

Looking back for words of wisdom on the subject of moving forward with the practice of higher-order mLearning, and new meaning making possibilities, Carroll does not disappoint. "Alice had got so much into the way of expecting nothing but out-of-the-way things to happen, that it seemed quite dull and stupid for life to go on in the common way." (p. 14). Once one has gone through the looking glass and becomes practiced at reframing challenges, steeped in double-loop learning, and familiar with flow, retreating back to reactive problem solving seems as silly as the Red Queen's policy of "jam to-morrow and jam yesterday— but never jam to-day." (Carroll, 2013, Kindle Location 524).

REFERENCES

Argyris, C. (1976). Single-loop and double-loop models in research on decision making. *Administrative Science Quarterly*, *21*(3), 363–375. doi:10.2307/2391848

Ashby, F. G., Isen, A. M., & Turken, A. U. (1999). A neuropsychological theory of positive affect and its influence on cognition. *Psychological Review*, *106*(3), 529–550. doi:10.1037/0033-295X.106.3.529 PMID:10467897

Brookfield, S. D. (2011). *Teaching for critical thinking: Tools and techniques to help students question their assumptions*. John Wiley & Sons.

Carroll, L. (1865). *Alice in Wonderland.* Academic Press.

Carroll, L. (1871). *Through the Looking Glass*. Academic Press.

Csikszentmihalyi, M. (1997). *Flow and the Psychology of Discovery and Invention*. Harper Perennial.

Csikszentmihalyi, M. (2014). *Applications of Flow in Human Development and Education.* Springer Netherlands. doi:10.1007/978-94-017-9094-9

Ellis, A. E. (2003). Personality Type and Participation in Networked Learning Environments. *Educational Media International, 40*(1/2), 101–114. doi:10.1080/0952398032000092152

Facione, P. A. (1990). *Critical thinking: A statement of expert consensus for purposes of educational assessment and instruction.* Newark, DE: American Philosophical Association.

Fiol, C. M., & Lyles, M. A. (1985). Organizational learning. *Academy of Management Review, 10*(4), 803–813. doi:10.5465/amr.1985.4279103

Fisher, M., & Baird, D. E. (2006). Making mLearning work: Utilizing mobile technology for active exploration, collaboration, assessment, and reflection in higher education. *Journal of Educational Technology Systems, 35*(1), 3–30. doi:10.2190/4T10-RX04-113N-8858

James, A., & Brookfield, S. D. (2014). *Engaging imagination: helping students become creative and reflective thinkers.* John Wiley & Sons.

Marquardt, M. J. (2000). Action Learning and Leadership. *The Learning Organization, 7*(5), 233–241. doi:10.1108/09696470010352990

Marsick, V., & O'Neil, J. (2007). *Understanding Action Learning (AMA Innovations in Adult Learning).* New York: AMACOM.

Mentor, D. (2016). EMxC3= e&mLearning Cultivating Connected Communities: Sustainable Workforce Talent Development. In *Handbook of Research on Mobile Learning in Contemporary Classrooms* (pp. 240–259). IGI Global. doi:10.4018/978-1-5225-0251-7.ch012

Mentor, D. (2018). Micro to macro social connectedness through mobile phone engagement. In *Encyclopedia of Information Science and Technology* (4th ed.; pp. 6184–6194). IGI Global. doi:10.4018/978-1-5225-2255-3.ch537

Mentor, D. J. (2011). *Exploring social connectedness via mobile phone texting* (Doctoral dissertation). Teachers College, Columbia University.

Mezirow, J. (1990). Fostering critical reflection in adulthood. San Francisco: Academic Press.

MoLeNET. (2011). Retrieved from http://www.molenet.org.uk

Paul, R., & Elder, L. (2003, Spring). Critical thinking: Teaching students how to study and learn (Part III). *Journal of Developmental Education*, *26*(3), 36–37.

Ryu, H., & Parsons, D. (2012). Risky business or sharing the load? Social flow in collaborative mobile learning. *Computers & Education*, *58*(2), 707–720. doi:10.1016/j.compedu.2011.09.019

Seo, M., & Creed, D. (2002). Institutional Contradictions, Praxis, and Institutional Change: A Dialectical Perspective. *Academy of Management Review*, *27*(2), 222–247. doi:10.5465/amr.2002.6588004

Shippee, M. & Keengwe, J. (2014). *mLearning: Anytime, anywhere learning transcending the boundaries of the educational box.* Academic Press.

Taylor, M. (1987). Self-directed learning: More than meets the observer's eye. In Appreciating adults learning: From the learner's perspective (pp. 179-96). London: Kogan Page.

Torbert, B., & ... (2004). *Action inquiry: The secret to timely and transforming leadership*. San Francisco: Barrett-Koehler.

Tuckman, B. W., & Jensen, M. A. C. (1977). Stages of small-group development revisited. *Group & Organization Management*, *2*(4), 419–427.

Voronov, M., & Yorks, L. (2015). "Did you notice that?" Theorizing differences in the capacity to apprehend institutional contradictions. *Academy of Management Review*, *40*(4), 563–586. doi:10.5465/amr.2013.0152

Chapter 2
Developing an "App-titude" for Cooking:
Helping Adults Promote Positive Cooking Skills With Children via Mobile Web-Based Accessible Media

Jamie Krenn
Columbia University, USA

ABSTRACT

Food preparation for preschool and elementary school children is described from the perspective of enhancing cognition and learning through the use of mobile devices. While most children are too young to use a stove or slice with a knife, with proper guidance and simulated experiences through mobile devices, children can actively contribute to food choice, preparation, and knowledge while educators contribute knowledge from various academic areas. The kitchen can be a positive learning environment for the educator, the student, and potentially, people in the student's household, as part of an effort to develop a positive attitude toward food preparation. Updated learning implications, more current apps, and cooking curriculum suggestions are provided, along with future research recommendations.

INTRODUCTION

The objective of this chapter is to provide theoretical guidelines and suggestions for mobile-media implementation within *culinary cognition* in the classroom. Several theoretical areas are to be examined around the term *culinary cognition*. Most relevant is the argument that feelings about cooking should be acknowledged and cooking

DOI: 10.4018/978-1-5225-9351-5.ch002

practices made possible by technology should be considered. Learning is discussed within the terms of observing and interacting with mobile media, along with suggested curriculum features and apps. The challenge beyond the classroom for researchers and mobile-media developers is finding creative and cognitively appropriate ways to combat dislike for food preparation, while increasing the probability of a higher culinary and food-appreciative skill set within the United States, where early childhood obesity is on the rise (Ng et al., 2014). This chapter also aims to present technological applications, and examine the multifaceted relationship between children's mobile technology and healthful food practices. It will address mobile technology in the context of learning from a cognitive perspective, for educators, parents, and educational consultants. Finally, the chapter seeks to define *culinary cognition* as the mental action or process of acquiring food-preparation knowledge and understanding through thoughts, experiences, and the senses as an instructional experience for educators inclusive of technology applications.

Considerable evidence supports the hypothesis that cooking can enhance one's health and well-being (Brown et al., 2006; Kruger et al., 2006; Larson et al., 2006; Olvera et al., 2008; Rosenkranz, 2008). Liquori et al. (1998) showed that when children cooked with vegetables and whole grains in the classroom, they consumed the same kind of food when it was served in school lunches and Children who received only nutritional education did not eat the new food. In addition, evidence supports the hypothesis that nutritional television programs increase public awareness and food knowledge (Chew et al., 1995, 1998; Smith & George, 2001), yet these research studies were conducted with adults and young adults (Larson et al., 2006) and not with preschool or elementary school students. Thus, a curriculum that includes apps targeted specifically to young children, about healthy eating choices will undoubtedly promote changes in cooking skills and overall healthful food choices. (Fildes et al., 2014; Martinez et al., 2014; Peters et al., 2014; Russell et al., 2015; Wiggins, 2014; Woodruff & Kirby, 2013 You indicated you have newer references, right?). Traditionally, many children learn their eating habits and to cook at home. In some homes there may be limited or no exposure to quality food. Additionally, school-based recipes or mobile-device food or cooking applications which may be useful for learning about healthy food habits (Norman et al., 2015) are also rarely present in children's lives. Therefore, much of this chapter offers valuable transferable mobile learning lessons to foster basic healthy eating habits, agency and cooking skills in young children.

Sadly, some families settle for fast food or other preprocessed alternatives, for convenience, to save money, marketing, or to have more time to spend with loved ones. Some parents are either not educated themselves around healthy food choices, some abdicate their responsibilities when it comes to healthy nutritional choices. However, children spend a lot of time in classrooms, so schools and educators can be the conduit between food appreciation and preparation while using mobile apps to help develop healthy cooking skills.

Mobile devices are ubiquitous enough for educators to leverage and share information that may support children in developing healthy cooking skills—the very skills that caregivers may not have time for with their children. A mobile device supports the user's ability to learn from multimedia presentations and illustrations (for example, visual and auditory representations) through instructional messages (Mayer & Moreno, 2003), such as nutritional ingredients, recipes, or videos lessons. These "anytime, anywhere learning" (Schuler, 2009) illustrations are usually enhanced by repeat, and re-do or tactile feedback and motion-detecting functions, thereby increasing feedback to the user on the basis of real-time responses (Attewell & Savill-Smith, 2004; Williams, 2003).

Computer tablets and smartphones can also freely enter the kitchen. They promote play or exploration. When young learners navigate a game or conduct a browser search, they are constructing new knowledge through symbolic representations (Ainsworth, 1999). Play, discovery, reward or incentive systems can help children stay interested in a cooking task, which some often put in the "mundane" or "chore" category. Having a positive play attitude about food and food preparation, as learned from an educator, can only help children build lasting, positive relationships to cooking. Children then have effective constructive attitudes, rather than negative ones, which they may bring into the home. By promoting discovery through unlimited access to recipes, ingredient information, cooking goals and attempts to model these, serve as an untapped resource for educators and developers. Mobile devices can also be used to harness, as well as apply to mobile cooking interaction and executions.

The design and production of downloadable apps for children move as swiftly as the technology changes. Correspondingly, theories of learning and cognition have recently evolved to incorporate these innovations [give examples here]. The transformations in mobile-media technology, and its learning theories, has led to changes in how learning is defined and how cognitive skills relate to multimedia processing. There are devices with applications that house numerous sophisticated animations and imagery, all of which allows educators to receive these devices and apps more positively. The ease of use and availability have made these applications less overwhelming for educators and young learners (Schuler, 2009). Many educators now use portable-device applications for supplementary classroom lessons. Thus, teachers can teach young children about a new area of learning by combining food preparation and cognitive skill sets and by making connections to various academic areas, such as mathematics and science. These inclusions seek to benefit the young learner, as mobile media are attractive tools for implementing valuable educational elements. In comparison to older education tools (such as lecture and demonstrations), modern learning tools have evolved into individual supplements for both the home and the classroom.

A few games that do involve food focus on unhealthy edibles, outside the kitchen (For example, Food Fight [Playdek, 2012] uses food as weapons, and food as targets to cut and waste as in Fruit Ninja. These are great examples. It would be great add newer ones to these examples. Another example is Cake Doodle, whose [Shoe the Goose, 2014] design and aesthetics does not focus on the baking process or show healthy ingredients.) Educators need guidance on properly selecting, using, and preparing food. Rethinking food practices and using positive media will support educators and provide students with agency, and positive life skills, such as the ability to learn about, and understand food preparation through thoughts, experiences, and the senses. This joint process of students building their own knowledge base while working on nutritional or cooking processes led by teachers in a mobile environment (Zurita & Nussbaum, 2004) will support both parties' *culinary cognition.*

Rationale

Food preparation can be an innovative way to flex a person's reasoning muscles. People can use their thinking skills to manipulate ingredients and create dishes to share with others. Food preparation may provide a sense of independence for someone who has felt uneasy in the kitchen. Overall, humans need food, and eating can be considered survival (i.e., solving the problem of being hungry and trying to find food) (Anderson, 2005; Johnson-Laird, 2005). However, some adults feel as if they could barely survive in the kitchen alone, not to mention with children.

Working with children in the kitchen can pose numerous challenges from keeping their attention, helping children understand a novel topic; having the openness, education and versatility to craft different dishes; unlearning poor cooking practices; familiarizing students with new food; managing safe techniques and tools; carving out cooking time in an already hectic school schedule. All these challenges are part of the *culinary cognition* concept. One reasoning skill useful to this task is mental model reasoning.

It is important to note that there are difficulties in learning, such as false understanding or limited experience. However, mental models can represent any number of people, things, or processes and underlie thinking with propositional content (Johnson-Laird, 1994, 2005). Likewise, each person can have models in place that assist in thinking. One particular way a person interacts with something may not be similar to another person. For that reason, *iconicity* states that "a mental model has a structure that corresponds to the known structure of what it represents" and how the person interprets it (Johnson-Laird & Byrne, 1991). This is similar to Sweller's Schema Theory which he espouses as he writes about cognitive load (Sweller, 1994) . Furthermore, Singaporan mathematic teaching models host the idea of divergent meaning making, and to incorporate it as a multiple perspective practice. This can lead to the understanding of new relationships (Schaeken et al., 1996). Mental models can be applied to logical reasoning situations and offer numerous solution possibilities. Which in turn, can have positive quality-of-life ramifications, such as consuming more healthful food and being more active through food preparation and shopping (Chen et al., 2012; Erlich et al., 2012).

From a cognitive perspective, a person who demonstrates good reasoning in the kitchen has developed a model and utilized it for some time. If an educator has a proper format to apply, like a parent or grandparent when it comes to cooking, there is an increased chance of passing on cooking lessons to young minds. Therefore, it is imperative that an instructor has culinary knowledge and experience, and that s/he can also teach young learners that hands-on experience is the key to developing cooking skills. Any degree of success in a kitchen is correlated with positive experiences (Erlich et al., 2012), and this is now the case with preschool children (Kihm & Singleton, 2012). Those with experience and stored mental models will have an easier and more successful time with any activity related to the initial learning experience or schema (Johnson-Laird & Wason, 1970; Sweller, 1994). Mental structures related to recipes and procedural kitchen activities can organize and simplify knowledge about the cooking environment—for example, knowing where tools are located, what combination of ingredients will make a successful dish, or how hot the oven should be set. Schemas are about people and places, as well as many life events (Craik, 1943; Hegarty, 1992; Schwartz & Black, 1996), and in many cultures a good majority of social interactions center on the preparation and conception of food (Kittler et al., 2012).

Any person familiar with cooking and the kitchen has a number of mental models for preparing different types of dishes. For example, if a person makes pancakes, he or she knows what ingredients are involved in preparing pancake batter, locating a skillet, turning on the stove to warm the skillet, pouring the batter onto a hot buttered skillet, finding a spatula, and flipping the pancakes when bubbles form. Changing some steps in the batter preparation, like adding blueberries for example, will modify the original model. Therefore, educators can help young learners vary a pancake recipe, for example, by adding bananas, walnuts, and cinnamon. Thus, one advantage of an educator's base knowledge and experience is that it allows him or her to build new schemas and yield new relations (Peirce, 1903). Tapping into mental models for event concepts from activities learned in the past, can be helpful in learning new skills with similar circumstances. By retrieving these memories and using cues, one can harness the knowledge of a stored category (Chase & Simon, 1973).

On the other side of the coin, someone who has little cooking experience is unlikely to have many mental models to bring into the kitchen. Recipes may be unfamiliar, particular cooking techniques may be baffling, and getting from a sack of flour, to muffins hot from the oven may seem impossible. From false beliefs arise false inferences. While preexisting knowledge is beneficial, educators in the kitchen should focus on children's false understandings about cooking, as these misunderstandings will affect first- and second-hand knowledge (Vosniadou & Brewer, 1989). Children may have spent little time in the kitchen at home, or a caregiver may be baffled by simple food-preparation activities (e.g., boiling an egg, or roasting a chicken). Educators may benefit from building on these misunderstandings and teach their students better ways to prepare food.

For example, when asked how an egg is prepared for breakfast, some people might reply, "In a frying pan, scrambled." Few people know that they can bake eggs in the oven, poach them, or boil them—the true result of mental models is that a person can learn to adjust. The same question can also yield this response: "I don't know how to bake an egg, but I can bake a cake." But educators will need to help people understand that baking eggs does not mean not putting the eggs, shells and all, in the oven at 425 °. Students may begin by reasoning ineffectively (Johnson-Laird, 2005), but eventually develop strategies through directed guidance. Self-directed learning can also be employed for trial and error, but that should come after a few successful first-time cooking experiences.

Mental models for making eggs, like making omelets, will help show students that variations are being added before tackling a more difficult task,. This will be less cumbersome to new cooks. It is more reassuring to say, "We are adding a few changes," rather than declaring, "We will begin from scratch"; the latter may be too intimidating. Research has shown that musicians with a higher number of mental models are able to improvise; this can likely be true for cooks, too (Johnson-Laird, 2005).

Novices with no mental models for activities in the kitchen will be some of the more difficult children to work with. Therefore, educators have to start from scratch and explain about particular appliances and gadgets. To overcome this hurdle, mental models of existing knowledge or simple recipes can be used. The more the person uses these recipes, the more likely it is that he or she will be able to reason and adjust dishes to taste without relying on recipes. Moreover, a recipe is a written mental model and provides steps, ingredients, and preparatory suggestions. One way to encourage "cook-type" thinking is to use Designs for Knowledge Evolution, modified to suit culinary cognition.

The Application of Designs for Knowledge Evolution in Food Preparation

Designs for Knowledge Evolution (DKE) is an explicit attempt to join first-hand [action] and second-hand [instruction] experiences into a prescription for learning. Knowledge acquisition begins with mechanisms that generate specific forms of learning. It proposes a framework in which multiple representations and mechanisms can interact to co-evolve a well-rounded understanding that supports future learning and adaptation. (Schwartz, et al., 2005, p. 32)

The DKE model adapts to learners' needs. The instructional sequence of DKE develops first-hand theories that focus on direct experience with the learner and second-hand theories that focus on descriptions of experience and communicated knowledge. Because cooking involves haptic experience, much as the manipulation of mobile devices does, by using DKE, parents or educators can formulate cooking methods for young learners to improve models for comparison and testing. In this way, children's models can evolve from understanding. Using this design, along with the student–teacher relationship, to structure the interplay between appropriation of technology and practice (Winters, 2007) will further student understanding of culinary concepts.

To begin, first-hand processing depends on students directly interacting with a topic. Before young children develop the ability to interpret and describe phenomena, engagement occurs. Engagement appears to occur more often with today's "learn on the go" mobile technology. Often, in today's graphic user face world, when a new device is obtained, an instruction manual is not consulted; device interaction is the first thing to happen. Most users, regardless of age, dive right into manipulating buttons, touching screens, or using voice-recognition systems. These initial interactions focus on perception and action, and tend to be more cued to the individual because their focus is direct and come from personal interactions. First-hand experiences are highly relevant to learning. The DKE model integrates the human experience into learning. The DKE model states that "embodied theories of cognition emphasize the mental simulation and organization of bodily experience" (Schwartz et al., 2005), and this is applicable as one uses mobile devices and moves through cooking spaces.

Second-hand theories are also extremely relevant to learning (Schwartz et al., 2005). Second-hand theories are the actual instructions that the learner must follow to complete a task. Through these experiences, learners take the context or instruction from the teacher and cognitively process it into a sequence of actions. It is the educator's and the designer's responsibility to clearly present and design these instructions for the young learners' developmental needs. An example of second-hand theory in play is a recipe. A recipe is a set of guidelines one must follow to prepare food. A recipe's instructional components can only be enhanced with mobile-device technology, which can present a combination of text, images, sound, and haptic interactions while taking the user's environment into consideration (Gellersen et al., 2002). Videos that illustrate a technique can only add to the vicarious experience of cooking through a well-formulated, second-hand experience design.

The DKE model also integrates scaffolding. With scaffolding, learners take on limited roles that allow them to learn "the form of the activity even though they may not fully understand its function" (Schwartz et al., 2005). Because young learners use recipes as they learn how to cook, it is highly likely that over time they may develop recipes on their own; they may even combine the original ingredients to form their own, unique dishes with the help of caregivers. By uniting the mechanisms of first-hand and second-hand experiences, along with scaffolding, to the DKE model, educators can provide a supportive learning environment that encourages learners to physically complete a cooking task and then process the task by preparing the dish at home, understanding the positive effects of nutrition through cooking, or teaching others how to prepare a dish. To complement second-hand theories, different types of instruction, such as audio, text, pictures, and digital interactions, are optimal for long-term learning. To complement first-hand theories, teachers can support student accomplishments by supporting good cooking practices or applying hands-on knowledge to other academic areas of learning.

Because learners are cooking, the DKE model's adaptability suits first-hand theories or experiences. Under the direction of educators, learners get hands-on experience from a lesson plan and a personal experience as they learn. Because first-hand experiences depend on people (students and educator) directly interacting with a phenomenon, and largely focus on perception and action, young learners must have a hand in the preparation of any dish (Schwartz et al., 2005). This enables processing to perfect second-hand theories, through synchronous or asynchronous feedback, positive or negative, from the instructor—perhaps a comment such as "Nice job washing the vegetables" or "Let's watch that video again about kneading dough." Mobile devices benefit this situation by allowing participants to access recipes and video instruction as they work in the kitchen. The mobile devices easily move with the cooks and are on hand for each new step in the recipe or when the teacher and students want to review a technique. To enhance second-hand experiences, text, video, and sequenced picture instructions for each step in preparing the meal are optimal for teaching young children positive cooking practices; the additional modes of communication only increase knowledge. Mobile devices can also mitigate the effects of an instructor who has had poor results in the kitchen or a negative attitude about cooking.

The design of mobile devices also permits scaffolding. Young learners can take the lessons and improve on them over time by repeating the steps and cooking a dish at home with their families. Young learners may be able to perfect the dish, depending on how well teachers maintain a balance between mobile devices and kitchen use. This is shown in Figure 1 and Table 1, which draw on evolutionary theory (Schwartz et al., 2005).

COOKING AND MOBILE DEVICES AS A MAIN FOCUS

Cognitive Design Pattern

Before working with preschoolers in any cooking activity, educators should prioritize safety and acknowledge any anxieties about food preparation, such as anxiety about producing a dish as expected, and the dangers or difficulties that can arise when working with this population (Knight et al., 2014). If a teacher is a poor cook, this should not prevent the process of preparation from entering the classroom kitchen. In fact, working through anxieties about cooking can be one motivation for achieving the end goal—a tasty meal. Identifying negative feelings about cooking can be the

Figure 1. Cooking model implementation flow chart based on DKE design (Schwartz et al., 2005)

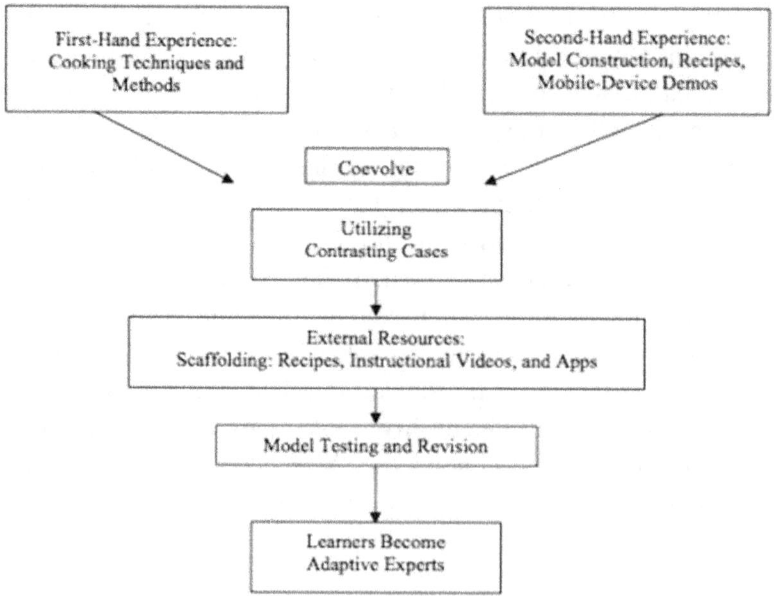

Table 1. Designs for knowledge evolution (DKE) (adapted from Schwartz et al., 2005, p. 35-36) How can the use of mobile devices be added to the table below.

Feature	Student Application
Students generate symbolic models to differentiate between contrasting cases.	This permits students to perceive what is significant about the cases and their contexts while developing a structured account of what they perceive as a quick response.
Students test their models across contexts of contrasting pairs.	As they confront new contexts, some models fail, and students notice properties of the new contexts and the models they select.
Students mutate new models that can survive in the context.	The new models evolve from the understanding developed from previous models, even if the students need to abandon the form of their earlier model and try new models.
Students juxtapose respective models to notice their "survival" value.	Whereas the contrasting cases introduce environmental variation, the juxtaposed models introduce species-level variation. Noticing the varying quality and useful features of different models introduces selective pressures to help students generate useful models.

starting point for a discussion with students about overcoming internal obstacles. Teachers, when starting any technology-supported culinary expedition, can benefit from reflecting on their own attitudes. This reflection can help teachers convey a positive attitude about cooking to their ready-to-be-molded students. Cooking in the classroom is a great foundation for future food habits. A cultural exercise or project where students bring their favorite recipe, and a collection to a multimedia link of images or videos of the recipe, and perhaps the result of the process for others to enjoy, would also be a great start. The latter idea can also be integrated into lessons related to Social Studies, Geography, History, Mathematics, Language or Literature themes.

Research suggests that a child's early years build the groundwork for future eating attitudes (Knight et al., 2014; Martinez et al., 2014; Norman et al., 2015; Peters et al., 2014). Food insecurities have been linked to many negative outcomes in childhood development (Cook & Frank, 2008; Morgan et al., 2014) and overall life quality (Casey et al., 2005). Most adults do not realize that eating is a learning process, not an automatic response. Classroom exposure to new food can be a tool for promoting good health and can give children the opportunity to try and like new kinds of food (Krenn & Hachey, 2015; Satter, 2008). And with the favorite food idea project supplied above, part of the project could also include the nutritional facts about the ingredients in their food project.

For many children, school is the primary source of meals, including breakfast, lunch, and afterschool snacks, during the week. This gives educators many opportunities to nurture in students' positive attitudes toward food and cooking. If these attitudes are not nurtured, disinterest in food or a lack of motivation about eating well and cooking can set in. Sadly, for some children school may be the only place where they are exposed to healthful food. As a result, young children may end up making unhealthful food choices. They may be indifferent to food preparation and be at a higher risk for nutrition-related health issues. The challenge is finding creative and cognitively appropriate ways to combat this debacle within our society, in which early childhood obesity is on the rise (Ng et al., 2014). Knowing about new food may go beyond the refrigerator to the mobile device.

For example, Super Abby vs. The Enchanted Sweet Stall (Forma Animada, 2013 Great apps), an app for children ages 6 and 8, weaves facts about nutrition and causes of obesity into its game. In the game's interactive story, Super Abby is curious about why her brother snores so loudly. She learns that obesity can cause snoring. Super Abby then discovers that there are other symptoms that come with being overweight. The game teaches children about healthful food and the benefits of exercise in a fun, developmentally appropriate way (Serf-Walls, 2013). Or Grow Garden (Common Sense Media, 2019) where children learn about the microsystem of growing food – farm to table). These are excellent supplemental game for educators to use before and after cooking lessons.

Equipment Needs for the Contemporary Classroom

The combination of play and learning about cooking skills through the meaningful use of mobile-media technologies can be challenging for teachers (Lee, 2015b). A simulated experience, which does not use actual kitchen tools or provide hands-on experience, will happen through the use of mobile devices. Thus, starting such a program may be difficult because a teacher will need to acquire both technology and kitchen equipment to give students simulated and hands-on experiences. Children's mathematics skills, especially, may greatly increase through the use of mobile media (Ginsburg, 2014). Many mobile-supported education programs cannot be possible without large investments of both time and money on the part of administrators. However, with the personal mobile device ownership, or cheaply available devices, the challenge of cost can be overcome. Because teachers will be doing the cooking, they will need to procure several items to prepare classroom kitchens. This can also be done through a cooking instructor or a volunteer parent who travels from classroom to classroom and leads weekly demonstrations for small groups. Funding, resources, and class time may limit administrators' ability to decide on implementing the recommendations included in this chapter. However, all listed before are necessary to the learning process. Many of the items listed in Table 2 are small and storable, which will allow educators to begin tackling simple recipes with their students.

Table 2. Suggestions for preparing the classroom kitchen

Implement Group	Supply List * *It is important to exercise caution and safety with regards to all utensils, knives, pots, pans, etc. with young children. Please take note of child's experience and developmental age when using sharp, heavy or heated objects.*
Tools	**Plastics knives, forks, and spoons; toothpicks; napkins; cutting boards; oven mitts; food thermometers; whisks; aluminum foil; plastic wrap; colander; wooden spoons; ladles; tongs; grater.**
Pots and pans	**Square baking pans, cookie sheets, skillets, pots of various sizes**
Containers	**Mixing bowls, measuring cups for dry and wet ingredients, tasting bowls, paper plates**
Small kitchen appliances	**Blender, toaster oven, hot plate, waffle iron, griddle**
Spices	**Salt, pepper, garlic powder, thyme, parsley, basil, oregano, cinnamon**
Oils	**Cooking spray, olive oil**
Equipment for clean-up and storage	**Refrigerator, sink, garbage vessel**

Educators can make the classroom valuable by organizing conversations, sharing beloved food experiences, and encouraging listening in the hopes that students might try a new suggestion, on the basis of classmate experiences, when using the tools in Table 2 for preparing food. Teachers can ask questions that seek clarification, extend students' thinking, and connect past and current cooking practices. While focusing on a simple snack is a great start, educators should also look into preparing meals with students. Teachers can discuss learning from mistakes. In addition to working with students in the classroom kitchen, educators can encourage parents to dedicate time to preparing food with their children at home. Or teachers can organize a monthly class dinner party and prepare homemade food, instead of having food delivered (Krenn & Hachey, 2015).

Nicholas' Garden (AppMatrix, Inc., 2014) is an app that can get children involved in planning a class meal. An 8-year-old boy, who wanted to promote healthful eating, inspired the development of this app. The app provides simple recipes that students can help make. (Some recipes are so simple that older elementary school students can make them on their own.) Children are able to try recipes, share photos of their dishes, and make recommendations, all while learning important life skills. The app focuses on the empowerment of children as they develop cooking, eating, and shopping skills. It benefits children by showing them recipes and teaching them what it takes to construct a shopping or gathering list. Educators can use this app to give students meaningful experiences and promote the development of skills that young learners can use to make food choices a valued part of their lives at home and in the classroom.

Setting up the Student for Success as a Competent Cook and Examining Current Cooking Practices

This step is about educators presenting food knowledge and promoting decision-making skills to create a positive experience in the classroom. It is not about making complex dishes with many ingredients and steps or using equipment that is difficult to manipulate. It is about using mobile devices to explore cooking apps and television programs (either cooking programs or programs in which food is an important part of the plot) to discover a handful of simple, efficient recipes organized by skill level and completion time that can be used in the classroom. Three apps that meet these criteria are Food Network in the Kitchen (Television Food Network, 2015), Whole Foods Market (Whole Foods Market, 2015), and Epicurious (CondeNet, 2015). If educators help students use these apps to discover new food options, as well as accept that failures are part of working in the kitchen the higher likelihood of success in cooking and enjoyment occurs. Just look to Julia Child, who often made a happy moment out of flouring a chicken too profusely, but who savored the finished product and knew that proficiency in the kitchen comes with practice (Public Broadcasting Service, n.d.). Finally, by taking the time to critically look at how negative feelings about cooking may be influencing what, and how, educators teach children about cooking can then help to assist in their pedagogy (Krenn & Hachey, 2015). To do this, educators should ask themselves these questions:

- Have I ever allowed children to perform simple, age-appropriate tasks related to food preparation?
- Do I have any recipes or cooking activities that can engage and inspire children?
- Do I seek opportunities to deepen my, as well my students', culinary understanding?
- Do I model for students the relevance and importance of best food practices to everyday life?
- Add questions about teachers' mobile technology knowledge and abilities.

These steps can occur independent of another and be used as stand-alone items. It simply depends on the skills of instructors, student and available supplies/materials.

Creating a Plan for What Good Food Practices at School Should Be and Ways to Measure Success

This step is about focusing on the purpose of cooking with children: to bring positive social experiences into the classroom and create confident, capable eaters, problem solvers and informed decision makers (Krenn & Hachey, 2015). To determine whether directives are effective, educators can keep in mind some cooking skill areas to assess success on the part of both the instructor and the students (Kihm & Singleton, 2012).

Preschool Cooking Skill Assessment Areas

The following are some areas that can be tested for content and skill. Classrooms can benefit from a digital presence of content by using Quizizz - https://quizizz.com/ (for content testing by having instructor input testing information) or Flipgrid - https://flipgrid.com/ (students can make their own cooking videos demonstrating skills for evaluations and sharing with others.

- Naming Ingredients: By sight, smell, taste, or touch
- Naming and Explaining Various Cooking Methods: Boiling, poaching, sautéing, warming, melting, mixing, blending, whisking, crushing, cracking, toasting
- Understanding Recipe Logic: Order is required when cooking, which can be reinforced by mobile technology with an educator's help.
- Completing Steps Hands On: Cutting (with safety knives); grating; weighing; measuring; sieving; cutting fat into flour; cracking eggs; separating eggs; beating eggs; mixing liquid into flour; kneading; rolling dough; using cookie cutters; brushing food (e.g., with egg wash or water); tearing; mashing; crushing; squeezing; shaping dough; peeling eggs; beating liquids (e.g., salad dressing); draining using a colander; garnishing
- Understanding Simple Tastes and Identifying Associated Foods: Bitter, salty, sweet, sour, umami
- Understanding the Concept of Healthful vs. Unhealthful, or "Sometimes," Food: For example, whole wheat toast vs. a doughnut, olive oil vs. butter, steamed veggies vs. French fries
- Understanding the Concept of Healthful vs. Unhealthful Food Preparation Methods: For example, baked potato vs. fried potatoes, grilled chicken vs. fried chicken (Food for Life, 2015).

Assessment will help an educator decide when to move forward with particular recipes, on the basis of skills met, and when to repeat a dish later in the school year for familiarity and fun. This will help remind the educator that it is important to see cooking as a process rather than a strict set of rules. While there are guidelines in a recipe, part of the inventiveness and fun of food preparation is adjusting recipes to taste and ability level. Educators can give children some pita bread to tear, not cut with a knife, for baked pita chips or have children include tomatoes in a salad if they are a favorite food. Educators should keep in mind, and reiterate to parents, that cooking is not a set of rules. Instead, the focus is on sensory stimulation (i.e., smells, textures, and tastes); problem solving; and MUNCHING! The progression through recipe steps, investigating and finding ways to include small hands in food preparation, is just as essential as the final result.

Viewing Cooking in School as Participant Play: A Way to Introduce Cooking Concepts, Methods, Enhance Language and Linguistic Skills

There are times when playing with food can be fun. Educators can try different cuisines and explore consistencies and colors to make food exciting. For example, play with different color safe jellies, offers children the chance to be motivated. It also offers a "learning by doing" environment, which has been shown to be a problem-solving arena that can benefit the student in future thinking practices (Anzai & Simon, 1979). Intrinsically motivated children often exhibit persistence and creativity in problem solving during play (Hachey, 2009). Play in the kitchen is about discovering the enjoyment of cooking, even if the teacher is not a fan of cooking. Experiencing new tastes can be gratifying for the teacher and children for the sake of genuine exploration. Engaging children in cooking themes coordinated with the season provides rewarding opportunities to discuss the function of time as it relates to food growth. Educators may want to place food-type toys in the classroom to create a training kitchen that students can play with and have fun in. Adding child-appropriate cooking tools, such as spoons, hand mixers, or cutting boards, along with a variety of reusable food containers, encourages play.

Educators should carefully observe children to determine methods for stimulating children's interest and desire to work in a kitchen. Children may select mobile applications that involve food or food games, or they may use the play kitchen more often. When young children engage in food-type play, there are many instances when direction following occurs, first by the educator and often among children during "play kitchen" interactions. This type of play can only be enhanced by quality cooking applications with interactive lessons that provide step-by-step directions through verbal and visual actions. A child's development and learning through play can positively evolve during a learning experience in which he or she is given the opportunity to practice new skills (National Association for the Education of Young Children, 2009). Educators who offer a quality, low-stress environment help children learn from mistakes (e.g., spilling something or tearing food into nonuniform pieces). Learning from mistakes, along with give-and-take verbal communication, is vital for young minds to learn (Lee, 2015b), especially when making the novel combination of technology and food preparation. Seeing and hearing directions through a device like a tablet computer can only augment play and learning. For example, when children use digital media for both cooking and play-cooking activities, they have more of a potential to develop vocabulary; knowledge (e.g., reading, spelling); writing (Lee, 2015b) and object recognition skills when compared with traditional education methods. Mobile devices offer a range of environments (Schuler, 2009) and because the cook needs to be moving as he or she prepares food, a portable device is optimal.

When inquisitiveness is roused, natural engagement in and enthusiasm for kitchen skills is a hands-on learning tool. Cooking should not involve an educator's drawn-out explanations and little child input. Educators should use simple activities that everyone can work on together. The fundamental aim is for children to cultivate knowledge and learn cooking techniques, along with the freedom and self-confidence to expand on the lessons they learn. Active participation encourages children to see cooking as an important daily activity in which they actively participate and have a hand in their own health choices.

Using Apps and Media to Capitalize on Children's Natural Curiosity of Steps and Processes Involved in Cooking

Each time a child follows the steps of a recipe—measures a quarter cup of flour, compares raw and cooked carrots, tears bread into small pieces for croutons—he or she is using literacy, mathematics, science, and motor skills (Lee et al., 2007). Children will use these skills later to make informed decisions about how they will choose and cook food. For educators in the kitchen, this means focusing on children's statements and ideas. A child may say, "I have *six* zucchinis," "That's about *three quarters* of the bread dough," or "He has *less* tomatoes." When teachers hear these statements, they can prompt children to mathematically evaluate their observations (Hachey, 2009).

Before encouraging food preparation though the use of mobile devices, an educator must understand the limitations and constraints of these devices, as well as the differences in each device and situation. However, because the goal is to give children lifelong skills and knowledge, it is important for educators to have some ideas that preschoolers can use in their own kitchens, if nothing more than ideas that children can take home with the recipes. The educator, as the directive center, can be both a guide and a thinking coach. Educators should keep in mind that students may benefit from a co-involvement app environment.

> *Because a teacher was always with them in the center, the children were encouraged to learn not only how to have divergent thinking to solve a problem with an iPad, but also, how to work together to solve it by thinking, discussing, and communicating with the teacher and their peers. (Lee, 2015a, p. 948).*

Over the years, there has been a considerable rise in the number of cooking shows on television (de Solier, 2005, as cited in Clifford et al., 2009) and in computer applications and interactive instructional videos and devices. Currently more efforts have been made to target children's cooking skills (or lack thereof) through platforms that teach basic skills (like Nomster Chef), subscription boxes (Raddish; Kidsitr; FoodStirs. Eat2Explore); or television programs (Gordon Ramsy's Master Chef Junior or Food Network's Star Kids) as a forum for providing entertainment through food and at home education. With assistance, educators and parents can connect mobile media to cognitive theory and determine how mobile media influence knowledge, attitudes, and behaviors related to cooking healthy food these offerings. While these are safe steps into culinary arts for young children, they often lack regular assessments and rigor similar to other academic subjects currently available in school. To make the culinary arts for young children more effective a turn toward a curriculum that is more educational, culinary, nutritional, and psychosocial based could have the potential to provide permanent ramifications on skills, lifestyle and independence.

Research studies have shown that nutrition-education television programs increase nutrition awareness and that results often augment nutritional knowledge (Chew et al., 1995, 1998; Puska & McAlister, 1981; Shannon et al., 1979; Smith & George, 2001, as cited in Clifford et al., 2009). However, most research in this area has investigated nutrition-oriented programs with adults (Chew et al. 1995, 1998; Puska & McAlister, 1981; Shannon et al., 1979; Smith & George, 2001, as cited in Clifford et al., 2009), while few research studies have focused on young children, who are too young to prepare meals and make informed decisions about selecting ingredients. Today, it is mostly adults who turn to Youtube or cooking shows for information about food preparation and nutrition, and children ought to do the same through portable devices, right in the kitchen (Clifford et al., 2009). Educators can capitalize on this by using mobile devices to access apps or show developmentally appropriate television programs focused on cooking. Quality children's television programs can be easily accessed on a mobile device and displayed in the cooking area, for a seamless transition from device to stove. Suggested children's television programs are listed in Table 3.

Apps listed in Table 4 will be great supplements to the cooking process, before and after cooking. The apps can help students build culinary cognition for each simple cooking method and promote problem solving and culinary vocabulary building. Each app is specially designed with easy-to-follow instructions.

Table 3. Food-focused preschool and elementary school television and web based programs

Program Season: Episode Title	Description	Educational Objective
Curious George Season 4: Episode 2 "Juicy George"	After one sip, George and the man with the yellow hat are so hooked on Juicy Jay's special blended juice that they buy their own juicer to recreate the drink at home. But operating a juicer is not as easy as it seems, and it's certainly not a mess-free process! With some fun-filled experimentation, George discovers the secret ingredient and creates a juice that impresses Juicy Jay! (Saric et al., 2009).	To learn about healthful snacking, specifically fruit and vegetable drinks; to introduce experimentation as a fun way to create new flavors
Daniel Tiger's Neighborhood Season 1: Episode 16 "Be a Vegetable Taster! / Daniel Tries a New Food"	Be a Vegetable Taster! Teacher Harriet introduces Daniel and his friends to the school's vegetable garden and asks them to help her pick some healthful vegetables for their snack. They explore the garden, picking and trying different vegetables…but not everyone is eager to taste the new food. (Friedman & Kapridov, 2012). Daniel Tries a New Food Miss Elaina is having dinner with the Tiger family, and she encourages Daniel to try some new food: vegetable spaghetti and banana swirl. Daniel is convinced he won't like vegetable spaghetti, even though he has never tried it. Miss Elaina encourages him to be adventurous and try the new food. She does the same, and even though she doesn't like all of it, she's proud of herself for trying. (Friedman & Kapridov, 2012).	To encourage children to try new food; the *strategy song* of this episode is "Try a new food. It might taste good!"
Super Why! Season 2: Episode 5 "Who Loved Spaghetti"	It's dinnertime at Whyatt's house, but things are not going well! He's perplexed when Mom and Dad say that what's on his plate isn't healthful. To find out why, the reading superheroes fly to King Eddie Who Loved Spaghetti and encounter a wacky king who wants to eat only one thing—spaghetti! As the super readers persuade King Eddie to taste new food, Whyatt learns that there's more to dinner than he had thought! (Davidson, 2011).	To learn about healthful eating; kids will discover the alphabet, practice the magic of spelling, learn what the word "ripe" means, and use the power of reading to change the story.
The Bow Girls https://www.youtube.com/watch?v=zXE3lGHCX1U	Briena and Britney, who make every kid-friendly, sparkly, and brightly colored treat you can imagine: cupcakes, milkshakes, cookies galore. But they also tackle fun savory recipes like green eggs and ham and homemade pizza, and make healthy lunches for school. The videos offer slow, step-by-step demonstrations that make the recipes easy for you and your kids to replicate. (Epicurious, 2017).	To encourage the kids by the viewing of other kids cooking on their own.
Charli's Crafty Kitchen https://www.youtube.com/watch?v=G_IsaJIDbp0	Charli, 10-year-old Australian host of this show, bakes with her little sister, Ashlee. These videos feature real, regular kids making approachable recipes that are easy to replicate at home. Since a lot of these kid-friendly channels focus on sweets, a refreshing part of Charli's Crafty Kitchen is that it includes healthy recipes (Epicurious, 2017).	To encourage the kids by the viewing of other kids cooking on their own.

Table 4. Food-focused preschool and elementary school apps

App	Description	Website
Issa's Edible Adventures (ages 4+)	Is your little one an aspiring chef? Issa's Edible Adventures explore the excitement of a working restaurant kitchen while learning about foods from around the world. (issasedibleadventures.com, 2017).	https://issasedibleadventures.com/
Nomster Chef (Ages 2+)	Illustrated recipes helps kids and parents break down cooking steps visually based on cognitive and developmental needs of children. They site also offers a web series to help parents provide proper safety measure when cooking with children in the kitchen.	H ttps://www.nomsterchef.com/
Jobi's Bistro (ages 6–8)	The games found on the "first floor" in this app contain no sweet treats. Instead, two chefs help users prepare meals. The cooking games operate through touch on the stove, pot, sink, and other places in the kitchen. Ingredients include healthful fruit and vegetables. Users can cook food in saucepans, chop vegetables, and clean up (including washing plates and putting cutlery back on wall brackets) (Yellephant, 2015).	http://www.yellephant.net
KidECook (ages 6–11)	This app allows users to measure and cook ingredients as a real chef does. There is a wide range of ingredients, including fruit, eggs, milk, butter, flour, vanilla, rice, and semolina. Kitchen equipment is available to use during food preparation (Chocolapps, 2013).	http://en.chocolapps.com

By devoting awareness through child-focused cooking apps and daily cooking conversations, educators can help children develop reasoning skills for cooking. Simple activities like adjusting preferred recipes, taste testing new food, experimenting with simple kitchen tools, and understanding the benefits of healthful food, allow educators to present "content and process knowledge, which is the heart of later cross-content academic thinking" (Krenn & Hachey, 2015).

Cooking Topics by Subject Area

Using a variety of games and basic learning tools, students can learn about food properties and cooking skills. Because most classrooms have access to mobile devices, and students are enthusiastic about using these devices, a mobile device is an excellent platform for knowledge acquisition, as well as for making connections to most general curriculum goals in science, social studies, problem solving, and mathematics. Moreover, the studies done by Ginsburg and his colleagues (2013, 2014) as well as Black, et al (2016) Incorporating Touch-Based Tablets into Classroom Activities: Fostering Children's Computational Thinking through iPad Integrated Instruction AS WELL AS Paek's (2012) *The impact of multimodal virtual manipulatives on young children's mathematics learning* clearly show the ways in which touch-screen technology and quality applications could support mathematics learning if money were not an issue, as "studies. . .have indicated that digital media use supported young children's mathematical and cognitive skills" (Clements & Sarama, 2003; Klibanoff et al., 2006). The same would be true for funding a preschool or elementary school culinary program.

One of the ways to promote informal learning in a classroom is by combining healthful food ideals with imaginative means of entertainment. Research and experience are required to help families understand the benefits of a healthful diet and to understand that cooking skills have both a positive physical benefit and a social–cognitive benefit throughout their lives (Bandura, 1986). This section seeks to augment content knowledge and thinking skills with app recommendations to provide educators with informal, yet valuable, teaching methods.

In the kitchen, children can learn about measuring, mixing; stirring; baking; broiling; safety; and, equally important, enjoying food. The following academic components can be included in each lesson but are not limited to the following areas. A list of topics for each area is provided, followed by ideas and questions educators can ask during lessons.

Science: Possible Food Growing and Preparation Topics Update below to show how mobile phones and iPads can be used with some or all of the below.

- Changes in temperature as they affect food growth
- Climate change as it affects yearly crop production
- Decomposition of food scraps as helpful compost agents
- Digestive system's breakdown of food within the human system
- Electrical currents in cooking devices
- Fertilization as helpful for growth
- Food structural changes when exposed to heat or cold
- Ingredients' chemical properties
- Photosynthesis of crops
- Plant growth as it varies from plant to plant
- Preservation of food
- Solar energy as used in farming
- Water cycles in nature

Sample Science Incorporation

Science-related suggestions focus on behavioral aspects of nutrition, development, and evaluation of curricula that link science education to food preparation as well as food knowledge. The suggestions examine food and food systems, development, evaluation, and ecological issues. The goal of these suggestions is to reduce children's risk of obesity by focusing on increasing the understanding of how the interactions among biology, personal behavior, and the food system affect eating patterns and weight and by enhancing skills for making competent choices (Contento et al., 2007; Islas et al., 2000; Koch et al., 2008; Liquori et al., 1998).

In the kitchen, there are plenty of opportunities to incorporate science lessons. By combining ingredients and adding heat, food undergoes certain changes. To continue with the theme of water, how does water change into ice? If water is boiled to cook rice, why does it sometimes turn into steam? What structural and chemical changes does the water undergo? This will teach children how food often changes its structure when heat or cold is applied. Table 5 lists two helpful apps that can get children involved with science education during a cooking lesson.

Children's television programs can also be helpful in this lesson with young chefs. Table 6 lists two *Sid the Science Kid* episodes that can help educators explain the property changes that food undergoes during ripening and cooling.

Table 5. Food-science-focused preschool and elementary school apps

App	Description	Website
Grow Garden (ages 4+)	Kids visit three: the garden plot, the house, and the compost bin. Plant seeds, shine the sun's rays, and drag rain clouds around to nurture crop growth in the garden. When the veggies are ready, bring them to the house, where hungry animals are waiting to eat. Gather the food scraps, feed them to microorganisms in the compost bin, and pick up the nutrient-rich compost. Bring the compost to the garden and use it to help your crops grow faster. Kids earn more seed types and some special food treats as their gardens grow. (Common Sense Media, 2019).	https://www.groplay.com/
Eat & Move-O-Matic (ages 9–11)	This app helps users recognize the relationship between food and exercise. It shows calories consumed and the time required to burn off the calories with activities such as finishing homework and dancing. It offers guidelines for healthful substitutions for high-calorie food, like burgers and fries (NMSU Board of Regents, 2015).	http://mediaproductions.nmsu.edu/eatmove/

Table 6. Food-science-focused preschool and elementary school television programs

Program Season: Episode Title	Description	Educational Objective
Sid the Science Kid Season 1: Episode 6: "My Mushy Banana"	Sid wants to know why his beloved yellow banana has turned brown and mushy. After exploring how fruit and vegetables slowly change, Sid learns that it's natural for his banana to ripen and turn brown and mushy. Over time, fruit and vegetables decay! (Zewig et al., 2008).	Plant decay
Sid the Science Kid Season 1: Episode 8: "My Ice Pops"	Sid wakes up and discovers that his ice pops have melted! This sticky situation leads him to ask, Why do things have to melt? After investigating at school, Sid learns that if frozen liquids don't stay in a really cold place, they slowly melt. He also discovers that liquids can freeze into solid ice and then melt right back into liquids. (Shaw et al., 2008).	Change caused by cold

Mathematics: Possible Food Measurement and Numerical Processing Topics

- Addition
- Deduction
- Division
- Law of averages
- Measurement skills
- Metric and standard conversions
- Multiplication
- Proportion and fraction lessons
- Subtraction

Sample Mathematics Incorporation

Food preparation is an opportunity to apply mathematics concepts to a real-life situation. For example, young children can learn about harvesting apples in fall and discover that only 15 apples fit in each basket. Children may realize that there are many trees in an apple orchard, and that each tree has many apples. To harvest all the apples, children need to figure out how many apples are on each tree and how many apples there are in total. Students can then figure out with educators how many baskets are needed. But do they need to count every single apple? Perhaps the law of averages can help the children estimate the number of apples. There are countless ways to study mathematics while cooking in the kitchen. Children can learn mathematics lessons by doubling recipes to feed a large group or converting ounces into cups. Say, for instance, that students are making squash soup. One soup recipe makes six bowls of soup. But maybe the students need to feed the entire class. Educators can help students realize that there may not be enough soup if the recipe is followed. Perhaps quadrupling the soup recipe will solve the problem. Mathematics can be used to multiply ingredients. Table 7 lists two useful apps that can encourage children to use mathematics during a cooking lesson. While these apps have great kitchen themes and developmentally appropriate mathematics topics, educators should refrain from using them as the primary method of teaching—rather educators should use the apps only as supplements. Because it has been demonstrated that cognitive load is reduced when redundant information is presented pictorially, repeating a lesson using a mobile app is more beneficial than continually presenting too much of a concept (Bobis, Sweller, & Cooper, 1993). An app can be an innovative visual-learning tool.

Table 7. Mathematics-focused preschool and elementary school apps

App	Description	Website
Tiggly Chef (preschool+)	With Tiggly Chef, young children can practice counting and addition in an entertaining and motivating setting. Users can experiment with putting groups of food together as they cook up imaginative culinary dishes. The app connects intuitive math to formalized math, reinforces simple addition concepts, introduces formal math symbols, allows users to create their own recipes, and encourages flexible thinking and playing with numbers. Users can put the ingredients into the bowl in groups, and an addition equation appears to show their progress. While the app does link food and mathematics, it is unable to provide palatable dishes. Users can put whatever they like into the bowl, and most of the combinations would not fare well at a real table (Tiggly, 2015).	https://www.tiggly.com/tiggly-chef
Grandma's Kitchen (preschool)	Grandma's Kitchen is an educational, quiz-type game covering prekindergarten subjects, such as counting by tens and basic addition. Grandma asks questions as she cooks and offers fun play breaks. Each activity joins the kitchen and mathematics themes, along with simple letter sounds and alphabetic organization (Fairlady Media, 2015).	http://fairladymedia.com

Social Studies: Possible Social–Emotional and Geographic Topics

- Basic living skills, including trying new food for nutrition's sake
- Cultural aspects of food
- History of local geographic areas
- Language skills—communication with cooking team members
- Sharing tasks, steps, and supplies
- Turn taking in the kitchen

Sample Social Studies Incorporation

Culinary lessons in the classroom are opportunities to apply lessons from a social studies curriculum. For example, perhaps a classroom is ready to harvest fresh tomatoes from a garden. It is a great time to learn about the history of the tomato. In this lesson, students discover that the tomato is native to western South and Central America and that the explorer Cortez discovered tomatoes in Montezuma's gardens and brought seeds back to Europe, where they were first planted for decoration and not consumed (Trowbridge Filippone, 2010). As children work in the kitchen, cooking shows and apps on mobile devices can explain how food is prepared, present helpful skills, and reiterate content. Two of these helpful apps can be found in Table 8. Say, for instance, that students want to make a sauce to complement a pasta dish. The educator can lead the students through several steps to create such a delicious topping. Every student may be eager to help, so how is the work divided? A teacher can present lessons related to social skills, history, and geography as children prepare the sauce, to help reiterate past lessons.

Table 8. Social-studies-focused preschool and elementary school apps

App/Site	Description	Website
Food Span	This free, downloadable curriculum provides high school students with a deep understanding of critical food system issues, empowers them to make healthy and responsible food choices, and encourages them to become advocates for food system change (foodspanlearning.org, 2011).	http://www.foodspanlearning.org
Veggie Circus Farm (preschool)	This app helps young children recognize vegetables. It provides animated vegetable performances that teach kids how to say the names of vegetables, and it visually shares basic nutritional benefits that even children who don't yet know how to read can understand (Brainster, 2014; Moore, 2015).	http://www.toddlerwords.com

Problem Solving: Possible Thinking and Processing Topics

- Critical thinking
- Complexity processing
- Goal directives
- Recipe adaptation for the home kitchen when listed ingredients are unavailable
- Cooking-tool substitution for when a kitchen utensil is unavailable
- Overcoming negative weather
- Growing and harvesting fruit and vegetables that have pest and fungal problems

Sample Problem-Solving Incorporation

Problem-solving suggestions focus on aspects of thinking and processing. For example, sometimes measuring tools are missing. In these situations, cooks often problem solve by using what is available. Perhaps a teacher incorporates making homemade pizza into a week's lesson. A half cup of fresh basil is listed as an ingredient. What if the teacher cannot find the half-cup measure? However, the 1-cup measure is available. To solve this problem, a teacher can instruct students to use available materials to complete the recipe. Students can use half of the 1-cup measure to figure out how much basil to use. Table 9 lists several mobile-device apps that can help educators incorporate problem solving into their cooking lessons.

Each day more apps become available. By the time this chapter is published, educators will have even more apps to choose from. Before choosing a cooking app for the classroom, educators should test it for quality, ingredient accuracy, curriculum goals, and fantasy elements. Is the academic content correct? Are there advertisements? Is the food used in a way that is true to its real-life properties if the educator is seeking a cooking concept? For example, if a *tomato* in an app is part of fresh spaghetti sauce, is lycopene (the antioxidant in tomatoes) discussed? Or is

Table 9. Problem-solving-focused preschool and elementary school apps

App/Book	Description	Website
Yummiloo Rainbow Power (ages 5 and under) App	In this educational app, playing with your food takes on a whole new meaning as kids learn the basics of nutrition and food identification through a fun and engaging game. Users are instructed to gather fruit and vegetables by dragging them into colored buckets to fuel the Rainbow Machine. Users are given a compost bin for spoiled food, and they make room for more fresh food to grow once the rotten food is removed. When all five carts are filled with food, the Rainbow Machine runs on full power, unleashing carnival rides and a Yum Yum party. The celebratory end of the game reinforces the importance of eating a rainbow of food, teaching kids about how eating a mix of brightly colored food helps their bodies grow (Night & Day Studios, 2016).	http://www.nightanddaystudios.com
Awesome Eats (ages 4–6) App	In this game, users sort, stack, pack, and plate fresh garden food. There are also lessons about recycling. At each level, users stack and sort fruit, vegetables, and whole grains to earn stars and score points. Thieving birds try to steal food, and if users shoo away the birds they win extra points. Healthful-eating tips, problem solving, obstacles, and surprises keep users interested (Whole Kids Foundation, 2013).	Presented by Whole Kids Foundation https://www.wholekidsfoundation.org Developed by Fun Machine http://www.fun-machine.com
The Prisoner of Carrot Castle (ages 6–8) Book	In The Prisoner of Carrot Castle, Aiden's imagination transports him to a faraway place, where he finds himself a prisoner in a cell made of carrots. What will he do? Must he face the angry king? Will his hate for vegetables get him into trouble? Users help Aiden as he tries to escape from Carrot Castle. Through creative storytelling, kids learn that eating vegetables is good—it can actually help them (Purple Carrot Books, 2013).	http://purplecarrotbooks.com
Food Truth (educators)	Through three levels of cooking, teachers can have food facts and recipes on hand as they teach. Most recipes are easy to prepare. For each of the 100 foods described in this educational app there are descriptions of nutritional values and at least three recipes. Users can select a level on the basis of their experience in the kitchen. This app can help teachers as they teach children about cooking with the best ingredients. Most recipes are sugar free and gluten free and comply with GAPS, SCD, and diabetic diets, which can be helpful for meeting every student's needs (Food Truth Online, 2015).	http://www.foodtruthonline.com/app

the tomato being used as a projectile, not as an ingredient? Entertainment may grab children's attention, but during a cooking lesson, mobile devices should help teach young child about health and nutrition.

CONCLUSION

It is anticipated that this chapter continues to make suggestions for mobile research having the potential to contribute to multimedia culinary education and cognitive and developmental psychology communities. The following questions can help school administrators decide whether to use mobile media and cooking curricula in their classrooms:

- Are educators incorporating young learners' food likes and dislikes into using mobile-media devices and including these interests in their classrooms?
- Do educators have sufficient food knowledge and materials about how and what to use to incorporate cooking into an academic curriculum?
- Are educators aware of their cooking limitations (e.g., negative attitudes or low skill levels)? And, if so, how can we support educators so that they pass along the message that food preparation can help students make healthful choices, be self-sufficient, and develop cognitive skills?
- If using a combination of mobile media and cooking practices will help elementary school and preschool students make learning more pleasant and exciting, is it not crucial for educators to incorporate cooking into play?
- What are the downsides of using an app/social media in assisting cooking? (Something that can be determined in a classroom by classroom, district by district basis).
- What are some guidelines for educator/caregiver on how we can use these apps in tandem with cooking? (Subject to be focused in future directions related to this topic within culinary cognition).

ACKNOWLEDGMENT

The author would like to acknowledge Emma Krenn (age 6) for her inquisitive and passionate approach in the kitchen. The author would also like to acknowledge Teachers College graduate student Xinyuan (Sonya) Zhang for her help in this project.

REFERENCES

Ainsworth, S. (1999). The functions of multiple representations. *Computers & Education*, *33*(2), 131–152. doi:10.1016/S0360-1315(99)00029-9

Amanda, C. (2014, June 4). Dr. Panda's Restaurant 2 gets kids cooking [Web log post]. Retrieved from http://www.geekswithjuniors.com/blog/2014/6/4/dr-pandas-restaurant-2-gets-kids-cooking.html

Anderson, J. R. (2005). *Cognitive Psychology and Its Implications* (6th ed.). Worth Publishing.

Anzai, Y., & Simon, H. A. (1979). The theory of learning by doing. *Psychological Review*, *86*(2), 124–140. doi:10.1037/0033-295X.86.2.124 PMID:493441

AppMatrix, Inc. (2014). Nicolas' Garden (Version 1.1.0) [Mobile application software]. Retrieved from http:// http://www.nicolasgarden.com

Attewell, J., & Savill-Smith, C. (2004). Mobile learning and social inclusion: focusing on learners and learning. In J. Attewell & C. Savill-Smith (Eds.), *Learning With Mobile Devices: Research and Development* (pp. 3–11). London: Learning and Skills Development Agency.

Bandura, A. (1986). *Social foundations of thought and action: A social cognitive theory*. Englewood Cliffs, NJ: Prentice Hall.

Bobis, J., Sweller, J., & Cooper, J. (1993). Cognitive load effects in primary-school geometry task. *Learning and Instruction*, *3*(1), 1–21. doi:10.1016/S0959-4752(09)80002-9

Bow Girls. (2017, May 17). *How to Make a Mystic Mermaid Shake* [Video File]. Retrieved from https://www.youtube.com/watch?v=zXE3lGHCX1U

Brainster Apps. (2014). Veggie Circus Farm (Version 1.1) [Mobile application software]. Retrieved from http://www.toddlerwords.com

Bransford, J. D., Franks, J. J., Vye, N. J., & Sherwood, R. D. (1989). New approaches to instruction: Because wisdom can't be told. In S. Vosniadou & A. Ortony (Eds.), *Similarity and analogical reasoning* (pp. 470–497). New York: Cambridge University Press. doi:10.1017/CBO9780511529863.022

Brown, S., Birch, D., Teufel, J., & Kancherla, V. (2006). Overweight in children: The perspectives of 9-13 year olds. *American Journal of Health Education*, *37*(5), 296–305. doi:10.1080/19325037.2006.10598918

Casey, P. H., Szeto, K. L., Robbins, J. M., Stuff, J. E., Connell, C., Gossett, J. M., & Simpson, P. M. (2005). Child health-related quality of life and household food security. *Archives of Pediatrics & Adolescent Medicine*, *159*(1), 51–56. doi:10.1001/archpedi.159.1.51 PMID:15630058

Charli's Crafty Kitchen. (2014, September 25). *Personalized Healthy Fruit Kebobs-Kababs Skewers – Educational how to DIY* [Video File]. Retrieved from https://www.youtube.com/watch?v=G_IsaJIDbp0

Chase, W. G., & Simon, H. A. (1973). The mind's eye in chess. In W. G. Chase (Ed.), *Visual information processing* (pp. 215–281). New York, NY: Academic Press. doi:10.1016/B978-0-12-170150-5.50011-1

Chen, R. C. Y., Lee, M. S., Chang, Y. H., & Wahlqvist, M. L. (2012). Cooking frequency may enhance survival in Taiwanese elderly. *Public Health Nutrition*, *15*(7), 1142–1149. doi:10.1017/S136898001200136X PMID:22578892

Chew, F., Palmer, S., & Kim, S. (1995). Sources of information and knowledge about health and nutrition: Can viewing one television programme make a difference? *Public Understanding of Science (Bristol, England)*, *4*(1), 17–29. doi:10.1088/0963-6625/4/1/002

Chew, F., Palmer, S., & Soohong, K. (1998). Testing the influence of the health belief model and a television program on nutrition behavior. *Health Communication*, *10*(3), 227–245. doi:10.120715327027hc1003_3 PMID:16370984

Chocolapps. (2013). *KidECook (Version 1)* [Mobile application software]. Retrieved from http://en.chocolapps.com

Clements, D. H., & Sarama, J. (2003). Strip mining for gold: Research and policy in educational technology—A response to "fool's gold." *AACE Journal*, *11*(1), 7–69.

Clifford, D., Anderson, J., Auld, G., & Champ, J. (2009). Good grubbin': Impact of a TV cooking show for students living off campus. *Journal of Nutrition Education and Behavior*, *41*(3), 194–200. doi:10.1016/j.jneb.2008.01.006 PMID:19411053

CondeNet. (2015). Epicurious (Version 4.5.2) [Mobile application software]. Retrieved from http://www.epicurious.com/services/mobile

Contento, I. R., Koch, P. A., Lee, H. W., Sauberli, W., & Calabrese-Barton, A. (2007). Enhancing personal agency and competence in eating and moving: Formative evaluation of a middle school curriculum—Choice, Control, and Change. *Journal of Nutrition Education and Behavior*, *39*(5), S179–S186. doi:10.1016/j.jneb.2007.02.006 PMID:17826699

Cook, J. T., & Frank, D. A. (2008). Food security, poverty, and human development in the United States. *Annals of the New York Academy of Sciences*, *1136*(1), 193–209. doi:10.1196/annals.1425.001 PMID:17954670

Craik, K. (1943). *The nature of explanation*. Cambridge, UK: Cambridge University Press.

Davidson, B. (Director). (2011). King Eddie who loved spaghetti [Television series episode]. In S. Freeman Alpert (Executive producer), Super Why! Toronto, Canada: Out of the Blue Productions & DHX Media.

de Solier, I. (2005). TV dinners: Culinary television, education and distinction. *Continuum (Perth)*, *19*(4), 465–481. doi:10.1080/10304310500322727

Dr. Panda Limited. (2014). Dr. Panda's Restaurant 2 (Version 1.2) [Mobile application software]. Retrieved from http://drpandagames.com/app/dr-panda-restaurant2/

Eat2Explore. (2019). Retrieved from https://eat2explore.com/

Erlich, R., Yngve, A., & Wahlqvist, M. L. (2012). Cooking as a healthy behaviour. *Public Health Nutrition*, *15*(7), 1139–1140. doi:10.1017/S1368980012002662 PMID:23062443

Fairlady Media. (2015). Grandma's Kitchen (Version 2.1) [Mobile application software]. Retrieved from http://fairladymedia.com

Fildes, A., van Jaarsveld, C. H., Wardle, J., & Cooke, L. (2014). Parent-administered exposure to increase children's vegetable acceptance: A randomized controlled trial. *Journal of the Academy of Nutrition and Dietetics*, *114*(6), 881–888. doi:10.1016/j.jand.2013.07.040 PMID:24091061

Food for Life Partnership. (2015). *Cooking in school: Your guide to planning and teaching cooking in primary and early years settings*. Retrieved from http://www.foodforlife.org.uk/~/media/files/resources/fof%20resources/fflp-cooking-in-school-checklist.pdf

Food Network's Star Kids. (2016). [Television Program]. Retrieved from https://www.foodnetwork.com/shows/food-network-star-kids

Food Span. (2011). Retrieved from http://www.foodspanlearning.org/

Food Truth Online. (2015). Food Truth (Version 2.0) [Mobile application software]. Retrieved from http://www.foodtruthonline.com/app

FoodStirs. (2019). Retrieved from https://foodstirs.com/

Forma Animada. (2013). Super Abby vs. The Enchanted Sweet Stall (Version 1.0) [Mobile application software]. Retrieved from http://www.superabbyworld.com

Friedman, B. (Writer), & Kapridov, V. (Director). (2012). Be a vegetable taster! / Daniel tries a new food [Television series episode]. In W. Harris (Supervising producer), & A. Santomero (Executive producer), Daniel Tiger's neighborhood. Pittsburgh, PA: The Fred Rogers Company.

Gellersen, H. W., Schmidt, A., & Beigl, M. (2002). Multi-sensor context-awareness in mobile devices and smart artifacts. *Mobile Networks and Applications*, *7*(5), 341–351. doi:10.1023/A:1016587515822

Gibson, J. J., & Gibson, E. J. (1955). Perceptual learning: Differentiation or enrichment. *Psychological Review*, *62*(1), 32–51. doi:10.1037/h0048826 PMID:14357525

Ginsburg, H. P. (2014). My entirely plausible fantasy: Early mathematics education in the age of the touchscreen computer. *Journal of Mathematics Education at Teachers College*, *5*(1), 9–17.

Ginsburg, H. P., Jamalian, A., & Creighan, S. (2013). Cognitive guidelines for the design and evaluation of early mathematics software: The example of MathemAntics. In L. D. English & J. T. Mulligan (Eds.), *Reconceptualizing early mathematics learning* (pp. 83–120). Springer Netherlands. doi:10.1007/978-94-007-6440-8_6

Gordon Ramsy's Master Chef Junior. (2019). Retrieved from https://www.fox.com/masterchef-junior/article/about-the-show-597bbb42ef528f0026dc0305/

Grow Garden. (2019). Grow Play (Version 1) [Mobile application software]. Retrieved from https://www.groplay.com/apps/grow-garden/

Hachey, A. C. (2009). I hate math: What we want young children NOT to learn. *Texas Child Care Quarterly*, 2-7.

Hegarty, M. (1992). Mental animation: Inferring motion from static diagrams of mechanical systems. *Journal of Experimental Psychology. Learning, Memory, and Cognition*, *18*(5), 1084–1102. doi:10.1037/0278-7393.18.5.1084 PMID:1402712

Huminah Huminah Interactive. (2013). Perfect Picnic (Version 1.2.3) [Mobile application software]. Retrieved from http://fightbac.org/kids

Islas, A., Contento, I. R., Koch, P. A., Hagiwara, S., & Barton, A. C. (2000). Food choice criteria in elementary school age children incorporating ecological concerns. *Appetite*, *34*, 125.

Issa's Edible Adventures. (2017). ALK Ventures (Version 1) [Mobile application software]. Retrieved from https://issasedibleadventures.com/

Johnson-Laird, P. N. (1994). Mental models and probabilistic thinking. *Cognition*, *50*(1), 189–209. doi:10.1016/0010-0277(94)90028-0 PMID:8039361

Johnson-Laird, P. N. (2005). Mental models and thoughts. In K. Holyoak & R. Morrison (Eds.), *The Cambridge handbook of thinking and reasoning* (pp. 185–208). New York, NY: Cambridge University Press.

Johnson-Laird, P. N., & Byrne, R. M. (1991). *Deduction*. Lawrence Erlbaum Associates, Inc.

Johnson-Laird, P. N., & Wason, P. C. (1970). A theoretical analysis of insight into a reasoning task. *Cognitive Psychology, 1*(2), 134–148.Jump App (2015). Cute Food (Version 1.5) [Mobile application software]. Retrieved from http://jumpapp.eu/?post_type=portfolio&p=2532

Kidstir. (2018). Retrieved from https://kidstir.com/

Kihm, H., & Singleton, K. (2012). Creative cooking with preschoolers: A service learning experience for FCS students. *Journal of Family and Consumer Sciences Education, 30*, 1.

Kittler, P. G., Sucher, K., & Nelms, M. (2012). *Food and culture*. Belmont, CA: Cengage Learning.

Klibanoff, R. S., Levine, S. C., Huttenlocher, J., Vasilyeva, M., & Hedges, L. V. (2006). Preschool children's mathematical knowledge: The effect of teacher "math talk." *Developmental Psychology*, *42*(1), 59–69. doi:10.1037/0012-1649.42.1.59 PMID:16420118

Knight, A., O'Connell, R., & Brannen, J. (2014). The temporality of food practices: Intergenerational relations, childhood memories and mothers' food practices in working families with young children. *Families, Relationships and Societies*, *3*(2), 303–318. doi:10.1332/204674313X669720

Koch, P., Calabrese Barton, A., Contento, I., & Crabtree, M. (2008). Farm to table and beyond: Helping students make sense of the global food system. *Science Scope*, *31*(9), 36–39.

Krenn, J. & Hachey, A. C. (2015). Cooking with attitude: How preschool children and early primary educators can promote positive lifelong skills in the kitchen. *Texas Childcare, 38*(4).

Kruger, J., Blanck, H. M., & Gillespie, C. (2006). Dietary and physical activity behaviors among adults successful at weight loss maintenance. *The International Journal of Behavioral Nutrition and Physical Activity*, *3*(3), 1–10. Retrieved from http://www.ijbnpa.org/content/pdf/1479-5868-3-17.pdf PMID:16854220

Larson, N., Perry, C., Story, M., & Neumark-Sztainer, D. (2006). Food preparation by young adults is associated with better diet quality. *Journal of the American Dietetic Association*, *106*(12), 2001–2007. doi:10.1016/j.jada.2006.09.008 PMID:17126631

Lee, H., Contento, I. R., Sauberli, W., Koch, P. A., & Calabrese Barton, A. (2007). Using science education to move middle schoolers toward more healthful food and activity choices: An outcome evaluation of Choice, Control, & Change (C3). *Journal of Nutrition Education and Behavior*, *39*(4S), S96. doi:10.1016/j.jneb.2007.04.270

Lee, L. (2015a). Digital media and young children's learning: A case study of using iPads in American preschools. *International Journal of Information and Education Technology (IJIET)*, *5*(12), 947–951. doi:10.7763/IJIET.2015.V5.643

Lee, L. (2015b). Technology play and learning. In D. P. Fromberg & D. Bergen (Eds.), *Play from birth to twelve: Contexts, perspectives, and meanings* (pp. 217–224). New York, NY: Routledge.

Liquori, T., Koch, P. D., Contento, I. R., & Castle, J. (1998). The Cookshop Program: Outcome evaluation of a nutrition education program linking lunchroom food experiences with classroom cooking experiences. *Journal of Nutrition Education*, *30*(5), 302–313. doi:10.1016/S0022-3182(98)70339-5

Mango Tree. (2015). WeCookit (Version 1.0.2) [Mobile application software]. Retrieved from http://www.we-cookit.com

Martinez, S. M., Rhee, K., Blanco, E., & Boutelle, K. (2014). Maternal attitudes and behaviors regarding feeding practices in elementary school–aged Latino children: A pilot qualitative study on the impact of the cultural role of mothers in the U.S.–Mexican border region of San Diego, California. *Journal of the Academy of Nutrition and Dietetics*, *114*(2), 230–237. doi:10.1016/j.jand.2013.09.028 PMID:24315129

Mayer, R. E., & Moreno, R. (2003). Nine ways to reduce cognitive load in multimedia learning. *Educational Psychologist*, *38*(1), 43–52. doi:10.1207/S15326985EP3801_6

Moore, M. (2015, February 2). 6 great apps to teach kids about nutrition [Web log post]. Retrieved from http://www.eatright.org/resource/food/nutrition/healthy-eating/6-great-apps-to-teach-kids-about-nutrition

MoreChinese. (2015). Dodo China Pro: the trip of experiencing Chinese culture, food and characters (Version 3.22) [Mobile application software]. Retrieved from https://itunes.apple.com/us/app/dodo-china-pro-trip-experiencing/id598851707?mt=8

Morgan, P. J., Collins, C. E., Plotnikoff, R. C., Callister, R., Burrows, T., Fletcher, R., & Lubans, D. R. (2014). The 'Healthy Dads, Healthy Kids' community randomized controlled trial: A community-based healthy lifestyle program for fathers and their children. *Preventive Medicine*, *61*, 90–99. doi:10.1016/j.ypmed.2013.12.019 PMID:24380796

Ng, M., Fleming, T., Robinson, M., Thomson, B., Graetz, N., Margono, C., ... Gupta, R. (2014). Global, regional, and national prevalence of overweight and obesity in children and adults during 1980–2013: A systematic analysis for the Global Burden of Disease Study 2013. *Lancet*, *384*(9945), 766–781. doi:10.1016/S0140-6736(14)60460-8 PMID:24880830

Night & Day Studios. (2015). Yummiloo Rainbow Power (Version 1.1.2) [Mobile application software]. Retrieved from http://www.nightanddaystudios.com

NMSU Board of Regents. (2015). Eat & Move-O-Matic (Version 2.2) [Mobile application software]. Retrieved from http://mediaproductions.nmsu.edu/eatmove/

Nomster Chef. (2019). Retrieved from http://nomsterchef.com

Norman, Å., Berlin, A., Sundblom, E., Elinder, L. S., & Nyberg, G. (2015). Stuck in a vicious circle of stress: Parental concerns and barriers to changing children's dietary and physical activity habits. *Appetite*, *87*, 137–142.

Octave Media International. (2012). Smash Your Food (Version 1.3) [Mobile application software]. Retrieved from http://smashyourfood.foodnme.com

Olvera, N., Knox, B., Scherer, R., Maldonado, G., Sharma, S., Alastuey, L., & Bush, J. (2008). A healthy lifestyle program for Latino daughters and mothers: The BOUNCE overview and process evaluation. *American Journal of Health Education*, *39*(5), 283–295. doi:10.1080/19325037.2008.10599052

Osherson, D. N. (1976). *Logical abilities in children: IV. Reasoning and concepts*. New York: Lawrence Erlbaum.

Peirce, C. S. (1903). Abduction and induction. In J. Buchler (Ed.), *Philosophical writings of Peirce* (pp. 150–156). New York, NY: Dover.

Peters, J., Parletta, N., Lynch, J., & Campbell, K. (2014). A comparison of parental views of their pre-school children's 'healthy' versus 'unhealthy' diets. A qualitative study. *Appetite*, *76*, 129–136. doi:10.1016/j.appet.2014.02.001 PMID:24524974

Playdek. (2012). Food Fight (Version 1.0.1) [Mobile application software]. Retrieved from http://www.playdekgames.com

Public Broadcasting Service. (n.d.). *Julia Child: PBS icon and cooking pioneer*. Retrieved from http://www.pbs.org/food/julia-child

Purple Carrot Books. (2013). The Prisoner of Carrot Castle (Version 1.1.2) [Mobile application software]. Retrieved from http://purplecarrotbooks.com

Puska, P., & McAlister, A. (1981). Television in health promotion: Evaluation of a national programme in Finland. *International Journal of Health Education*, *24*, 238–250. PMID:7344372

Radish Kids. (2019). Retrieved from https://www.raddishkids.com

Rosenkranz, R. R. (2008). *The impact of family mealtime on human nutrition*. Paper presented at the Food & Youth: Finding the Tipping Point for Wellness, Kansas Nutrition Conference, Junction City, KS.

Russell, C. G., Worsley, A., & Liem, D. G. (2015). Parents' food choice motives and their associations with children's food preferences. *Public Health Nutrition*, *18*(06), 1018–1027. doi:10.1017/S1368980014001128 PMID:24887485

Saric, L. (Writer), Bakalian, P. (Writer), Rey, H.A. (Writer) & Malkasian, C. (Director). (2009). Juicy George/The Big Picture [Television series episode]. In B. Howard, B. Grazer, D. Kirschner, & J. Shapiro (Executive producers), Curious George. Image Entertainment & Universal Animation Studios WGBH-TV.

Satter, E. (2008). *Secrets of feeding a healthy family: How to eat, how to raise good eaters, how to cook*. Kelcy Press.

Schaeken, W., Johnson-Laird, P. N., & d'Ydewalle, G. (1996). Mental models and temporal reasoning. *Cognition*, *60*(3), 205–234. doi:10.1016/0010-0277(96)00708-1 PMID:8870513

Schuler, C. (2009). *Pockets of potential: Using mobile technologies to promote children's learning*. New York: The Joan Ganz Cooney Center at the Sesame Workshop.

Schwartz, D., Martin, T., & Nasir, N. (2005). Design for knowledge evolution: Towards a prescriptive theory for integrating first- and second-hand knowledge. In P. Gardenfors, P. Johanssson, & N. J. Mahwah (Eds.), *Cognition, education, and communication technology* (pp. 21–54). Mahwah, NJ: Lawrence Erlbaum Associate, Inc.

Schwartz, D. L., & Black, J. B. (1996). Shuttling between depictive models and abstract rules: Induction and fallback. *Cognitive Science*, *20*(4), 457–497. doi:10.120715516709cog2004_1

Serf-Walls, S. (2013, December 7). Super Abby vs. the enchanted sweet stall [Web log post]. Retrieved from http://www.bestappsforkids.com/2013/super-abby-vs-the-enchanted-sweet-stall

Shannon, B., Thurman, G., & Schiff, W. (1979). Foodsense: A pilot TV show on nutrition issues. *Journal of Nutrition Education*, *11*(1), 15–18. doi:10.1016/S0022-3182(79)80057-6

Shaw, E. (Writer), & Martin, H. (Director). (2008). My ice pops [Television series episode]. In B. Henson, L. Henson, H. Stanford, & B. Zweig (Executive producers), Sid the science kid. Hollywood, CA: The Jim Henson Company.

Shoe the Goose. (2014). Cake Doodle (Version 1.15) [Mobile application software]. Retrieved from http://www.shoethegoose.com

Smith, J., George, V., & Easton, P. S. (2001). Home-grown television: A way to promote better nutrition. *Journal of Nutrition Education*, *33*(1), 59–60. doi:10.1016/S1499-4046(06)60013-4 PMID:12031209

Television Food Network. (2015). Food Network In the Kitchen (Version 4.8.1) [Mobile application software]. Retrieved from http://www.foodnetwork.com/site/mobile.html

Tiggly. (2015). Tiggy Chef (Version 1.1) [Mobile application software]. Retrieved from https://www.tiggly.com

Toca Boca. (2015). Toca Kitchen 2 (Version 1.2) [Mobile application software]. Retrieved from http://tocaboca.com/app/toca-kitchen-2

Trowbridge Filippone, P. (2010). *Tomato history: The history of tomatoes as food*. Retrieved from http://homecooking.about.com/od/foodhistory/a/tomatohistory.htm

Vosniadou, S., & Brewer, W. F. (1989). *The concept of the earth's shape: A study of conceptual change in childhood* (Technical Report No. 467). Retrieved from Illinois Digital Environment for Access to Learning and Scholarship website: https://www.ideals.illinois.edu/bitstream/handle/2142/17996/ctrstreadtechrepv01989i00467_opt.pdf

Wei, R. (2013). Mobile media: Coming of age with a big splash. *Mobile Media & Communication*, *1*(1), 50–56. doi:10.1177/2050157912459494

Whole Foods Market. (2015). Whole Foods Market (Version 5.2.0) [Mobile application software]. Retrieved from http://www.wholefoodsmarketcom/apps

Whole Kids Foundation. (2013). Awesome Eats (Version 1.3) [Mobile application software]. Retrieved from http://www.fun-machine.com

Wiggins, S. (2014). Adult and child use of love, like, don't like and hate during family mealtimes: Subjective category assessments as food preference talk. *Appetite*, *80*, 7–15. doi:10.1016/j.appet.2014.04.024 PMID:24793063

Williams, V. (2003). Designing simulations for learning. *E-journal of Instructional Science and Technology*, *6*(1), 50–71.

Winters, N. (2007). What is mobile learning. *Big Issues in Mobile Learning*, 7–11.

Woodruff, S. J., & Kirby, A. R. (2013). The associations among family meal frequency, food preparation frequency, self-efficacy for cooking, and food preparation techniques in children and adolescents. *Journal of Nutrition Education and Behavior*, *45*(4), 296–303. doi:10.1016/j.jneb.2012.11.006 PMID:23481894

Yellephant Co. (2015). Jobi's Bistro (Version 1.2) [Mobile application software]. Retrieved from http://www.yellephant.net

Yogome. (2015a). Healthy Heroes 1: Nutrition for Kids (Version 1.2.1) [Mobile application software]. Retrieved from http://yogome.com/

Yogome. (2015b). Healthy Heroes 2: My Plate for Kids (Version 1.0.3) [Mobile application software]. Retrieved from http://yogome.com/

Zeiger, S. (2015, January 12). Toca Kitchen 2 [Web log post]. Retrieved from http://www.bestappsforkids.com/2015/toca-kitchen-2

Zewig, B. (Writer), Henson, B. (Director) & Calamandrei, C. (Director). (2008). My mushy banana [Television series episode]. In B. Henson, L. Henson, H. Stanford, & B. Zweig (Executive producers), Sid the science kid. Hollywood, CA: The Jim Henson Company.

Zurita, G., & Nussbaum, M. (2004). A constructivist mobile learning environment supported by a wireless handheld network. *Journal of Computer Assisted Learning*, *20*(4), 235–243. doi:10.1111/j.1365-2729.2004.00089.x

KEY TERMS AND DEFINITIONS

App: A self-contained program or piece of software designed to fulfill a particular purpose; an application, especially as downloaded by a user to a mobile device.

Culinary Cognition: The mental action or process of acquiring food-preparation knowledge and understanding through thoughts, experiences, and the senses.

Media Artifact: Television programs, web-based platforms, computer games, and other advanced technologies.

Mental Model: An explanation of someone's thought process about how something works in the real world.

Chapter 3
Mobile-Assisted Language Learning From Language Instructors' Perspectives

Brenda I. Lopez Ortiz
Florida International University, USA

Mara J. Luna
University of Puerto Rico in Río Piedras, USA

ABSTRACT

This chapter explores instructor use of MALL in EFL/ESL classrooms. It begins with a discussion of instructor ownership as well as personal use of mobile technology. The chapter includes a discussion of overall perceptions and/or reported educational use and benefits and barriers (not tied to specific language skills.) The chapter goes on to describe how instructors use mobile devices while teaching reading, writing, listening, and speaking skills. This is followed by a brief discussion of mobile devices in situations that integrate all four skills. For each language skill, the authors also discuss perceived and/or reported benefits and barriers. Thereafter follows a discussion of trends in the focus of research questions, research methodologies, geographic location of studies, and a brief comparison in trends between earlier and later studies. The review ends with a conclusion that provides directions for future research and reminds the reader of the utility of this literature review.

DOI: 10.4018/978-1-5225-9351-5.ch003

INTRODUCTION

This review of literature examines the past and current state of research on English as a second/foreign language (ESL/EFL) instructor use of mobile-assisted language learning (MALL) along with associated experiences, perceptions, and attitudes. The search for research on this topic yielded limited results. On the other hand, student learning using mobile devices monopolized search findings (Jung, 2015). Characterizing instructor use of MALL does not appear to attract the attention of researchers in the field. Yet since their role in adopting mobile technologies is so instrumental, this chapter will shine light on this subject. Researchers may use the knowledge from this review as a foundation for fostering mobile device use in the EFL/ESL classroom.

The authors queried several databases in the process of identifying articles for this literature review. The researchers conducted searches using ERIC and all databases that the Ebscohost vendor provides (approximately 114). They queried the ProQuest database for dissertations on the topic. Google Scholar was also included in the search. The authors examined the reference lists of several articles to identify additional articles that would be useful. Finally, they explored the table of contents of several mLearning Conference proceedings from 2015 – 2018 (Mobile Learning International Conference, 2019).

There are several issues that complicate the search for literature on this matter. The phrase "mobile-assisted language learning" is not yet a key word in most popular research databases (ERIC, 2019). In order to search for articles that explore the instructor perspective on using MALL for teaching and learning, the authors used several combinations of search terms. Table 1 lists the various search terms used and their purposes.

Table 1. Search terms used in databases

Search Terms Used	To Search for Articles Related to
"mobile-assisted language learning"	the use of mobile devices in language learning
Handheld devices, iPad, iPhone, tablet	mobile devices even if the complete MALL phrase did not appear in the text
Faculty, Adjunct Faculty, College Faculty, Teacher, Instructor; instructor's perspectives, instructor's perceptions, instructor's opinions, and instructor's attitudes (in singular and plural forms)	the instructor perspective
language arts, language learning, linguistics	English as a foreign language (EFL) or English as a second language (ESL)

The resulting literature set includes articles that describe instructor perceptions, interest, opinions, and attitudes as well as usage of mobile devices in ESL (8 studies) and EFL (10 studies). The authors of three research studies did not specify the type of language learning that they focused on. Furthermore, the survey contains 5 articles with K-12 teachers, 2 articles with adult education teachers, and 11 studies with university faculty as participants. Lastly, the authors included studies in which researchers compared the perspective of the instructor to the perspective of the student (3 studies) if the instructor's perspective could be isolated in the findings. The purpose of the inclusion of these types of studies was to borrow from sister subjects and alternative contexts to provide a strong description of faculty perspectives on MALL.

The literature review that follows first examines ESL/EFL instructor ownership and personal use of mobile devices. Then, the chapter describes general uses of mobile devices in teaching and learning. This literature survey continues with a summary of general benefits and common issues found in the literature. Next, the review addresses the use and perceptions of MALL in teaching reading, writing, listening and speaking skills. This review includes a discussion of the instructor's perception of benefits and barriers of using MALL to teach each type of skill when such data is available. The review presents a brief discussion of MALL in teaching contexts that integrate the four skills. Then, it presents a discussion of trends in areas of the literature related to the focus of research questions, research methodologies, and geographic location of studies. The review also presents a brief comparison of trends between earlier and later studies. The review ends with a conclusion that provides directions for future research and reminds the reader of the utility of this literature review.

FACULTY OWNERSHIP AND USE OF MOBILE DEVICES

Ownership

Research in the field of MALL has considered personal ownership and use of mobile devices. For example, Bakhsh (2015), found that 95.7% of the instructors surveyed in his research owned a smartphone. Among these instructors, 59.4% owned Android devices whereas 26.1% owned iOS devices. This trend was also visible in additional research in which most participants owned an Android device (Dogan & Akbarov, 2016). Half of the respondents in Bakhsh (2015) indicated that they owned a tablet as well. Other research confirmed the trend of multiple device ownership (Grimshaw,

Cardoso, & Collins, 2017; Martiz, 2015; Saidouni & Bahloul, 2016; Stauffer, 2014). These findings seem encouraging because they lead one to believe faculty should possess basic knowledge of mobile devices.

Personal Use

In addition to determining whether faculty own mobile devices, researchers have explored whether, and how faculty use them. Grimshaw, Cardoso and Collins (2017) reported that 13 of 21 faculty in their study ranked mobile devices as first, second, or third device most frequently used. These instructors were comfortable with this technology and attributed their comfort partly to ease of use of their devices. In Bakhsh's (2015) study, instructors frequently reported that it was easy for them to use mobile features and applications (apps). Instructors rarely needed assistance using mobile devices. Male instructors rarely reported facing difficulties using mobile devices whereas female instructors sometimes reported facing difficulties (Bakhsh, 2015). Males were reportedly able to deal with mobile software, and they were sometimes able to deal with hardware issues and fixing common technical problems. The instructors in Stauffer (2014) had mixed levels of frustration with technology. However, 82% of them somewhat agreed or completely agreed they liked learning new things by using technology. This data is also encouraging because it minimizes the role of familiarity with the tools as an adoption challenge.

Several studies looked at the kinds of mobile device features instructors frequently used. Park and Slater (2014) reported that instructors from the Midwest in the United States used the call and text messaging features. Consistent with this previous study, two other studies observed that the most common type of use for smartphones was to make phone calls (Karnjanapun, 2015; Saidouni & Bahloul, 2016). In Karnjanapun (2015), a study contextualized in Thailand, approximately 52% of teachers indicated that they used smart phones for texting while all Algerian instructors in Saidouni & Bahloul's (2016) admitted to doing so. Ghrieb (2015) surveyed 7 language instructors at a university in Algeria. Like the studies previously discussed, 85% of faculty used mobile phones to talk and 57% claimed they used them to send text messages.

Authors also reported on the popularity of various other mobile device functions. Stauffer (2014), for example, noted that all participants in their study indicated using their mobile devices for email, Internet browsing, and photo capturing. Karnjanapun (2015) confirmed that photography and web search were the two other most popular categories of common activities for faculty in this research. Ghrieb (2015) reported that other teachers widely used these types of devices to set up alarms (85%). Reading PDF files was another activity that instructors in other research reported engaging in when using mobile devices (Saidouni & Bahloul, 2016).

Some reports of mobile device use presented contrasting results regarding video production/consumption and social networking. For example, all participants in Stauffer's (2014) study indicated that they used their mobile devices for video capturing. Slightly more than half of instructors in Saidouni and Bahloul's (2016) study told researchers that they used mobile devices to download and watch videos. By contrast, Park and Slater (2014) found that a large percentage of faculty did not use the video recorder features. Ghrieb (2015) indicated that teachers reported seldom or never watching videos with their mobile phones. In this study only one of the seven instructors reported using the phone for social networking. This contrasts with the 50% of faculty who indicated that they accesed social networks via mobile devices in a study by Saidouni and Bahloul (2016).

Regarding features that were unpopular among instructors, Park and Slater (2014) found that a large percentage of faculty did not play games in their devices. Faculty reported gaming with mobile devices was one of the least common activities in Karnjanapun (2015), Ghrieb (2015) and Saidouni and Bahloul (2016). Likewise, Stauffer (2014) reported that faculty did not use their devices as frequently for gaming as for other activities. However, this statistic was more applicable to the older members of the sample population.

The authors of this review have identified a few common trends in personal use of mobile devices. First, communication via phone calls or text messaging is common among most faculty. This is not surprising given these features have been available since earlier versions of mobile devices. Second, some faculty use mobile devices for Internet browsing and sending email. Third, playing video games is not a popular choice for faculty except among the younger generation.

USE IN TEACHING AND LEARNING

This section begins with a discussion of general uses of MALL. The authors describe benefits and barriers of MALL in addition to a discussion of faculty ambivalence towards MALL. Next, they present studies that focus on the use of mobile devices to facilitate learning language skills and learning in contexts that integrate all four skills (i.e. reading, writing, speaking and listening). They also describe benefits and barriers instructors have identified in using them for each language skill.

General Uses

The research literature explored whether instructors have used mobile devices for teaching and learning in general, and instructional ways they have used it. When instructors did not use MALL, research explored issues including their interest in using MALL or their readiness to use it. The findings vary depending on the participants in each study, but they suggest little use of MALL. For example, instructors in UK and Spain reported infrequent use of smartphones and tablets, although UK instructors reportedly used tablets more frequently than instructors in Spain (Pérez-Paredes, Ordoñana Guillamón, & Aguado Jiménez, 2018). In this study, about half of both UK and Spain instructors confessed they either never used mobile devices or used them a few times a year in the classroom. In Stauffer's (2014) study, faculty admitted that they did not frequently use their mobile devices for teaching purposes. In Park and Slater's (2014) study, 85% of teachers reported never using MALL in teaching, and 97.61% of faculty reported never using MALL in the study conducted by Dashtestani (2013). In another study, faculty were almost equally divided between using or not using mobile devices for teaching (Dogan & Akbarov, 2016). In other studies, faculty have shown moderately positive attitudes towards the implementation of MALL (Dashtestani, 2013). Grimshaw et al., (2017) conducted a study where 62% of instructors indicated that they were familiar with language learning applications, 67% indicated that they would recommend language learning apps to students and 43% expressed that they had made students aware of the apps and/or trained them. Chen (2017), on the other hand, found that almost all instructors in his study reported having some experience with mobile learning for teaching and learning. However, they were doubtful about adopting mobile learning on a regular basis and thought adoption would be superficial. In Tayan's (2017) research, instructors demonstrated more positive attitudes, and they viewed mobile learning as a tool to effectively support language learning, favored its implementation in the classroom, highlighted its positive influence in language learning, and reported that they would recommend it to others.

Researchers were also curious about whether faculty asked students to search the Internet for course-related information on their mobile devices. Some instructors reported doing so. Only 14% of faculty in Stauffer's (2014) study reported asking students to find information on the Internet. Other researchers also asked faculty about using mobile phones for educational purposes, such as accessing the Internet, taking notes, reading PDF's, writing notes, translating words, and media recording (Ghrieb, 2015; Saidouni & Bahloul, 2016). Faculty reported accessing the Internet and translating words more often than other activities (Ghrieb, 2015). More than half of faculty reported browsing the Internet and taking notes using their mobile devices (Saidouni & Bahloul, 2016). Grimshaw et al. (2017) found that 67% of faculty in their study used mobile devices to search the web. Finally, Bakhsh's (2015) reported that instructors sometimes encouraged students to search course information through browser apps on their mobile devices.

Another focus of examination was whether instructors used the Internet and specifically, whether they used cloud storage to share course-related materials. In at least one study (Grimshaw et al., 2017), downloading course content was low on the list of activities that instructors reported encouraging students to do with mobile devices (Grimshaw et al., 2017). In a separate study (Stauffer, 2014), only 32% of faculty reported sharing documents with their students on cloud storage. In Ghrieb's (2015) study, 57.14% of faculty agreed that they would appreciate sending learning material to their mobile phone through the Internet or through Bluetooth. In Martiz's study (2015), a cloud storage repository allowed the teacher and students to share supplementary material. Additionally, mobile phones delivered content that could be used as a foundation for group reports. Bakhsh's (2015) also surveyed instructors about using mobile devices for course-related communication and file storage/ exchange. The instructors in this study reported frequently using mobile devices to send course assignments, educational content, or other files or documents. They also sometimes reported encouraging students to download educational apps to view course content and apps that facilitate learning English and for cloud storage. Most instructors in Dogan & Akbarov's (2016) study specified using free mobile content with students. In fact, instructors indicated they would pay a reasonable price for high-quality teaching and learning content.

An additional area of focus in researching instructors' perceptions and use of MALL is its application for communication. This is an area that several researchers have chosen to focus on. This is not surprising because language learning occurs for the purpose of communication. The instructors who participated in Ghrieb's (2015) study claimed they thought that mobile devices are good tools to keep in touch with students outside of the classroom. This finding is shared by participants in other studies (Bakhsh, 2015; Pegrum, Oakley, & Faulkner, 2013). Bakhsh (2015) specifically indicated that close to 80% of instructors expressed they frequently thought that using mobile learning helps language instructors to contact students and colleagues easily. They also reported frequent use of these devices for communication (e.g., sending course alerts and emails). However, they noted only sometimes using these devices for discussion, question and answers, assignment feedback, or collaborative work. While instructors sometimes encouraged students to use the text messaging feature to contact others in English, they frequently encouraged students to use WhatsApp to keep in contact with them (Bakhsh, 2015). In Chen's (2017) study, instructors revealed they recognized the role of mobile devices in improving communication between instructors and students. The instructors also reported seeing mobile learning as ways to receive quicker feedback, which is a very important piece of language learning. Finally, Tayan (2017) showed that teachers recognized device affordances for communication and collaboration.

At least three studies examined the use of MALL for multimedia creation/consumption (Bakhsh, 2015; Dashtestani, 2013; Saidouni & Bahloul, 2016). These are not very popular applications of mobile devices for the faculty who participated in studies in this review. Bakhsh (2015) inquired about specific apps faculty encouraged students to use for class purposes. Instructors reported sometimes encouraging students to use their camera to take course-related pictures or videos. Likewise, they sometimes encouraged students to use YouTube to upload or download course videos. In a study by Saidouni & Bahloul (2016), only 14.30% of instructors declared using mobile devices for recording audio. Dashtestani (2013) found that in instances in which faculty did not yet use mobile learning for multimedia creation/consumption, they reported having positive attitudes towards the affordances of mobile devices to perform these tasks, and they indicated they believed that the audio-visual properties of mobile devices can facilitate learning.

Furthermore, Jung (2015) examined the factors influencing EFL teachers' adoption of mobile learning. This author studied the following factors:

1. Instant connectivity (i.e. capability to provide anytime, anywhere connectivity to teaching resources),
2. Compatibility with teaching environment and style,
3. Interaction (i.e. the unit of activity of the teacher in establishing a better and useful English learning environment),
4. Content enrichment (i.e. the variety of teaching materials that can be accessible), and
5. Computer self-efficacy (i.e. teachers' confidence in their knowledge and skills to successfully teach English using m-learning technology) (p. 227).

This researcher concluded that these factors influenced teachers' reasoning for choosing to implement m-learning in their English teaching.

Two of the factors listed by Jung (2015) proved to be important for teacher appropriation of mobile learning in the study by Martiz (2015). In her research, Martiz reported that the teacher was concerned with issues of connectivity and self-efficacy. The issue of self-efficacy applied to her students in this particular case. During the study, the teacher started to consider the use of cell phones to teach English given her positive assessment of three other adoption factors: compatibility with teaching environment and style, interactivity, and content enrichment. The research by Oh, Lee, Park, and Cho (2014) also found that instructors appreciated the use of mobile devices because of their content enrichment.

One final area of general MALL uses in teaching ESL/EFL is readiness factors that influence adoption and training. Instructors have expressed some level of comfort in using mobile phones for teaching and learning in a study by Chen (2017). About half of the faculty in a study conducted by Ghrieb (2015) agreed or strongly agreed that they were ready to adopt mobile devices for learning purposes. However, 71% of instructors indicated that they needed training to do so. Similarly, Dashtestani (2013) found that 47.27% of faculty indicated a need for training. In another study, 68.69% of faculty in the UK and 51.15% of faculty in Spain indicated that their institutions fostered the use of mobile devices (Pérez-Paredes et al., 2018). However, 65.66% of faculty in the UK and 76.34% of faculty in Spain indicated they lacked training in the use of mobile devices. These authors indicated that fostering use and training is associated with frequency of use. Oh, et al. (2014) claimed that both faculty and students agreed that instructor proficiency in the use of the tablets (i.e., Jung's (2015) self-efficacy factor) was satisfactory. Bakhsh's (2015) research found that many instructors had positive attitudes towards mobile learning adoption. Furthermore, as Jung (2015) described above, faculty had positive reactions towards questions related to the category of self-efficacy.

When it comes to the use of mobile devices for teaching and learning, faculty are in the beginning stages of exploration. Therefore, their use is basic. On one hand, faculty have barely used mobile devices. On the other hand, they are hopeful about MALL, but they are not confident in its use. Key in efforts to facilitate adoption of MALL would be adequate infrastructure to support adoption and adequate training to increase self-efficacy.

Faculty Ambivalence Towards MALL

Research that considers faculty use and perceptions of use of MALL has found that faculty are still at odds between the benefits and challenges this technology can bring. Research describes that faculty are somewhat skeptical of using MALL because of technological and pedagogical limitations, but at the same time, they are somewhat eager to incorporate these technologies into their teaching (Park & Slater, 2014). Only 15% of teachers in this study by Park and Slater reported using MALL in their teaching. Skepticism about the capabilities of mobile devices (i.e. iPads) and its applications to support the English language learning classroom was voiced in a study of faculty perspectives in an iPad integration project (Waller, Balanyk, & Thornquist, 2015). Lin, Hsu and Lee (2015) found contradictions between teachers' interest in using mobile devices in teaching and their concerns. These authors reported that

100% of teachers reported seeing mobile devices as capable of motivating students to learn. However, 87.5% of teachers reported that using these devices would result in problems, such as workload increase. Ghrieb (2015) found conflicting perceptions about mobile devices. While 57.14% of instructors agreed or strongly agreed that they could use mobile devices for learning purposes, only 14.29% agreed or strongly agreed that they should be encouraged and 28.57% agreed that they are a tool for distraction. In the same study, 57.14% of faculty agreed that mobile devices should be allowed as learning materials and that they are more suitable for learning than printed material. However, 85.71% of instructors disagreed or strongly disagreed that these devices are better than computers. The same percentage of instructors agreed that mobile devices can be used as supplementary to printed materials.

Instructor ambivalence towards MALL is an important issue to address because faculty attitudes towards technology can be transmitted to the students. This was evidenced in the account of a specific faculty member who reported that student comfort with technology increased as he felt more comfortable with it (Waller, et al., 2015). Moreover, Pegrum et al. (2013) reported that iPads had been motivating for teachers. Ideally, institutions can take advantage of motivation and self-efficacy to lead faculty to increased use of MALL for activities in which the teaching of the four skills is significantly improved.

Benefits

Increased student engagement is a finding that is frequently noted in MALL research focused on instructor use or perceptions and attitudes (Al-Okaily, 2013; Ghrieb, 2015; Rashid, Cunningham, & Watson, 2017; Waller et al., 2015). In the eyes of faculty, this might be because they believe mobile learning allows instructors to prepare more interesting activities (Bakhsh, 2015). In the Waller et al. (2015) study, one instructor reported that student socialization, which would otherwise be distracting, still occurred in the context of iPad integration. The difference was that socialization occurred for the purpose of sharing resources that were useful for the academic task at hand. Another instructor in the same study described the flip side of this type of socialization: a course that was supposed to help students learn to speak English now had students "hiding" behind their iPad screens. Notwithstanding, the third instructor in the study recognized that those screens did provide students with a voice. In his words, "The screens in front of them offered a new kind of engagement, one that offered students time to compose and refine their ideas. It was a little more thoughtful and reflective than, say, a large-group class discussion" (Waller et al., 2015, p. 10). Ghrieb (2015) reported that 57.15% of instructors in his study thought that students would be motivated and would participate more if they could use mobile devices. Perception of higher student engagement was also reported in research on the effects of tablet PC-based English learning on instructors' and students' perceptions of MALL (Oh et al., 2014). One of the instructors in the study by Waller et al., (2015) warned that the novelty effect wore off as the use of electronic books started to become mainstream. This means that MALL researchers and practitioners should not exclusively rely on student engagement as a driver of MALL adoption and use.

Ubiquitous learning was another advantage reported by one of the instructors in the studies by Bakhsh (2015), Tayan (2017), Dashtestani (2013), and Waller et al. (2015). This type of learning is powered by the portability of mobile devices. Additionally, the combination of hardware and software, such as learning management system, and its capabilities for delivering learning content and tracking usage enables faculty to monitor those ubiquitous learning experiences (Waller et al., 2015).

Another advantage reported in the literature is the opportunity to use mobile devices to provide content that meets the needs of students. Waller et al. (2015) reported faculty were able to provide specific content to meet the needs of students who lagged behind as well as of those who need an extra challenge.

Yet another advantage of the integration of iPads in the English classroom, as reported by two participants in the study by Waller et al. (2015), was related to the availability of student performance report tools. One of the instructors declared that this reporting capability allowed him to flip his classroom: he would assign work to students and clearly see who had completed it and who had not. A second instructor in the Waller et al. study specifically mentioned that instantly accessing information on quiz questions commonly answered incorrectly and instantly accessing information on students with low scores were benefits of MALL. This information allowed the instructor to identify material that needed revision as well as students who needed additional help. This finding has been confirmed by at least one other study (Oh et al., 2014).

In addition to these findings, another benefit of incorporating MALL is its affordance as time-saving and time-management tools (Martiz, 2015). In this study, the teacher reported using mobile phones to keep track of time either by monitoring time or by using timers. Additionally, the mobile phone made some tasks more efficient. An example was the use of a mobile-friendly testing platform. This platform provided automatic grading, which allowed students to see the results immediately instead of waiting for the instructor to grade each submission individually. Enhancements in productivity were also seen as benefits in another investigation (Bakhsh, 2015). Likewise, lesson delivery and student assessment were other advantages reported in studies (Oh et al., 2014). Furthermore, Pegrum et al. (2013) described the organizational capabilities of mobile devices to support teaching. The features of mobile devices that enable faculty to track student performance also double as agents to enhance efficiency in the teaching and learning process.

Rashid, Cunningham, and Watson (2017) stated another benefit of MALL in their research. These authors highlight the potential of mobile devices to become an alternative to a combination of technologies that would otherwise be needed for language teaching in developing countries.

Barriers

Recent MALL research has stressed several troublesome issues. Some of the barriers are related to factors that influence the adoption of MALL (Jung, 2015). Faculty's level of mobile literacy was one reported barrier. In several studies, even if some faculty were comfortable learning independently, the authors recognized mobile literacy as a key factor in sustained use of mobile learning for English teaching (Grimshaw et al., 2017; Lin et al., 2015; Waller et al., 2015). For example, instructors in the study by Stauffer (2014) stated they had received training through their departments, programs, institutions, colleagues or on their own. However, they also indicated they would integrate more technology if they received more training. Pedagogical training on mobile integration is one of the supporting factors that researchers have listed to enable more wide-spread adoption of MALL (Martiz, 2015). In another study, lack of training was the highest-ranked obstacle to using mobile devices for teaching (Dogan & Akbarov, 2016).

There were two other factors that researchers have found to influence teacher acceptance and adoption of mobile technology for teaching English (Lin et al., 2015; Waller et al., 2015), The first factor is the pedagogical relevance of applications. The second factor is the compatibility of technology features with the characteristics of needed teaching materials and activities.

Authors in the literature of MALL have also considered the impact of mobile devices in classroom management. The research conducted by Ghrieb (2015) considered the first of these challenges to the use of mobile devices for language learning, as seen by faculty. 100% of faculty indicated that mobile devices cannot be controlled inside the classroom. The introduction of distraction is an issue that other authors list in their research (Grimshaw et al., 2017; Martiz, 2015; Oh et al., 2014; Rashid et al., 2017). Due to their potential to distract students, mobile phones have been banned in some educational contexts. The teacher in Martiz's study (2015) set up etiquette guidelines for the use of mobile phones in the classroom to counteract this problem. Easy access to content is another issue that faculty are reported to face. Easy access means accessibility to content that students were not skilled enough to evaluate for quality or appropriateness (Waller et al., 2015).

Research studies also investigate issues related to supporting the diversity of mobile devices, cost of Internet and devices, and time investment. In some research, faculty expressed concerns because of the diversity of students' devices (Dogan & Akbarov, 2016). Ghrieb (2015) stated that 85.71% of instructors indicated that one of the challenges in using MALL was the diversity of students' devices. For faculty with incipient knowledge of mobile technology, the diversity of device brands and operating systems can seem like an unsurmountable obstacle in their attempts to incorporate MALL into their classrooms. For 57.14% of faculty in this study, the cost of Internet access and mobile devices was also a challenge. While male instructors rarely reported problems with the cost of mobile fees in Bakhsh's (2015) study, female faculty were more prone to report problems with cost. Bakhsh's (2015) studied additional possible challenges to implementation of mobile devices in language learning. This author reported that faculty sometimes thought that using different mobile features and apps would require extra time and effort.

Technical support, connectivity, and harware issues are additional areas of interest in the literature of MALL. Lin et al. (2015) mentioned the availability of technical support as another key factor in mobile device adoption and persistence in use. Half of the instructors in the study by Stauffer (2014) declared that they received enough support to teach with technology. However, a study by Waller et al. (2015) reported technical difficulties as factors that inhibited faculty work on iPad integration. In this study, one faculty member reported that while there was technical support during the first year of implementation, it was withdrawn at the begininng of the second year. Faculty were left to be the providers of support. Waller et al. (2015) reported difficulties with finding technology mentors and presenters for discovery sharing sessions because of faculty reluctance towards the integration of these devices. Another faculty member in the same study complained about the disruptions in learning because of interruptions on wi-fi service or because of wait time for technical support to solve problems. However, over time,, the problem of connectivity became less of an issue in his institution, thereby enabling him to confidently plan MALL lessons. Issues of Wi-Fi connectivity, screen size, hardware and software, and poor battery life or lack of charging facilities have also been discussed in the literature (Dashtestani, 2013; Oh et al., 2014; Pegrum et al., 2013; Rashid et al., 2017; Tayan, 2017). In contrast, male instructors in the study by Bakhsh (2015) rarely had difficulty using mobile learning because of lack of technical support, while female instructors sometimes did. Finally, Dogan & Akbarov (2016) reported connectivity issues as a moderate area of concern.

Reading

The use of mobile devices for teaching reading and vocabulary has been one of the most favored applications of this type of technology (Burston, 2014, 2015; Duman, Orhon, & Gedik, 2015). This section examines literature on using mobile devices in teaching reading skills and vocabulary. It also discusses research findings related to perceived or reported benefits and barriers of using mobile devices for these purposes.

When examining reading tasks that could be performed via mobile devices, Park and Slater (2014) reported that faculty ranked reading course materials highly in a list of tasks that could be done in MALL. Saidouni and Bahloul (2016) revealed that 78.60% of faculty would use mobile devices in teaching reading. In a study that compared instructors in Japan with instructors from Europe and countries in the Asia-Pacific Rim, Pagel, Lambacher, Kikuchi, and Mori (2018) indicated that 67% of faculty preferred using MALL to develop reading skills. Ghrieb (2015) reported that 57.14% of instructors thought mobile devices are good for reading activities. In another study that sought to understand factors influencing EFL teachers' adoption of mobile learning, Jung (2015) found that 35.45% of teachers reported using mobile devices for reading activities. Nonetheless, the use of mobile devices for teaching reading was the most frequently cited use of mobile devices in Jung's study. In contrast, participants in other studies reported less frequently or rarely using mobile devices to read text or teach reading skills (Bakhsh, 2015; Grimshaw et al., 2017).

Teachers have also reported seeing mobile devices as alternatives to computers for displaying books because of their capabilities to store interactive eBooks (Pegrum et al., 2013). In contrast, in a different study, smartphones as devices for reading eBooks were not at the top of the list for English teachers (Karnjanapun, 2015). Finally, Dogan & Akbarov (2016) found instructors views', split on using eBooks with students.

The study of mobile devices in the teaching of vocabulary has also caught the attention of several researchers. Park and Slater (2014) reported that teachers listed looking up vocabulary more often than others as tasks that could be performed via mobile devices. Pagel et al. (2018) stated that vocabulary was at the top of their list of applications that faculty recommended students to use. Martiz (2015) also expressed that the teacher in her study used mobile phones in the teaching of vocabulary words via a dictionary app. Ghrieb (2015) indicated that 85.71% of participants in his study thought mobile devices are good for vocabulary building activities. A total of 71% of instructors in another examination indicated that they had allowed students to look up words or definitions (Grimshaw et al., 2017). Instructors in the study of Bakhsh (2015), reported sometimes using these devices to teach vocabulary. In contrast, 16.36% of instructors in one study indicated that students used dictionary apps to look up the meanings of words (Dashtestani, 2013). Waller, et al., (2015) reported on their perspectives as faculty at three different entry points of an iPad integration project. In their study, one of the authors reported using the devices to have students look up the meanings of vocabulary words. Another instructor reported taking advantage of the capabilities of these tools to facilitate the study of vocabulary.

Perceived/Reported Benefits and Barriers

The research from 2013 to 2018 reveals a tendency to favor MALL for teaching reading and vocabulary. Researchers can consult additional research to delve deeper into the subject of mobile devices as providers of eBooks. Meanwhile, the studies examined highlighted several benefits associated with MALL for teaching reading. A teacher who used Edmodo to post articles for students to access and read using their mobile devices reported student interest in mobile devices as one benefit (Al-Okaily, 2013). According to an instructor in the Waller, et al., study (2015), looking up words in mobile devices was consistent with students' preferences of finding information online. According to another instructor, iPads could also deliver adaptive and customizable quizzes and games, which goes beyond recreating traditional strategies of memorization of vocabulary words (e.g. the use of flashcards). The research studies in MALL for reading that were included in this review did not reveal specific barriers.

Writing

Research has also considered the use of mobile devices in teaching writing skills. In general, the percentages of use of these devices for teaching writing are usually lower than the percentages related to teaching reading. Pagel et al. (2018) reported that 36% of participants preferred to use mobile devices for writing. A lower percentage of faculty (28.58%) agreed or strongly agreed that mobile devices were good for writing activities, according to another study (Ghrieb, 2015). In research by Jung (2015), 16.4% of teachers reported using mobile devices for teaching writing. Saidouni and Bahloul (2016) found similar results. In their study, 14.30% of participants were open to the idea of using mobile devices for teaching writing. Similarly, instructors reported rarely using mobile devices to teach writing skills, according to findings reported by Bakhsh (2015). This included the use of the "Notes" feature to take notes and posting status updates to social apps.

When considering writing tasks that could be performed via mobile devices, Park and Slater (2014) reported that teachers listed writing short messages, academic papers, and emails as well as taking notesmore often than online chatting and posting on social networking sites. In a closed-ended response format, teachers ranked writing formal and informal emails as the highest writing-related MALL task. Grimshaw et al. (2017) reported that a 38% of their participants allowed students to use mobile devices to take notes. The teacher in Rashid, et al. study (2017) had students use smartphones to write blogs.

Perceived/Reported Benefits and Barriers

Despite the low popularity of mobile devices for teaching writing, there is a wide range of writing applications that researchers have discussed in the literature. Researchers have also reported on related benefits. The instructor observations reported by Al-Okaily (2013) highlighted two benefits of using mobile devices for student writing: (1) mobile devices allowed students to read each other's blogs, thus providing more authentic audiences for each other, and (2) mobile devices facilitated teachers' access to student writing, eliminating the need for hard copies of entries. Rashid, et al. (2017) also reported that they saved time by providing feedback to students with mobile devices due to their ease of use.

Not surprisingly, research has also examined the limitations of MALL in teaching writing. Waller et al. (2015) reported that one faculty member witnessed students struggle with writing on the iPads. The students resorted to bringing notebooks, taking notes, and writing on their notebooks instead of their mobile devices. The same faculty member also described a writing problem that resulted from the easy access to content: there was an increase in plagiarism in writing assignments. This finding is not exclusive to the use of MALL in the English language learning classroom.

Listening

Some authors of the articles included in this review also investigated the use of mobile devices for developing listening skills. The range of activities for which MALL can help is wide. In one study, authors indicated that faculty see MALL as opportunities for students to listen to lectures, news, and music (Park & Slater, 2014). Interestingly, they reported that teachers did not list listening to pronunciation with MALL as often as other tasks included in the study. However, in a closed-ended response format questions, teachers highly ranked listening to language podcasts and voice messages, as well as practicing listening comprehension. The findings in Ghrieb's (2015) study support teachers' use of MALL for listening activities: 85.71% of faculty agreed or strongly agreed that mobile devices were good for listening activities. Pagel et al. (2018) found that 77% of participants in their study preferred mobile devices for teaching listening. Saidouni and Bahloul (2016) reported that 57.10% of their participants were open to the idea of using mobile devices for teaching listening. In the case of the Karnjanapun's (2015) study, approximately 46% of English teachers thought audio lectures should be made available for students on smartphones. Jung (2015) found that only 20.63% of teachers reported using mobile devices for teaching listening. On a similar note, instructors rarely used mobile devices to teach listening skills, according to the findings reported by Bakhsh (2015).

The teacher in the study of Martiz (2015) was able to use her mobile phone to teach listening skills despite connectivity issues. She used it offline to play audio files that she had previously downloaded to her mobile phone. Because of institutional limitations in resources, these audio files would have not been accessible to students otherwise. In doing this, she saw mobile devices as tools for content enrichment. In this case, overcoming the connectivity factor that Jung (2015) discussed made possible the adoption of mobile devices for the teacher in Martiz's study.

Perceived/Reported Benefits and Barriers

There is a 65% gap in instructor's perceptions and/or use of MALL to teach listening skills (between more positive and more apprehensive), according to the observations of the authors of this review. However, the literature discovered a few benefits of using MALL to teach listening skills. Student interest and engagement are two of the benefits of using MALL to teach listening skills. Student engagement while practicing listening via audio files posted in Edmodo and accessed via mobile devices was a finding of the study by Al-Okaily (2013). Martiz (2015) reported that the use of cell phones for listening comprehension enabled the teacher in his study to find real world uses of target verbs in popular songs. This technique maintained student interest in a way that is different from using listening exercises from textbooks. The research that was consulted for this review did not uncover any specific barriers of the use of MALL to teach listening skills.

Speaking

Research on MALL has also addressed the use of this technology for teaching speaking skills. Formal and casual phone calls were opportunities for MALL that faculty listed often in a study that examined MALL tasks in the four language skills (Park & Slater, 2014). In a closed-ended response format survey, teachers ranked speaking tasks that included participating in informal daily conversation highest than other speaking tasks listed, followed by speaking in the context of voicing opinions related to classwork or social problems. Making reservations was another task that they ranked highly. The participant in Martiz (2015) reported using an offline dictionary app to learn the pronunciation of words.

Karnjanapun (2015) studied English teachers' openness to the use of smartphones in teaching English pronunciation. This researcher found that 76.3% of female teachers and 68.4% of male teachers were open to using smartphones for this purpose. Similarly, Ghrieb (2015) found that 71.43% of faculty agreed or strongly agreed that mobile devices were good for speaking activities. Saidouni and Bahloul (2016) noted that 57.10% of their participants were amenable to the idea of using mobile devices for teaching speaking skills. The numbers of faculty who saw value in the use of MALL for teaching speaking and pronunciation were much lower in another study (Pagel et al., 2018). The authors showed that 44% preferred MALL for pronunciation, while 28% felt that way about speaking.

In contrast, other studies have found little use of MALL in teaching speaking skills. In one case, only 7.41% of teachers reported using mobile devices for teaching speaking skills (Jung, 2015). Likewise, the research by Bakhsh (2015) reported that instructors rarely used mobile devices to teach speaking skills. Instructors in this study only sometimes used them to teach pronunciation.

Perceived/Reported Benefits and Barriers

The studies considered in this literature review seemed to lean towards using MALL for teaching pronunciation more often than for teaching speaking. Perhaps this is due to the fact that pronunciation is a skill that can be drilled and practiced using mobile apps whereas speaking requires the intervention of others even if speaking can be mediated by mobile technologies. Regardless, there are a few benefits worth noting. Al-Okaily (2013) reported that the teacher in his study thought mobile devices gave students the extra opportunity to access resources to practice speaking (e.g. VoiceThread). Additionally, the availability of VoiceThread, according to the instructor observations, meant that students who would otherwise miss a class when presentations were due, were now able to post their presentations, thus fulfilling class requirements.

Martiz (2015) reported on several benefits of using MALL for speaking skills. The author described an offline dictionary app that the participant used to make up for the lack of connectivity during pronunciation lessons. Mobile phones also proved to be beneficial in the evaluation of students' speaking skills, according to this researcher. In this case, the instructor was not limited by class time constraints in assessing students. Using cell phones, students could record their conversations. They would then post their conversations to a cloud repository. The teacher had the flexibility to listen to them by accessing the files from students' mobile phones, her own, or the computer. The recordings also allowed for self- and peer assessment.

One of the studies included in this review reported a specific barrier for the use of MALL to teach speaking skills. Lin, et al. (2015) found that speaking practice in group contexts resulted in loud students impacting quieter students' practice. Teachers were concerned with losing class control with the kind of exercise that they performed during the research. However, they were able to find ways to keep classes under control because of discussion of this issue.

MALL Integrating the Four Skills

Research has also focused on the use of mobile devices in teaching that integrates the four skills (Waller et al., 2015). These authors reported that faculty encouraged students to use movie editing apps in English classes for project-based learning, which integrated all four skills.

Perceived/Reported Benefits and Barriers

The research within this context uncovered several benefits. One of the co-authors in the study by Waller et al. (2015) indicated that using MALL for project-based learning was an example of a task that was empowered by new technology, and that it would have not been possible without its use. Another faculty member in this study described how iPads contributed to an increase in the scope and quality of student products generated in this project. The same faculty member also reported on the positive impact of iPads on the process as well. According to a third faculty member in this study, the introduction of the iPad as a tool for project-based learning also brought about more student-led experiences in the classroom. This study did not list specific challenges related to using MALL to teach all four skills in an integrated way.

DISCUSSION OF TRENDS IN THE LITERATURE

Upon analysis of the research literature on usage of mobile devices for language learning, several trends become apparent. Trends in areas including the focus of research questions, common research methods, and geographic location of research studies will be described. The section ends with a general comparison of findings in earlier studies compared to findings in more recent studies.

Trends in the Focus of Research

Upon inspection of the studies in this review, it is clear that research related to the adoption of MALL has focused on general adoption issues (including personal and professional ownership and use), general benefits, and barriers. Adoption of MALL for vocabulary and reading purposes is the next common focus of the studies included in this review. This is also a common subject of research when studying MALL and student learning (Burston, 2014, 2015; Duman et al., 2015). The areas of research focus that follow are speaking, listening and writing skills. These common areas of research focus are not surprising. Technologies are seen as a means for access to content to practice when they are first introduced, hence the popular emphasis on reading and vocabulary. Many of the mobile devices that are available for teaching and learning are mobile phones. Since their primary purpose is voice communication, phones have many features that support teaching and learning of speaking. Writing via the small screen and perhaps keyboard-less mobile devices is uncomfortable for students. That may be the reason that writing is not the main focus of MALL research. Although many modern devices include dictation capabilities, it appears researchers have not yet attempted to harness this feature to facilitate writing. The research community would be well served if more recent research attempts to explore whether there is a dearth of research in listening and writing skills or whether that research is just not as easily available. More studies on those two types of skills can help strengthen the subject by filling the gaps.

Research Methods Used in the Studies

This section identifies the trends in research methods (Figure 1). Research in instructor use, perceptions, and attitudes of/towards MALL predominantly uses quantitative or mixed methods of data collection. About a third of the studies conducted interviews frequently in addition to surveys. The studies in this review are characterized by the use of one data collection method and one analysis method. This finding is troublesome because it means there was no triangulation of data especially in cases where samples were not randomly selected. It is understandable, though, since some of the areas of focus of the research do not necessarily lend themselves to the collection of observable data. Researchers must rely on instructor self-reports (e.g., instructor opinions, attitudes, predictions of use, etc.). All studies that used qualitative methods performed some sort of coding to classify findings and identify themes. Although a few studies used advanced statistical methods to analyze data, most used frequencies and calculated percentages and/or other descriptive statistics to describe data.

The studies that explored the research issues from the perspective of one instructor used qualitative methods and explored the issue from the perspective of students. The reports of research that had less than 10 instructors as participants incorporated qualitative methods, which may account for the small number of participants. Finally, the papers with 100 or more instructor participants relied on surveys to collect data. This characterization of studies warrants the investigation of faculty and MALL using larger and more systematically selected samples.

Geographic Location

Curiosity about instructors' proclivities, towards MALL is highest in the Middle East and Asia, as inferred by the geographic location of the studies examined in this literature review. Table 2 presents the number of studies per world region. Table 3 (Appendix A) presents the actual studies (identified by author(s) and organized

Figure 1. Research methods trends

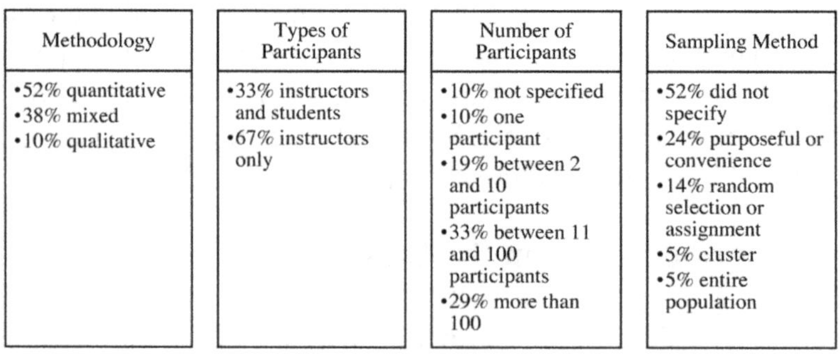

Table 2. Number of Studies per World Region

World Region	Number of Studies
Asia	6
Australia	1
Caribbean	1
Europe	1
Europe/Asia	1
Middle East	6
North Africa	2
North America	3

by world region). Figure 2 presents a map as visual representation of the number of studies per world region. More studies in diverse areas around the globe could contribute to strengthen our knowledge in this area of research. Furthermore, future research could investigate whether researchers in the areas with fewer studies simply do not consider this as an important area of research.

Comparison Between Earlier and Recent Studies

The studies considered in this survey of literature span six years. Technology evolves so quickly that devices or software made more than two years ago (sometimes even earlier) are considered outdated. Because of this, the reader may think that research in the earlier studies would yield vastly different findings than those of later studies. This is not evidenced in this review. The studies reviewed in this study present findings related to little use and not so positive perceptions, widespread use and positive attitudes, etc. or mixed results across focus areas. However, both earlier and recent research studies present similar findings in most of these areas. The only difference between earlier and more recent studies is that MALL is slowly becoming more popular over time. Interest in the use of MALL for pronunciation seems to be dwindling, according to more recent studies. This seems to indicate that the research findings related to MALL issues are more stable than the technologies that enable MALL.

Figure 2. Map of number of studies per world region

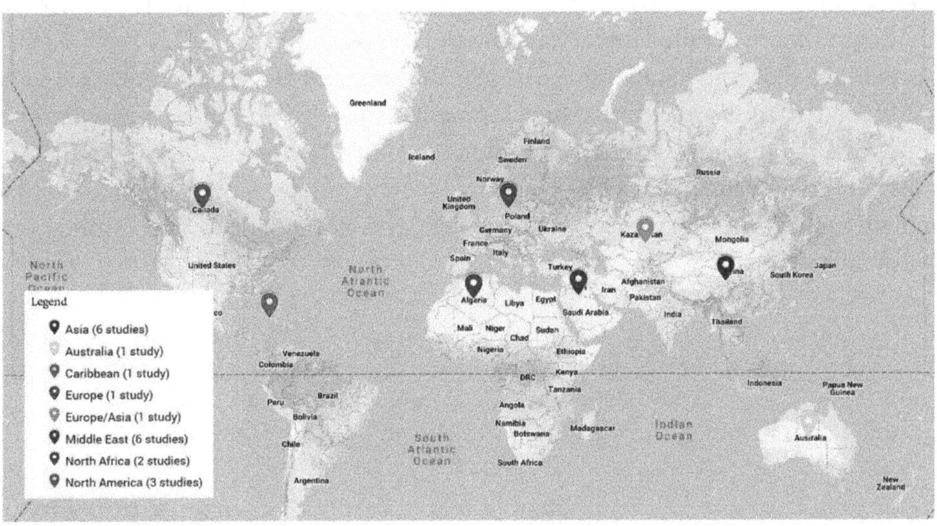

CONCLUSION

Future examinations of MALL from the perspective of instructors can mirror several aspects of the trends related to research methods and improve upon others. The studies can rely on several sources of data collection to strengthen their design. This research should aim to collect data from large numbers of instructors using appropriate methods of sampling that can help ensure wider generalizability of results. Finally, just as the majority of surveys in this review, future research can rely on frequency counts and descriptive statistics to summarize the data collected. However, studies can also benefit from more advanced statistics when appropriate.

New research endeavors can benefit from knowledge from the literature review in two ways. First, literature findings regarding benefits and barriers to MALL can inform the design of surveys and coding or interpreting responses to survey questions related to both MALL aspects. Second, Jung's (2015) issues related to technology adoption can aid in the design of responses that seek explanations of why faculty have used/not used mobile devices in their teaching. They can also provide ideas to design questions that inquire into circumstances that could encourage those faculty who have not used these devices to adopt them for teaching.

Given the state of research in instructor adoption of MALL, future research has the potential to add to the current body of knowledge in at least two ways. First, it can help to characterize areas of adoption that are not equally represented in literature (e.g. listening, writing and, to some extent, speaking). Second, it can also help to characterize the adoption of MALL in regions of the world that are underrepresented in this literature.

According to Savas (2014), the key to facilitate adoption of MALL may be in providing "a more detailed understanding of the possibilities" (p. 221). A first step in providing that understanding is to learn where instructors currently stand in order to better focus professional development. This is what future investigation of faculty use, perceptions, attitudes, opinions, etc. of MALL can seek. In achieving this, future studies can become a springboard to recommend ways to provide English language educators everywhere with the understanding of the possibilities that Savas referred to above.

REFERENCES

Al-Okaily, R. (2013). Mobile learning and BYOD: Implementations in an intensive English program. *Learning & Teaching in Higher Education: Gulf Perspectives*, *10*(2), 1–17.

Bakhsh, S. S. H. (2015). Mobile learning adoption by language instructors in Taibah University. *Education Research International*, *4*(5), 117–150.

Burston, J. (2014). The reality of MALL: Still on the fringes. *CALICO Journal*, *31*(1), 103–125. doi:10.11139/cj.31.1.103-125

Burston, J. (2015). Twenty years of MALL project implementation: A meta-analysis of learning outcomes. *ReCALL*, *27*(1), 4–20. doi:10.1017/S0958344014000159

Chen, K. (2017). Examining EFL instructors' and students' perceptions and acceptance toward M-learning in higher education. *International Journal (Toronto, Ont.)*, *16*(4), 967–976. doi:10.100710209-016-0494-8

Dashtestani, R. (2013). Implementing mobile-assisted language learning (MALL) in an EFL context: Iranian EFL teachers' perspectives on challenges and affordances. *JALT CALL Journal*, *9*(2), 149–168.

Dogan, A., & Akbarov, A. (2016). Teachers' attitudes toward the usage of mobile devices in EFL classroom. *European Journal of Educational Research*, *5*(1), 11–17. doi:10.12973/eu-jer.5.1.11

Duman, G., Orhon, G., & Gedik, N. (2015). Research trends in Mobile Assisted Language Learning from 2000 to 2012. *ReCALL*, *27*(2), 197–216. doi:10.1017/S0958344014000287

ERIC. (2019). *ERIC thesaurus*. Retrieved from https://eric.ed.gov/?ti=M

Ghrieb, E. B. (2015). *Teachers' and students' attitudes towards the use of mobile assisted language learning* (Master's thesis). University of Biskra, Biskra, Algeria.

Grimshaw, J., Cardoso, W., & Collins, L. (2017). Teacher perspectives on the integration of Mobile-Assisted Language Learning. In K. Borthwick, L. Bradley, & S. Thouësny (Eds.), CALL in a climate of change: A dapting to turbulent global conditions – short papers from EUROCALL 2017 (pp. 135-139). Research-publishing.net.

Jung, H.-J. (2015). Fostering an English teaching environment: Factors influencing English as a foreign language teachers' doption of mobile learning. *Informatics in Education, 14*(2), 219–241. doi:10.15388/infedu.2015.13

Karnjanapun, S. (2015). *Nakhon Ratchasima teachers' using of smartphones in English Languageteaching*. Paper Presented at the EdMedia: *World Conference on Educational Media and Technology*, Montreal, Quebec, Canada.

Lin, J., Hsu, H.-Y., & Lee, G. (2015). *Inducing language teachers to use tablets in the classroom: Lessons learned*. Paper Presented at the E-Learn: World Conference on E-Learning in Corporate, Government, Healthcare, and Higher Education 2015, Kona, HI.

Martiz, G. (2015). *A qualitative case study on cell phone appropriation for language learning purposes in a Dominican context* (M. Recker, K. DeJonge-Kannan, A. Diekema, V. Lee, & J. Rogers, Eds.). ProQuest Dissertations Publishing.

Mobile Learning International Conference. (2019). *Previous Editions: From 2005-2018 | ML 2019*. Retrieved from https://www.mlearning-conf.org/previous-editions-2005-2015/

Oh, J., Lee, H.-K., Park, M., & Cho, Y. (2014). Exploring the effects of tablet PC-based English learning. *English Teaching, 69*(3), 151–176. doi:10.15858/engtea.69.3.201409.151

Pagel, J., Lambacher, S., Kikuchi, H., & Mori, S. (2018). *Student and instructor attitudes toward CALL and MALL in the L2 classroom*. Editorial Universitat Politècnica de València. doi:10.4995/HEAD18.2018.8108

Park, M., & Slater, T. (2014). A typology of tasks for mobile-assisted language learning: Recommendations from a small-scale needs analysis. *TESL Canada Journal, 31*, 93–115. doi:10.18806/tesl.v31i0.1188

Pegrum, M., Oakley, G., & Faulkner, R. (2013). Schools going mobile: A study of the adoption of mobile handheld Technologies in Western Australian independent schools. *Australasian Journal of Educational Technology, 29*(1), 66–81. doi:10.14742/ajet.64

Pérez-Paredes, P., Ordoñana Guillamón, C., & Aguado Jiménez, P. (2018). Language teachers' perceptions on the use of OER language processing technologies in MALL. *Computer Assisted Language Learning, 31*(5-6), 522-526. doi:10.1080/09588221.2017.1418754

Rashid, S., Cunningham, U., & Watson, K. (2017). Task-based language teaching with smartphones: A case study in Pakistan. *Teachers and Curriculum*, *17*(2), 33–40. doi:10.15663/tandc.v17i2.167

Saidouni, K., & Bahloul, A. (2016). Teachers and students' attitudes towards using mobile-assisted language learning in higher education. *Arab World English Journal*, *3*, 123–140.

Savas, P. (2014). Tablet PCs as instructional tools in English as a foreign language education. *Turkish Online Journal of Educational Technology*, *13*(1), 217–222.

Stauffer, K. (2014). *Technology use in post secondary language education* (M. L. Thomas-Ruzic, K. Hall, & P. Mosele, Eds.). ProQuest Dissertations Publishing.

Tayan, B. M. (2017). Students and teachers' perceptions into the viability of mobile technology implementation to support language learning for first year business students in a Middle Eastern university. *International Journal of Education and Literacy Studies*, *5*(2), 74–83. doi:10.7575/aiac.ijels.v.5n.2p.74

Waller, T., Balanyk, J., & Thornquist, E. (2015). *Three perspectives of mobile learning integration at a tertiary institution in the UAE*. Paper presented at the 9th International Technology, Education and Development Conference, Madrid, Spain.

KEY TERMS AND DEFINITIONS

Higher Education Professors of English: College professors who have taught at least one English language course in a college or university.

Language Skills: Skills that enable a person to communicate orally or in writing in English. These include reading, writing, listening, and speaking.

Listening Skills: Skills that enable a person to derive meaning out of oral communication.

Mobile Devices: Portable electronic equipment that allows the user to communicate, access, and/or share data at any time and/or any place. Examples of these include any kind of handheld device, such as cell phones, personal digital assistants (PDAs), smartphones, iPads, iPods, tablets, or laptops.

Mobile Learning (mLearning): Educational experiences in which mobile devices are used to facilitate learning across multiple contexts, through social interactions and by accessing and/or sharing different types of content.

Mobile-Assisted Language Learning (MALL): A branch of mobile learning, referring to educational experiences in which mobile devices facilitate language learning.

Reading Skills: Skills that enable a person to derive meaning out of written communication.

Speaking Skills: Skills that enable a person to orally express ideas accurately and coherently.

Writing Skills: Skills that enable a person to express ideas accurately and coherently in writing.

APPENDIX

Table 3. Studies organized by world region

World Region	Studies
Asia	• Chen, 2017 • Jung, 2015 • Karnjanapun, 2015 • Lin, et al., 2015 • Oh, et al., 2014 • Rashid, et al., 2017
Middle East	• Al-Okaily, 2013 • Bakhsh, 2015 • Dashtestani, 2013 • Dogan, & Akbarov, 2016 • Tayan, 2017 • Waller, et al., 2015
North America	• Grimshaw, et al., 2017; • Park & Slater, 2014 • Stauffer , 2014
North Africa	• Saidouni & Bahloul, 2016 • Ghrieb, 2015
Europe	• Pérez-Paredes, et al., 2018
Europe/Asia	• Pagel, et al., 2018
Australia	• Pegrum, et al., 2013
Caribbean	• Martiz, 2015

Chapter 4
mHealth Interventions for Self-Management of Chronic Disease

Cristina A. Sumilang
Columbia University, USA

ABSTRACT

Chronic diseases are long-term conditions that require ongoing medical treatment and are among the leading causes of morbidity and mortality worldwide, imposing a high socioeconomic burden. Self-management is a critical element of chronic disease management and requires considerable effort from patients. The pervasive adoption of mobile health (mHealth) technology has the potential to revolutionize healthcare delivery and support patients in the self-management of their chronic conditions – thereby improving health outcomes and reducing healthcare expenditures. Although a variety of mHealth interventions are widely used by diverse populations for chronic disease self-management, there is limited evidence regarding their efficacy. This chapter explores the literature regarding the effectiveness of mHealth interventions in the self-management of hypertension, diabetes, chronic obstructive pulmonary disorder (COPD), asthma, and cancer. The mHealth issues pertinent to each disease process are addressed, as are the areas that warrant further research.

DOI: 10.4018/978-1-5225-9351-5.ch004

INTRODUCTION

A chronic disease is broadly defined as a physical or mental health condition that lasts more than one year and requires ongoing medical attention and/or which limits activities of daily living. Chronic diseases can be managed through medications and lifestyle modifications, but generally cannot be cured. They are among the most prevalent and costly health conditions in the United States. According to the Centers for Disease Control and Prevention (CDC), six in 10 American adults suffer from at least one chronic disease and four in 10 adults have two chronic conditions or more (Centers for Disease Control and Prevention [CDC], 2019).

Chronic diseases can lead to hospitalization, long-term disability, reduced quality of life, and death. In fact, chronic conditions are the primary causes of death and disability in the United States and are the leading drivers of the nation's $3.3 trillion in annual health care expenditures (CDC, 2019). Chronic diseases are responsible for seven out of 10 deaths in the U.S., killing more than 1.7 million Americans each year. In the U.S. alone, chronic diseases account for nearly 75 percent of aggregate health care spending, or an estimated $5,300 per person annually. In terms of public insurance, treatment of chronic diseases comprises an even larger proportion of spending: 96 cents per dollar for Medicare and 83 cents per dollar for Medicaid (Raghupathi & Raghupathi, 2018).

Current trends demonstrate that the number of individuals living with a chronic disease is only increasing. The nation's aging population—coupled with existing risk factors (poor nutrition, lack of physical activity, and/or tobacco/alcohol use) and medical advances that improve health and extend longevity—suggests these problems are only going to magnify if not effectively addressed now. Thus, the understanding, management, and prevention of chronic diseases are important objectives in order to provide patients with higher quality health care and enhance their overall quality of life (Raghupathi & Raghupathi, 2018).

Along with the rise in chronic disease prevalence is the unprecedented rise in mobile phone use. Approximately 77% of adults in the U.S. and 36% of the global population currently own a smartphone (Pew Research Center, 2018; The Statistics Portal, 2019). Smartphone use has rapidly increased over the past 10 years (18% in the U.S. and 15% globally in 2009), and the number of worldwide users is expected to hit 3.8 billion by 2021 (The Statistics Portal 2019; Takahashi, 2018). The widespread adoption of mobile technologies is opening new and innovative ways to improve health and health care delivery. The rapid growth in mobile phone use has been accompanied by a huge expansion in health and health-related behavior apps, and more than 100,000 of these are used by over 500 million people (Alessa, Abdi, Hawley, & de Witte, 2018; Food and Drug Administration [FDA], 2018).

In this way, a growing number of patients affected by chronic diseases are turning to smartphone applications in order to manage and improve their condition. It is important to note that 44% of American adults age 65+ currently use a smartphone compared to only 18% in 2013. This suggests that even the senior population, in which chronic diseases are most prevalent, have the ability to self-manage their condition with mobile technology (Pew Research Center, 2017). Moreover, research has shown that although rates of smartphone adoption may be lower in older populations, their willingness to use and engage with mobile devices for chronic disease management is equal to that of younger populations (Silva, Lawler, & Langbecker, 2019).

Self-management is considered to be an essential element in the management of chronic disease. Self-management demands that patients take an active role in managing their symptoms, treatments, psychological and physical effects, and changing lifestyle. However, achieving an optimum level of self-management is difficult and requires considerable effort from patients. Mobile health (mHealth) technology, defined as the use of mobile devices to deliver care, has the potential to facilitate and optimize patients' self-management of their conditions (Alessa et al., 2018). Interventions that incorporate mHealth can employ different features of mobile technologies, including (but not limited to) Short Message Service (SMS) text messages, interactive-voice response, telephone calls, Bluetooth, and mobile applications (apps) (Xiong et al., 2018). It is estimated that 50% percent of the more than 3.4 billion smartphone users have downloaded mobile health applications (FDA, 2018). Many health apps are targeted to support people with their self-management of chronic diseases by offering self-monitoring activities, reminders, tailored information, and feedback (Alessa et al., 2018). These apps are free or of low cost and have the potential to reduce morbidity and mortality and improve quality of life. mHealth interventions can play a crucial role in meeting National Health Service (NHS) policy goals by empowering patients to self-manage their chronic conditions—providing them with better access to personalized information and support for active involvement in treatment as well as producing significant savings in healthcare costs (McClean et al., 2016a).

There are a variety of innovative mHealth interventions that currently exist in order to support individuals with self-management of their chronic conditions; and such interventions are widely utilized by diverse populations throughout the world. However, the evidence concerning their efficacy is limited. This chapter synthesizes the literature regarding the effectiveness of mHealth interventions in the self-management of hypertension, diabetes, lung diseases, and cancer, which are among the most prevalent chronic conditions and leading causes of death. It addresses mHealth issues pertinent to each disease process, and it identifies areas needed for further research.

HYPERTENSION

The American College of Cardiology and the American Heart Association recently updated their guidelines to newly define Hypertension (HTN) as a blood pressure greater than 130/80 (American College of Cardiology [ACC], 2017). HTN is the principal risk factor for the development of cardiovascular disease and stroke, which are the first and fifth leading causes of U.S. deaths, respectively. It affects approximately 103 million American adults, which is nearly half of the U.S. adult population (American Heart Association [AHA], 2018). The World Health Organization (WHO) refers to HTN as the "silent killer" and acknowledges it as being the top modifiable risk factor causing the largest amount of disease burden worldwide. High blood pressure affects nearly a third of the adult world population and contributes to 8.88% of global disability-adjusted life years and almost one fifth of global deaths. Despite the massive efforts of heath care providers to control patients' blood pressure as well as the similarly strenuous attempts of public health campaigns to bring HTN awareness to the community, studies show that most hypertensive patients do not have their blood pressure at optimal levels (Xiong et al., 2018). Clinically significant effects and improvements in mortality can be achieved with relatively small reductions in blood pressure (McClean et al., 2016a). In 2015, the AHA issued a statement about the use of mHealth for cardiovascular prevention, and it included a section on effective blood pressure control through mobile apps. Many healthcare organizations are now incorporating mHealth strategies for HTN self-management due to their potential benefit of lowering blood pressure (Alessa et al., 2018). However, there is still a lack of reliable evidence that validates the effectiveness of their use (Parati, Torlasco, Ombroni & Pellegrini, 2017).

Effects of mHealth Interventions on Blood Pressure

Alessa et al. (2018) conducted a systematic review to assess the effectiveness of mHealth apps in lowering blood pressure (BP), as well as usability and patients' satisfaction with their use. This review included 21 studies that were conducted throughout the world. Together, the studies included a total of 3,112 participants with a mean age that ranged from 42.4 to 69.5 years old. The studies included 16 different apps that all had some similar characteristics and functionalities (see Figure 1). All the apps had self-monitoring capabilities for BP and other health data (medication adherence, physical activity, healthy eating and drinking, maintaining weight, sleep, and managing stress and coping). The second most common functionality was a reminder and alert component—one that prompts self-monitoring by reminding patients about their medication time, BP measurements, hospital visits, or personal

Figure 1.

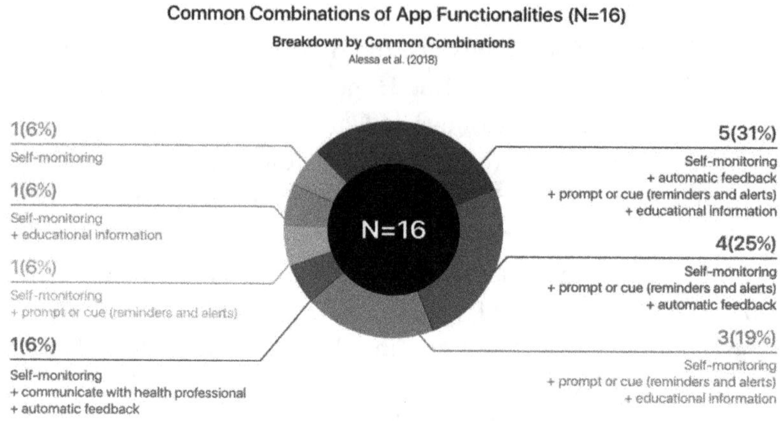

goals; or which alerts another person (e.g., a health professional) when a medication dose is missed or when BP is higher than the normal level. Educational information and automatic feedback were the next most common features (see Figure 2). The follow-up period of the studies ranged from 2-12 months. Of the 14 studies that assessed the effectiveness of mHealth apps in lowering blood pressure, 10 studies (71.4%) reported significant BP reduction (a systolic BP reduction range of -5.7 to -34.8 mmHg; and a diastolic BP reduction range of -4.9 to -12 mmHg). Due to variation in the studies' quality, the results are inconclusive as to which functionality combinations are most effective in lowering blood pressure. However, the data suggests that apps incorporating more comprehensive functionalities are likely to be more effective.

Figure 2.

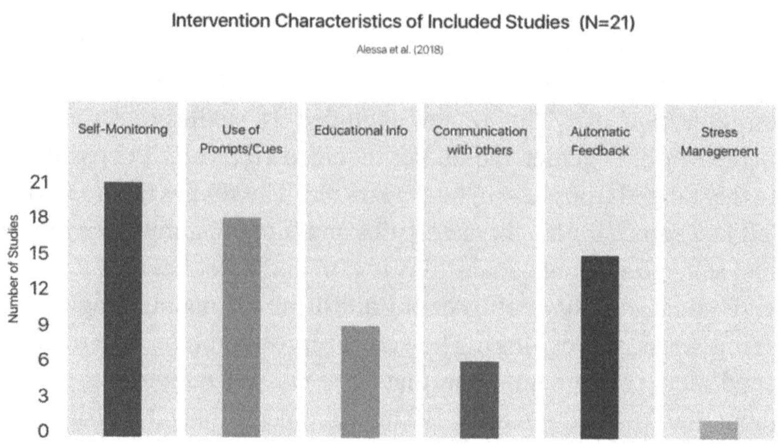

Two of the 21 studies explored app usability, and 9 studies assessed user satisfaction and attitudes towards the apps. Satisfaction questionnaires and interviews showed that the apps were easy to use, convenient, helpful in effectively communicating with health care professionals, and beneficial in hypertension management. Most participants stated both that they would continue using the app after the study and that they would recommend it to their friends (Alessa et al., 2018).

McClean et al. (2016a) conducted a systematic review to evaluate a variety of mHealth interventions that targeted BP. The review comprised of seven studies with 35-387 individuals per study and a total of 1,259 participants. The studies varied considerably in the nature and delivery of the intervention, but they all included mHealth interventions that provided self-management information and could be used either online or offline. The intervention had to function without the need for direct input from a health professional and must have been interactive and personally relevant to the users. Mobile apps were the most common mode of delivery (3/7 studies). One study used a website in which participants could upload BP readings through their smartphone and view trends and automated rules-based messages. One trial used a series of e-mails to deliver the intervention, whereas another offered the choice to participants of using either the study website or an interactive voice response system to monitor their progress. The last study used a telephone-linked computer system which spoke to patients using computer-controlled speech; and where patients communicated using the touch-tone keypad on their phones. The duration of the included studies ranged from 6 weeks to 10 months. Four of the seven (57%) studies demonstrated a significant reduction in systolic BP (MD -3.74 mmHg). Of the 6 studies that measured diastolic BP, four (67%) showed a considerable reduction (MD -2.37mmHg). In addition to BP, two studies reported a reduction in weight; and two studies indicated significant improvements in glucose, total cholesterol, and triglyceride levels. Though this review revealed that mHealth interventions can significantly lower BP, it could not determine the frequency, duration, or method of mHealth delivery that was most effective (McClean et al., 2016a).

mHealth Targeted to Medication Adherence

Patients' nonadherence to medications is widely acknowledged to contribute to the suboptimal control of hypertension (Xiong et al., 2018). Various factors have been shown to be associated with medication non-compliance; and mHealth-based interventions have the potential to target a number of these factors. For example, non-compliance resulting from a lack of information regarding the benefits of medication; a lack of information about how they work and how to take them;

misconceptions about side effects; complex or unclear advice; or poor recall of information may be addressed through simply worded SMS texts. Lack of social support has also been linked to poor medication adherence; and previous research has found that the receipt of text message-based interventions fills this void. mHealth interventions can be designed to target psychological factors such as a lack of motivation and low self-efficacy. The concept of social connectedness—a concept put into practice as part of these initiatives – can aid in faster healing (Mentor, 2011) and create a sense of a supportive affiliated community (Mentor, 2018). Existing interventions targeting medication adherence have employed mobile technologies to encourage self-monitoring of medication intake; encourage habit formation relating to medication-taking behaviors; provide medication information; and facilitate links to healthcare services where required.

Systematic reviews evaluating the effect of mHealth interventions have reported promising but inconclusive results relating to improved medication adherence (Palmer, Barnard, Perel & Free, 2018). Xiong et al. (2018) conducted a review involving mHealth interventions related to medication adherence and their potential to lower blood pressure. The review included 21 studies with sample sizes ranging from 10 to 1,372 participants, ages 40 to 69 years old. The studies were conducted in 12 different countries throughout the world. The platforms used for delivering mHealth interventions varied. Fifteen studies applied SMS texts for sending out reminders and/or health education and/or lifestyle-related recommendations. Nine studies utilized smartphone-based applications. Other mHealth intervention components included voice calls, emails, digital medicines, wireless BP monitoring, and electronic medication trays. The intensity and frequency of mHealth delivery varied between studies. The length of intervention ranged from 1 month to 1 year, though most lasted less than 6 months. Of the 21 studies, 12 (57%) found significant improvement in the patients' medication adherence levels. An additional 5 studies suggested that their mHealth interventions improved medication adherence, but the results were not statistically significant. Sixteen of the studies included blood pressure outcomes, among which 12 (75%) studies reported significantly greater BP reduction in the mHealth intervention group. However, the level of BP reduction varied widely between studies, with a systolic BP reduction ranging from 2.06 mmHg to 47.2 mmHg and a diastolic BP reduction ranging from 1.28 mmHg to 19.91 mmHg. Although the strength of the evidence is limited due to the variation across studies in their mHealth intervention design, outcome measurements, and study quality, Xiong et al. (2018) found that mHealth strategies offer promising potential in improving medication adherence and hypertension self-management.

Limitations, Conclusions, and Implications for Practice

The included systematic reviews demonstrate that the use of mHealth interventions for HTN self-management could lead to a greater reduction in blood pressure as compared to usual care, which entails medication administration combined with lifestyle modifications (diet and exercise) without the use of mHealth tools. Evidence also reveals that mHealth interventions targeting BP can concurrently improve weight and cholesterol levels. Each review contained a multitude of studies consisting of large sample sizes—studies that were conducted in various countries throughout the world. This indicates that these results are generalizable across diverse ethnic and gender populations, suggesting that mHealth can influence a wide range of hypertensive patients in different regions (Xiong et al., 2018). However, the effect size of blood pressure reduction varied across studies. This is likely due to the multifarious functionality and nature of app delivery and the differences in study duration and follow-up. Yet, it should be noted that only a 5-mmHg reduction in systolic BP is generally considered to result in a clinically significant reduction in the relative risk of stroke and heart disease (Palmer et al., 2018).

Further research needs to be performed in order to determine which functional combinations of mHealth delivery are most effective (Alessa et al., 2018). There is also a strong need for long-term clinical trials that evaluate user adherence and sustainability of mHealth interventions (Parati et al., 2017). Additionally, patient perspectives regarding usability and attitudes towards mHealth in BP self-management warrant consideration during development. The literature demonstrates that evidence is still insufficient for making definitive recommendations regarding clinical practice or policy. However, the benefits of employing mHealth tools in conjunction with routine care for BP self-management outweigh the possible risks. mHealth technologies may be a cost-effective method of health care delivery and have enormous potential to reduce HTN-related illness and death.

DIABETES

Diabetes is a chronic condition in which the body's ability to produce or respond to the hormone insulin is impaired, resulting in an abnormal metabolism of carbohydrates and thus, in prolonged periods of elevated glucose. Diabetes is considered a global public health problem that adversely impacts an individual's clinical, social, and economic aspects of life, ultimately leading to morbidity and mortality (Bonoto et al., 2017). More than 30 million Americans have some form of diabetes, which includes type 1, type 2, and gestational diabetes. An additional 86 million American adults have a condition called prediabetes, which puts them at risk for type 2 diabetes. In the United States, type 2 (insulin-resistant) diabetes comprises approximately 90-95% of all patients with diabetes, while type 1 (insulin-insufficient) accounts for 5%. Uncontrolled blood glucose elevation can lead to complications such as kidney failure, adult onset blindness, lower limb amputations, and heart disease. The CDC reports that diabetes costs the U.S. health care system and employers $245 billion each year in medical expenses and lost wages. The American Diabetes Association (ADA) recommends using a glycosylated hemoglobin (HbA1c) blood level as a measurement of glycemic control for both the diagnosis and the treatment of diabetes. A 0.5% reduction in HbA1c is considered to be clinically significant, as diabetes-related complications are directly proportional to HbA1c (Veazie et al., 2018).

Diabetes Self-Management

Management of diabetes varies, depending on the type and severity. Type 1 diabetes, an autoimmune condition usually diagnosed in childhood or adolescence, requires insulin administration as a life-long treatment. Type 2 diabetes management is often initially focused on lifestyle changes (diet and physical activity) and may include the use of oral hypoglycemic medications. Patients with type 2 diabetes may also need insulin in addition to, or in place of, other medications. Optimal glycemic control entails self-monitoring glucose levels multiple times per day and modifying insulin regimen, diet, and physical activity as needed. Diabetes self-management is considered to be a cornerstone of diabetes care and plays a crucial role in preventing diabetes-related complications. Components of self-management include diabetes education; healthy eating; physical activity; medication and device utilization; monitoring data to adjust behaviors and medication doses; preventing, detecting, and treating acute and chronic complications; and coping with psychosocial issues and problem solving (Veazie et al., 2018).

Similar to hypertension, clinicians and patients have started to use mHealth technologies to support diabetes self-management. In 2017, diabetes apps accounted for 16% of disease-specific apps available to consumers, second only to mental health apps. There are more than 1,100 different diabetes-related apps available for download in the iTunes App store for iOs and Google Play for Android. Diabetes apps vary in the functions they provide, including monitoring blood glucose measurements; nutrition database and carbohydrate tracking; physical activity and weight tracking; sharing data with clinicians or peers; social support; and messaging and reminders. The use of these features could theoretically enhance self-management plans, which in turn could lead to improved diabetes-related outcomes (Veazie et al., 2018). Unlike mHealth interventions for hypertension, practitioners are inundated with a vast amount of research that explores the efficacy of mHealth in diabetes management. Therefore, synthesis of this evidence is needed in order to provide clinicians and decision makers with reliable and accessible information. Greenwood et al. (2017) and Kitsiou et al. (2017) conducted a "review of reviews," in which they combined reports that had pooled evidence from multiple studies. This presents a higher level of evidence than a single systematic review alone, thus achieving more accurate results.

mHealth Interventions on Glycemic Control

Greenwood et al. (2017) analyzed 25 high quality systematic reviews that collected research from throughout the world. Each review consisted of 12 to 107 individual studies, and the number of participants ranged from 422 to 73,579. (Four reviews did not include the number of participants). The mean age reported was from 1 to 80 years old. (Nine studies did not report age ranges). Most of the included studies included outcomes in both type 1 and type 2 diabetes. Nine of the 25 reviews focused on app-based interventions, 9 included text messaging, 8 included wireless devices, 5 included e-mail or video conferencing, and 4 reviews identified decision support for diabetes education. The interventions in the reviews primarily featured four key concepts: 2-way communication between the participant and the health care team; patient-generated heath data (PGHD) analysis; general or customized education content; and automated and/or live feedback. Of the 22 reviews that measured HbA1c as an outcome, 18 (82%) reported significant improvements in HbA1c— with reductions ranging from 0.1% to 0.8%. In general, participants with type 2 diabetes improved HbA1c more than participants with type 1 diabetes. Nonetheless, this review demonstrates that utilizing mHealth technology for diabetes self-management notably improves HbA1c, which has a tremendous impact on health outcomes (Greenwood, Gee, Fatkin & Peeples, 2017).

Kitsiou et al. (2017) also combined results of systematic reviews to determine the effects of mHealth delivery on HbA1c. Their study included 15 reviews that were conducted across the world, with the number of participants ranging from 132 to 1,840. The number of studies per review varied from 2 to 21 (with a median of 10) and included 52 unique studies focusing on both type 1 and type 2 diabetes. The mHealth interventions were diverse in nature and utilized a wide range of technological innovations, including SMS text messaging; mobile applications; Bluetooth-enabled glucose meters; and secure websites/web-portals that could be accessed through the participants' mobile device for data entry and support. The most prevalent method for sending blood glucose measurements and/or receiving self-management support involved the use of SMS text messages (34 out of 52 studies). Twelve studies involved the use of web-portals that patient accessed through their cell phones in order to enter diabetes self-care data and feedback. In 19 studies, data entry and the transmission of glucose data were performed automatically via Bluetooth-enabled devices or other connectivity methods. In the remaining studies, data entry was performed manually. The majority of studies (38/52) provided educational support to patients via text and/or the Internet. The duration of follow-up ranged from 3 to 12 months. Glycemic control measured by HbA1c was the primary outcome of interest in all the included reviews. The pooled results demonstrated that mHealth interventions improve HbA1c as compared to standard care by as much as 0.8% for patients with type 2 diabetes and 0.3% for patients with type 1 diabetes—attesting that mHealth interventions are effective in diabetes self-management (Kitsiou, Pare, Jaana, & Gerber, 2017).

Most Effective mHealth Functions

Greenwood et al. (2017) observed that the incorporation of multiple modalities—such as text messaging or other mobile interventions and web-based tools in combination—was more effective at improving outcomes as compared to singular interventions. (See Figure 3 for intervention features associated with significant HbA1c outcomes). Similarly, Kitsiou et al. (2017) found that mean reductions in HbA1c appear to be more pronounced in interventions that combine text-messaging with Internet functionalities for self-monitoring of blood glucose and delivery of clinical feedback, including health education. Stand-alone text messaging interventions (a single media approach) had a smaller effect on HbA1c by 0.4%. Furthermore, mobile apps

Figure 3.

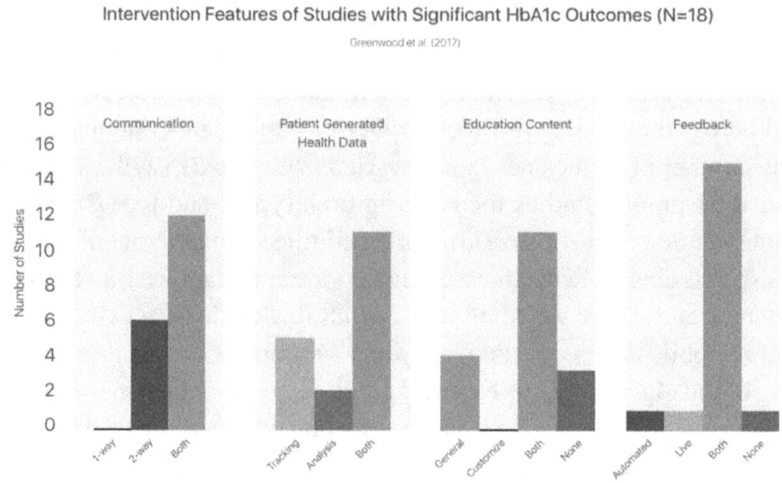

containing more than 2 features (e.g., medication reminders; assistance with diet and exercise; and blood glucose monitoring) show more favorable HbA1c results than apps with just one or two features. The frequency of engagement is also associated with glucose levels. Studies with daily interventions found a greater reduction in HbA1c as compared to those with weekly interventions (Bonoto et al., 2017).

A systematic review done by Garabedian et al. (2015) consisting of 20 randomized trials found that mHealth interventions that actively engage patients (via self-measurement, self-help education tutorials, or remote monitoring) and provide real time feedback or reminders via text messaging have a greater impact on HbA1c reduction. Those that provide an avenue of communication with the patients' health care provider are also found to be more effective (Garabedian, Ross-Degnan, & Wharam, 2015). Consequently, Wu et al. (2018) suggested that clinical decision-making features—defined as treatment recommendations made by algorithms alone without the participation of a health care provider—have limited efficacy. This report also demonstrated that apps which compromise complication prevention modalities (such as modalities associated with smoking cessation and hypoglycemia prevention) are associated with a greater HbA1c reduction (Wu, Yao, X & Li, 2018).

Other Outcomes Measured

The most critical adversity related to diabetes self-management is hypoglycemia (low blood sugar). Bonoto et al. (2017) analyzed 5 trials that reported hypoglycemic episodes and found variable results among the mHealth intervention and control groups. Three studies showed no difference in hypoglycemic episodes. Two studies determined that mHealth users had a lower risk, but only one of these two trials was statistically significant. An additional review done by Wu et al. (2018) showed similar results. Although these outcomes may not be favorable, it is crucial to note that mHealth utilization is not associated with an increased risk of hypoglycemic episodes—and that users reported that these tools actually decrease fear of hypoglycemia. Other improvements expressed by participants were awareness of hyperglycemic (high blood sugar) episodes, enhanced social relationships, and the perception that mHealth apps aid in the treatment of diabetes and in the adoption of healthier dietary habits (Bonoto et al., 2017).

Limitations

Though evidence reveals that patient utilization of mHealth tools improves glycemic control, the majority of the trials included were of a short-term duration (less than 12 months); thus, long-term benefits cannot be determined. Past reviews, unfortunately, show a decline of user engagement over the course of the intervention. Therefore, further research is warranted— in order to understand how patients engage with these mHealth interventions, and in order to learn how to adopt strategies for boosting adherence and sustainability (Nelson, Coston, Cherrington & Osborn, 2016).

An additional limitation is the reduced effect on HbA1c that mHealth had on individuals with type 1 diabetes as compared to those with type 2 diabetes. This may be due to a variety of factors ranging from higher initial A1c levels, less familiarity with glucose monitoring and the associated treatment, and perhaps less experience with self-management (Greenwood et al., 2017). Studies involving adolescents and young adults with type 1 diabetes frequently have higher HbA1c levels due to hormonal fluctuations, difficulty with increasing demands of diabetes self-care, competing priorities with school or work, and personal life/social stressors. Under these circumstances, mHealth interventions may have limited impact on glycemic control. However, as younger age groups demonstrate higher utilization of mobile devices, it is crucial for researchers to investigate incentive-driven mechanisms (e.g., social connectedness and gamification strategies) for this population (Kitsiou et al., 2017). The literature also illustrates that evidenced-based guidelines for diabetes self-management are usually neglected during app development due to a lack of expert involvement. For example, the current guidelines state that complication prevention strategies are critical components of diabetes management. Wu et al. (2018) found that such strategies are associated with greater HbA1c reduction but are uncommon app-based interventions. Additionally, prior research has shown that mHealth interventions are more successful when they are designed with a theoretical basis, and when the content of the intervention is dynamic and adaptive to the needs of each individual (Mentor, 2011; 2018). However, there is insufficient evidence to determine whether mHealth interventions that are adaptive in nature or based on behavioral change theories are more effective than others. The use or lack of theoretical models may explain variations in the effectiveness of mHealth tools for diabetes (Greenwood et al, 2017).

CONCLUSION AND IMPLICATIONS FOR PRACTICE

Current evidence establishes that the use of mHealth delivery for the self-management of diabetes improves glycemic control as compared to usual care. The studies analyzed in this review included participants of all age groups, education levels, ethnicities, and socioeconomic statues, which increases the generalizability of the findings. This suggests that mHealth tools have the potential to reduce HbA1c across diverse populations, thereby reducing diabetes-related complications and improving health outcomes worldwide. Although there has been tremendous progress in the use of technology in diabetes self-management, there remains a lack of empirical knowledge of how to integrate technology into new models of care, into the daily life of a person with diabetes, and into the workflow of the healthcare team. However, diabetes self-management support services are required during 4 key times—at diagnosis, during an annual health assessment, when complications arise, and during transitions in care (Greenwood et al., 2017).

Though an immense number of studies included in the reviews done by Greenwood et al. (2017) and Kitsiou et al. (2017) demonstrate that mHealth interventions could lead a significant improvement in HbA1c with minimal adverse events, further research is needed to investigate the impact of mHealth on health care utilization and expenditures (Kitsiou et al., 2017). If robust evidence shows that mHealth users have reduced hospitalizations, emergency room visits, and unanticipated doctor visits, then health care providers and society would be compelled to integrate these technologies into routine care and recommend them to patients as part of their practice or service. Despite limited research regarding the long-term effectiveness, economic advantages, and behavioral theories associated with mHealth delivery and variable HbA1c reduction among users, mHealth technologies have shown to significantly improve glucose levels, and may therefore reduce diabetes-related morbidity, enhance quality of life, and prolong life expectancy.

CHRONIC LUNG DISEASE: CHRONIC OBSTRUCTIVE PULMONARY DISEASE (COPD)

The World Health Organization (WHO) anticipates that COPD will become the third cause of mortality and the seventh cause of morbidity worldwide by 2030, imposing a high socioeconomic burden. COPD is an irreversible and progressive condition characterized by persistent airflow obstruction due to a chronic inflammatory response in the airways and lungs. The deterioration in lung function is associated with long-term exposure to noxious particles or gas, such as cigarette smoke, dust, fumes, or poor air quality (Sobnath et al., 2017). The natural course of COPD is interrupted by exacerbations—which are defined as a sustained worsening of patients' respiratory symptoms that are beyond normal day-to-day variability and may warrant medical treatment. Exacerbations are now a leading cause of hospitalizations, and an increased frequency of episodes leads to a faster decline in lung function, reduced quality of life, greater numbers of hospital readmissions, and increased health care costs (Shah, Velardo, Farmer, & Tarassenko, 2017). Therefore, treatment goals for

COPD aim to control symptoms, maintain normal activities, and minimize risks of exacerbation. However, research demonstrates low achievement of these goals (Morrison, Mair, Yardley, Kirby & Thomas, 2016).

COPD Self-Management

As with all chronic diseases, self-management is a critical aspect of evidenced-based care for those with COPD. The current guidelines advise that optimum self-management focus on smoking cessation, reducing exposure to indoor and outdoor pollution, promoting physical activity and a healthy diet, adhering to pharmacological therapies, and improving uptake of vaccinations (Morrison et al., 2016). mHealth interventions could address these complex challenges by delivering comprehensive, dynamic and individualized strategies that aim at motivating and engaging patients to positively adapt such behaviors. mHealth delivery may empower patients to play a more active role in managing their health; and the integration of these interventions in COPD self-management has the potential to improve respiratory symptoms, delay disease progression, reduce acute exacerbations, minimize hospital admissions, and enhance health-related quality of life (Alwashmi et al., 2016). Research over the past decade has focused on innovative methods and assistive technologies that facilitate self-management; and there are over 200 COPD management-related apps readily available to consumers (McCabe, McCann, & Brady, 2017; Sobnath et al., 2017). Therefore, mHealth is now being widely used in the self-management of COPD; but current evidence is still unclear as to whether these interventions are beneficial to patients (Yang, Wang, Yang, Hu, and Xiong, 2018).

Effects of mHealth Technology in COPD

Unlike mHealth interventions for hypertension and diabetes, the efficacy of mHealth interventions for COPD self-management is not primarily determined by numerical measurements, such as blood pressure or HbA1c level. Although objective measurements such as spirometry (lung function) and pulse oximetry (oxygen saturation) can reflect disease status, the effectiveness of mHealth delivery in the self-management of COPD is gauged by the severity of respiratory symptoms, the ability to do daily activities, the frequency of exacerbations, and hospital admissions (McCabe et al., 2017).

McCabe et al. (2017) conducted a systematic review in order to evaluate the effectiveness of mobile interventions aimed at facilitating, supporting, and sustaining self-management in individuals with COPD as compared to face-to-face or hard copy/digital documentary-delivered interventions. This review included three studies with a total of 557 randomized participants who had a clinical diagnosis of COPD at any stage of the illness. The mean age of participants was 64 years. The mHealth interventions were broadly similar across all studies, which focused on individualized education, support, and guidance towards goal achievement. The primary outcomes measured were hospital admissions, acute exacerbations requiring healthcare provider visits and/or additional treatment, and health-related quality of life. Health-related quality of life was measured using well recognized, valid, and reliable questionnaire tools. Data was reported by all three included studies at four weeks, at four months, and at six months. At these time points, the effect of mHealth technology on self-management and subsequent quality of life was significantly better as compared to that of the control group. This reflects improvements in symptoms, functional state, and mental state. However, there was no significant difference in hospitalizations or acute exacerbation between the two groups. Two of the three studies measured daily step count and indicated that the activity levels of those in the mHealth intervention group were significantly better at 4 weeks and 4 months. Thus, the evidence from this review suggests that mHealth delivery is effective in improving health-related quality of life and activity levels in COPD patients but does not have a significant impact on hospitalizations and acute exacerbations (McCabe et al., 2017). However, the following reviews do not coincide with these findings and show more favorable results.

A systematic review done by Yang et al. (2018) compared the efficacy of mHealth interventions with the usual care in the self-management of COPD by measuring hospital admissions and length of hospitalization. This review included eight studies with sample sizes ranging from 24 to 99 and a total of 391 participants with COPD. The lung function of the participants varied and their ages ranged from 63.5 to 81.0 years. The intervention included a mobile app to support patients in recording and monitoring their own physiological status (such as oxygen saturation levels, pulse rates, or blood pressure); or health behaviors (such as medication adherence, dietary intake, and exercise levels). Patients uploaded the data to their phones, which was sent to the networking/monitoring systems in order for healthcare providers to follow-up and personalize feedback. The aim of the intervention was to train patients and to promote self-monitoring and healthy lifestyle behaviors. The duration of follow-up ranged from 1 to 12 months. Of the 6 studies that that assessed the effects of mobile apps on hospital admission, the results revealed a significantly lower risk for hospital admission among patients using mHealth tools than that of the control group. Six studies measured the average days of hospital stay, but no significant difference was found between the two groups. However, five studies reported that mHealth systems could significantly improve exercise capacity and activity. Therefore, the results of this review show that mHealth utilization for self-management could reduce hospital admission and improve exercise and activity levels in patients with COPD, but that it might not reduce the average days of hospitalization (Yang et al,.2018).

Alwashmi et al. (2016) conducted a review in order to determine if smartphone interventions could reduce the frequency of exacerbations in individuals with COPD. There were 6 studies included in this review, with sample sizes ranging from 30 to 99 participants and a mean age of 70.5 years. The studies varied according to the type of mHealth intervention, frequency of data collection from the participants, and feedback strategy. However, smartphones in each study were primarily used to collect data about the daily symptoms of the patient. Participants used the phone to report physical activity level and daily symptoms, as well as heart rate and oxygen saturation. When an exacerbation was detected, patients were contacted in order to confirm the worsening conditions. One of the included studies used an automated feedback mechanism that advised to start medication in case of an exacerbation. The duration of follow-up ranged from 4 to 9 months. The rates of COPD exacerbations among participants receiving a smartphone intervention proved to be less as compared to those treated with the usual forms of care. Three of the six studies were included in a statistical analysis due to the similarities of the study designs. The pooled results indicate that those who received mHealth delivery had reduced COPD exacerbations by 80% as compared to usual care. The main objective of these mHealth interventions was the early identification of COPD exacerbations. Early identification allows the patient and the health care team to intervene successfully, thus improving the management of COPD and reducing exacerbations (Alwashmi et al., 2016).

Patient and Healthcare Provider Perspectives

In order to allow for the development of effective mHealth interventions, and in order to optimize engagement with these interventions, the patients' needs and the current healthcare context should be thoroughly investigated. A deeper understanding of both the patients' and the healthcare providers' perspectives towards mHealth delivery for COPD and exacerbation-related self-management will greatly enhance developing strategies with optimal usability and feasibility. Although mHealth technology is of emerging interest in health care, patient and clinician perspectives towards using mHealth for self-management are relatively unexplored. Therefore, Korpershoek et al. (2018) performed a qualitative study in order to survey both COPD patients' and healthcare providers' willingness to use mHealth for self-management of exacerbations; to identify potential benefits and barriers of using mHealth; and to uncover their needs and preferences regarding the potential content of mHealth interventions. A total of 13 COPD patients and 6 healthcare providers participated in focus group interviews. Maximum variation in COPD patients was reached in terms of disease severity, age, sex, and mobile technology usage (Korpershoek, Vervoort, Trappenburg & Schuurmans, 2018).

Benefits and Barriers

Patient and healthcare providers (HCPs) had similar perspectives regarding the potential benefits and barriers of mHealth delivery for COPD self-management. Most importantly, they thought that mHealth may contribute to an awareness of symptom deterioration and the identification of possible causes. In their opinion, mHealth interventions could determine the patients' current health status, which could in turn increase patients' self-empowerment towards communication with HCPs. Additionally, some patients felt that using mHealth strategies could support them in taking the necessary actions—thereby eliminating feelings of insecurity and reducing the threshold for contacting an HCP. According to both patients and HCPs, potential barriers to using mHealth tools were avoidance of confrontation with the disease, preference for personal contact with a heath care provider, difficulties with displaying feelings in an application leading to invalid patient measures, and lack of trust in advising characteristics of an mHealth intervention. Furthermore, an important potential barrier according to HCPs is a limited group of eligible patients. HCPs expressed doubts regarding eligibility based on disease severity, limited health literacy, and a lack of patient motivation and enthusiasm for self-management and mHealth utilization (Korpershoek et al., 2018).

Preferences Regarding Content of an mHealth Intervention

Both patients and HCPs had varied preferences regarding the content of an mHealth intervention designed to support self-management behavior and the intensity of its use. However, they all believed that an mHealth intervention should be a multi-component and tailored tool that improves patients' self-management skills over time by adequately detecting exacerbations, providing decision support to take prompt actions, and offering feedback on self-management behavior Korpershoek et al., 2018).

HCPs' Perspectives Towards Their Role Regarding mHealth Use

Most HCPs believe their role regarding patient utilization of mHealth strategies for self-management of COPD is to estimate patient capabilities for mHealth use based on the patient's motivation, technical and self-management skills, and cognitive level. However, it is ultimately the patient's responsibility to be proactive in using mHealth tools in order to achieve positive outcomes. From the HCPs' perspective, an mHealth intervention should be suitable for a wide range of patients and could support them in developing self-management skills over time. The HCPs role should then focus on monitoring self-management behavior, providing feedback, and setting future goals (Korpershoek et al., 2018).

Limitations, Conclusions, and Implications for mHealth in COPD

Although the literature demonstrates that the utilization of mHealth tools in the self-management of COPD achieves better outcomes as compared to usual care, there are several limitations of the featured reviews. There was high variability of the results due to the small number of studies and the relatively small numbers of participants within each review. Furthermore, each of the three reviews included studies of short-term duration and follow-up (most lasting less than one year). Therefore, patient adherence and long-term benefits of mHealth interventions cannot be determined. Nonetheless, the evidence suggests that mHealth strategies for the self-management of COPD has the potential to reduce the frequency of exacerbations and hospitalizations; enhance activity levels and exercise capacity; and improve health-related quality of life. The studies were conducted in different countries throughout the world and included patients of a wide range of ages and at various stages of their disease, indicating generalizability of the findings. In other words, mHealth tools can assist diverse populations of COPD patients in optimizing outcomes. The mean age of participants was that of older adults, suggesting that the senior population has the ability to utilize mobile technology.

It is of note that the literature search primarily presented studies that used mHealth technology supported by healthcare monitoring—but that few studies focused on the sole use of mHealth tools in promoting self-management for people with COPD, suggesting that this idea is still in its infancy and that a great deal of additional research is required for definitive confirmation of its true effects (McCabe et al., 2017). In addition, the evidence suggests that mHealth interventions are most effective in COPD self-management when they provide some type of feedback mechanism (either automated feedback or feedback provided by a health care provider). Most of the available studies only include a feedback component as part of the mHealth strategy, since early detection of exacerbations is key to reducing hospitalizations, healthcare costs, and morbidity. However, the implementation of a mixed methods research design to investigate the validity and clinical utility of mHealth interventions could establish other successful components, such as personalized education or social networking (Alwashmi et al., 2016).

Additional rigorous experimental research and clinical trials are crucial for determining the sustainability and long-term effectiveness of mHealth tools in the self-management of COPD. These trials should also analyze the economic benefit of mHealth strategies by calculating the cost of the intervention, lost wages from work, expenses from hospitalizations, and unexpected HCP visits (McCabe et al., 2017). It is also essential for researchers to evaluate patients' perspectives of mHealth tools for self-management of their COPD in order to optimize engagement, feasibility, and usability as well as to incorporate opportunities and motivational strategies designed to achieve behavioral change (Kopershoek et al., 2018). Although stronger evidence is required for health care providers and policy makers to implement mHealth tools into practice for the self-management of COPD, research demonstrates the potential benefit in improving outcomes without any indication of harm. Therefore, mHealth interventions could serve as a useful complementary treatment approach to standardized care. Health care providers can take an active role in assessing patients' eligibility for mHealth use for self-management; recommending such interventions if deemed eligible; discussing self-management behavior during consults; and providing necessary feedback (Kopershoek et al., 2018).

CHRONIC LUNG DISEASE: ASTHMA

Asthma is a common chronic inflammatory disease of the airways characterized by respiratory symptoms, such as coughing, breathlessness, and wheezing. Asthma affects more than 334 million people worldwide and it is predicted that the number of individuals with asthma will increase by 100 million or more by 2025 (Farzandipour, Nabovati, Sharif, Arani & Anvari, 2017). In the United States alone, 24 million people suffer from asthma (8.6% of children and 7.4% of adults) (Tinschert, Jakob, Barata, Kramer, & Kowatsch, 2017). According to the WHO, asthma is both the most common non-communicable disease and the third-ranking cause of hospitalization among children. A number of cost analysis studies have reported that the annual economic cost of asthma due to direct medical expenditures, as well as indirect expenses from lost school and workdays, amounts to more than $56 billion in the United States. Combined with the increasing cost of health care services, its high prevalence rate, and the need for ongoing treatment throughout a patient's lifetime, asthma represents a major societal burden as well as substantial challenge to the traditional models of healthcare (Al-Durra, Torio, & Cafazzo, 2015).

Asthma Self-Management

Asthma is characterized by reversible airflow obstruction and variable symptoms that are often underestimated by patients. Therefore, optimizing symptom control is a key treatment goal, with the aim that patients lead a full, active and impeded lifestyle (Morrison et al., 2016). There are well-developed, evidenced-based asthma management guidelines for healthcare providers, which recommend comprehensive patient education on asthma self-management. Adequate self-management has beneficial outcomes such as reduced emergency department visits and hospitalizations, healthier lung function, few symptom-related disturbances (e.g., nocturnal awakening and restricted activity), and less use of rescue medications. Despite effective patient education and pharmacological therapies, asthma control remains suboptimal, and patient adherence to preventative medication is reported to be as low as 30%. Therefore, asthma continues to be a growing cause of morbidity and mortality (Simpson et al. 2017; McClean et al., 2016b). Due to poor implementation and under-utilization of asthma self-management programs, there is a need for innovative self-management methods (Farzandipour, 2017).

Since smartphones have become an integral part of life, mHealth systems are promising tools that could revolutionize asthma self-management through automating and personalizing routine aspects of education, monitoring, and support – while simultaneously providing patients with constant and convenient access to detailed, personalized feedback (McClean et al., 2016b). mHealth applications, in particular, could enable low-cost and clinically efficacious interventions for asthma—and there are over 200 available mobile apps available for asthma patients. With virtually unlimited scalability and availability, apps could empower patients in self-management by delivering evidenced-based interventions. Moreover, data obtained through mobile phone sensors and connected medical devices (e.g., smart inhalers) can be used to deliver self-management interventions tailored to the specific needs of patients, thus increasing the intervention's efficacy (Tinshert et al., 2017). However, current literature indicates that the effects of mHealth utility for asthma self-management are still inconclusive (Simpson et al., 2017).

Effects of mHealth Technology in Asthma

Farzandipour et al. (2017) conducted a systematic review in order to evaluate the outcomes of patients who used mHealth apps for asthma self-management and to assess the functionalities of effective interventions. This review included 10 studies that were done in various countries throughout the world. The sample sizes of the studies ranged from 20 to 288 (with a median of 42.5), and the participants included adolescents and adults with varying severity of asthma. The functionalities of the apps were categorized into seven groups: inform, instruct, record, display, guide, remind/alert, and communicate. Most studies (80%) employed more than one functionality. Nearly all (90%) interventions had the functionality of recording user-entered data; 50% of the apps provided education information and reminders to patients; 40% graphically displayed user-entered data and used apps to communicate between health care providers and patients; and 30% of them provided instructions and guidance based on user-entered information. The duration of interventions varied from 7 days to 6 months (with a median of 3.5 months). These studies evaluated several outcomes, which were classified into three main categories: clinical (e.g., asthma control; exacerbations; lung function), patient-reported (e.g., medication adherence; quality of life), and economic (e.g., healthcare visits; hospitalizations; days off work). Nine of the 10 studies (90%) demonstrated that mHealth apps had a positive impact on most of the outcomes measured. Seven of the 10 studies assessed the impact of mHealth apps on clinical outcomes (lung function; asthma control; exacerbations). Of the 7 studies, there was a statistically significant improvement for all clinical outcomes, especially asthma/symptom control. Regarding patient-reported outcomes, 3 of the 4 studies (75%) that evaluated mHealth effect on quality of life showed a significant positive impact. However, the effect on medication adherence had mixed results. In general, the results of this review showed that the number of healthcare visits and hospital admissions (economic outcomes) improved in some studies but did not improve in others. According to the evidence observed in this review, mHealth app-based interventions are effective in the self-management of asthma by improving clinical outcomes (asthma control; lung function; exacerbations) and enhancing quality of life. Although mHealth apps have the potential to improve medication adherence in chronic disease by providing reminders, their effects on asthma is mixed; and consequently, there is a need for further studies in this area. Furthermore, it was reported that studies of a short duration and of a small sample size may be the cause of insignificant economic-related outcomes (Farzandipour et al.,2017).

Hui et al. (2016) reviewed the evidence in order to identify which information and communication technology features implemented in mobile apps to support asthma self-management are associated with adoption, adherence to usage, and clinical effectiveness. The review included 12 randomized controlled trials done in various countries. The number of participants for each intervention ranged from 16 to 300. Most studies included both teenagers and adults with varied severity of asthma. Of the 12 interventions, there were 3 mobile phone apps, 4 web applications, 3 SMS texts, 1 electronic inhaler reminder system connected with a web app, and 1 customized asthma monitoring system. Within the 12 studies, there were 10 common features grouped into 7 themes: education, asthma diary, action plan, medication adherence, facilitating professional support, raising patients' awareness of asthma control, and decision support for the healthcare professional. Eleven of the 12 interventions included more than 1 feature. Four interventions included 5 or more features. Eight included an asthma diary, 9 an action plan, and 11 professional support. The duration of trials ranged from 3 to 30 months. The measured clinical outcomes included asthma control, quality of life, and exacerbations. In 6 of 11 studies (55%), researchers reported improved asthma control over time scales of 3 to 30 months in the intervention groups. Of the 8 studies that reported asthma-related quality of life, only 50% found that quality of life improved over 6 to 30 months. Five trials reported outcomes relevant to exacerbations (e.g., emergency department visits), but none of the interventions were associated with a significant clinical impact (either positive or negative).

Only 1 included study reported adoption of an asthma action plan. A significant increase in asthma action plan adoption was reported in both intervention groups (web-based monitoring, from 2% to 88%; web-based specialist support, from 3% to 55%) compared to the usual care group (0% to 6%). There were no interventions that explicitly reported adoption of, or adherence to, an mHealth intervention; since there was no long-term follow-up after the end of the trials. However, data usage and drop-out rates may be indicative of the general level of interest in the mHealth tool. Only 2 studies reported data transmission, which showed inconsistent results. However, there was a low rate of participant withdrawal and loss to follow-up, indicating interest in app engagement. The outcomes of this review are in accordance with the review done by Farzandipour et al. (2017), with respect to its finding that mHealth strategies to support asthma self-management can be effective with regard to asthma control and quality of life. However, the degree of impact is variable. The mHealth interventions had no significant effect on exacerbations, which conflicts with the evidence synthesized by Farzandipour et al. (2017) which showed fewer exacerbations with mHealth use. The featured studies suggest that mHealth interventions may empower users to establish an asthma action plan as well as to adopt mHealth tools for overall self-management. Moreover, this review cannot provide clear evidence as to which combinations of mHealth features are most essential for effectiveness due to inconsistent results of the trials (Hui et al., 2016).

Perspectives on mHealth for Asthma Self-Management

Asthma is a lifelong condition that requires continuous attention from both healthcare professionals and patients. If mHealth systems are to be utilized in routine practice, it is essential to understand patients' and healthcare professionals' perspectives regarding these interventions. Simpson et al. (2017) conducted a mixed method qualitative study using focus groups and questionnaires in order to determine: 1) experiences and perceived uses of mHealth systems for asthma self-management; 2) views on what measurements would be useful in managing asthma; and 3) perspectives on the acceptability of and barriers to using mHealth systems for asthma self-management. The focus group included 18 individuals with asthma (9 men and 9 women) and five healthcare providers (two doctors, two asthma nurses, and a physiologist). The questionnaire was completed by 186 individuals with asthma (who had a mean age 40) and 63 healthcare professionals (including general practitioners, hospital doctors, asthma nurses, practitioners of other healthcare disciplines) (Simpson et al., 2017).

Experiences and Perceived Uses of mHealth for Asthma Self-Management

Individuals with asthma had considerably varied experience with mHealth; and healthcare professionals' experience with mHealth delivery was limited to the use of such systems during research projects. When those with asthma were asked what they would like from mHealth, patients most requested an mHealth system to monitor asthma over time (72%) and to collect data to present to healthcare teams (70%). Healthcare professionals most frequently selected functions that alert patients to deteriorating asthma control (86%) and advise them when to seek medical attention (87%). Furthermore, 46% of patients and 79% of healthcare professionals stated that they would find it useful to have an asthma action plan that is incorporated into an mHealth tool (Simpson et al., 2017).

Useful Measurements for Managing Asthma

Participants were questioned regarding measurements they believe are useful for the support of asthma self-management. Lung function measurements (71%) were commonly identified as being helpful in maintaining asthma control. This was linked with patients' perceptions of a connection between asthma control and lung function, and their perception of when to take appropriate action. On the contrary, healthcare professionals believed that measuring inhaler technique (87%) and medication adherence (89%) would be comparatively more helpful for asthma control. Measurements regarding environmental conditions were believed to be helpful by 71% of patients and 68% of healthcare professionals (air pollution and allergen exposure are associated with increased frequency of asthma attacks) (Simpson et al., 2017).

Acceptability and Barriers to Using mHealth

The ability to personalize mHealth systems is a vital component influencing participant acceptance of its use in the self-management of asthma. Discussions highlighted that different populations (e.g. children and the elderly and those with differing asthma severity) have different user requirements. Additionally, the interpretation of the data, either by automated systems or by the individuals themselves, was highlighted as being a potential source of error that deserves important consideration. Consequently, only 12% of patients indicated that they would accept recommendations to change their medication based on feedback from mHealth. Data security is also a potential barrier. Patients expressed opposing views with regards to data security, with some patients being unconcerned how their data was managed and others being insistent that data security is of the utmost importance (Simpson et al., 2017).

Limitations, Conclusions, and Implications For mHealth in Asthma

The literature regarding the effectiveness of mHealth systems in asthma self-management reveals inconsistent results; and thus, it has a more variable impact as compared to those used for the self-management of the chronic diseases previously discussed. The small treatment effects of mHealth interventions in asthma are similar to those seen in diabetes type 1, indicating that adolescents may have a substantial effect on unbalanced results. Unfortunately, these reviews did not conduct a subgroup analysis in order to determine if adult populations had more favorable outcomes than adolescents. In addition, most of the studies included small sample sizes and were of short duration, which may have contributed to the insignificant results.

Although impact is variable, the literature shows that mHealth interventions have the potential to be effective in supporting asthma self-management, especially with regard to symptom control, lung function, and quality of life. More rigorous, extended trials need to be done in order to establish long-term effectiveness and to resolve unclear effects of mHealth, such as those related to medication adherence and economic benefits. It is also necessary for further research to determine the specific combination of features that are associated with adoption, adherence, and improved outcomes. It was established that both healthcare providers and patients advocate for mHealth use in asthma self-management, though their expectations from the interventions differ. Numerous factors have been identified that would need consideration during development; and developers should incorporate these opinions, as well as include theoretical frameworks for behavioral change when creating new mHealth devices (Simpson et al. 2017; Al-Durra et al., 2015). Due to the inconsistency of the evidence, the potential for implementation of mHealth strategies into routine practice for asthma self-management trails far behind those used for other chronic diseases. However, with there being over 200 asthma-related applications available through the iTunes store and Google Play, a healthcare provider can encourage a patient to engage with an app that tailors to the specific needs of the user in order to support asthma self-management (Tinschert et al., 2017).

CANCER

Cancer is a collection of related diseases in which abnormal cells divide uncontrollably and spread into surrounding tissues (National Cancer Institute, 2015). Cancer is a leading cause of death worldwide, and the global burden is increasing in both developed and developing nations as the world population ages and as established risk factors associated with economic development and urbanization become more prevalent, including smoking, obesity, sedentary lifestyle, and lower parity. In the United States alone, more than 1.7 million people are diagnosed with cancer each year, and more than 600,000 affected individuals die of the disease (Uy et al., 2017). By 2020, the number of new cancer cases is expected to increase to nearly 2 million per year. Although mortality rates are declining, cancer is still the second leading cause of death in the United States (CDC, 2018).

Effects of Text Messaging Interventions on Cancer Screening Rates

The cancer mortality rate in the United States has dropped 23% since 1991 due to population-level screening tests, which detect early cancers and facilitate treatment before the disease becomes clinically apparent. Currently, there are only four screening tests that have shown to reduce mortality: 1) Mammography reduces breast cancer mortality; 2) Papanicolaou (Pap) testing, combined with human papillomavirus (HPV) testing, reduces the risk of death from cervical cancer; 3) Colonoscopy, sigmoidoscopy, and fecal occult blood test (FOBT) are the three tests that reduce colorectal cancer mortality; and 4) Low-dose computed tomography (CT) has shown to reduce lung cancer deaths in heavy smokers. The evidenced-based guidelines regarding the timing and frequency of screening depends on which organization guidelines the healthcare provider and patients choose to follow (Uy et al., 2017).

Several studies have demonstrated both patient interest in SMS text reminders for cancer screening appointments and a positive impact on screening rates. Uy et al. (2017) conducted a systematic review to assess the effect of text messaging interventions on increasing patient adherence to screening recommendations for breast, cervical, lung and colorectal cancer. However, no data was found regarding lung cancer screening reminders. The review consisted of 9 studies with sample sizes ranging from 500 to 48,901 with a total of 77,099 participants. Seven of these studies used a single or set of text messages delivered in 1 day. One study delivered several text messages over 7 days, and the other delivered individual texts at 1-month intervals until screening occurred (up to three texts maximum). The control group received usual care, such as letters, brochures, and phone call reminders. The length of follow-up from text delivery to outcome measurement ranged from 2 days to 6 months. The results of the review found that SMS texts can moderately increase screening rates for breast and cervical cancers and may improve colorectal cancer screening to a lesser degree. Across all studies, text message interventions led to increases in absolute screening rates of 0.6% to 15% and relative screening rates of 4% to 63%. Specifically, SMS texts increased breast and cervical cancer screening rates from 4.5% to 15% and colorectal screening rates from 0.6% to 3.3% (see Figure 4). Text reminders for colorectal cancer screening were thought to be less effective, due to the complexity of the tests (in comparison to those for breast or cervical cancer screening). Furthermore, the included studies had a relatively short duration of follow-up, which may have resulted in an underestimation of the SMS text effect, especially if there were substantial delays in scheduling the test.

Figure 4.

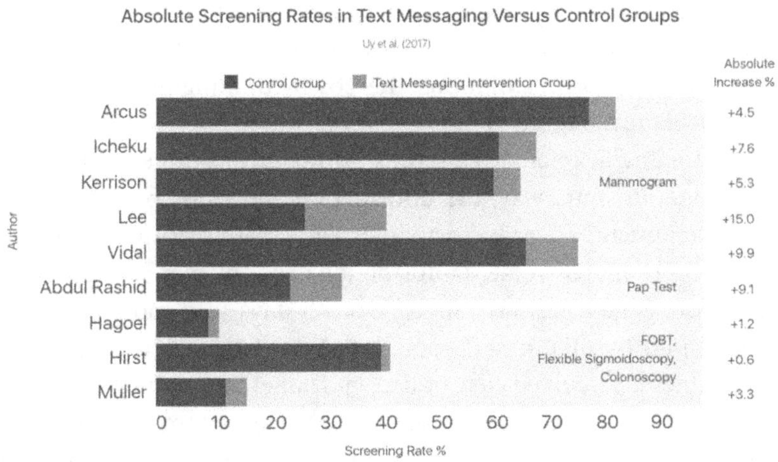

mHealth in Cancer Self-Management

Through earlier detection and improved treatments, the cancer survival rate has dramatically increased. Subsequently, there has been a shift from inpatient to outpatient and community cancer care, requiring patients to self-manage their condition at home. This change in care compels patients to take a more active role in their own treatment and survivorship. Patients are often faced with an uncertain future, unfamiliar tests and procedures, complex decisions about treatment options, treatment-related side effects, and lifestyle changes. While many people with cancer want as much information as possible about their condition and related issues, studies across the world have reported very high rates of unmet information needs. In addition to limiting patients' ability to participate in their care, there is evidence that unmet information needs are associated with a lower quality of life, loss of control over one's life, increased anxiety and depression, and dissatisfaction with care. Information-based mHealth interventions have the potential to support cancer self-management due to its ability to reach (and meet) a wide variety of patients at the point of need; lower cost, as compared to traditional healthcare interventions; and enable access to tailored healthcare to those in resource-poor settings or to those facing barriers to traditional health care. Similar to mHealth strategies for self-management of other chronic conditions, there has been substantial development of mobile interventions to support patients in the self-management of cancer (Richards et al., 2018).

A systematic review done by Richards et al. (2018) described the use of mHealth interventions designed to enable patients to meet their cancer-related information needs. A total of 20 studies were analyzed and sample sizes ranged from 4 to 125 participants. Two interventions were primarily designed to directly increase patients' knowledge of their upcoming surgical operations and coping with cancer-related pain. One intervention primarily aimed to improve patients' communication of symptoms to clinicians in consultations, thereby facilitating information exchange. The primary aim of the remaining 17 interventions was to improve the monitoring and management of treatment-related symptoms. Although these interventions attempted to educate patients in some way, the information and skills provided were solely related to their treatment. There were no interventions that primarily aimed to meet patients' full range of illness-related information needs by increasing understanding of their condition. Study periods ranged from 5 days to 6 months. The findings were organized into two themes: 1) acceptability of the interventions, including perceived usefulness, perceived ease of use, and adherence to mobile interventions; and 2) benefits of the interventions, which included symptom management, patient empowerment, reassurance and reduced anxiety, patient-clinician communication, and health-related quality of life

Acceptability

The mobile interventions were perceived useful by most patients, particularly the self-care advice provided in response to symptom reports. Patients reported that the information provided them with expectations for their treatment, reminded them to watch for symptoms, and suggested helpful home remedies. However, it was conveyed that patients with lower education and chemotherapy-naïve patients rated the intervention as being significantly more useful than did those with higher education. Almost all patients found the mobile interventions easy to use regardless of age, cancer type, and experience with technology. Studies indicated that daily use of an mHealth tool was nonintrusive on daily routines; and demonstrated that participants encountered none or very few technical problems. Adherence to interventions was generally high, regardless of the length of the study; however, there was considerable variation in the usage of the different interventional components within and between studies (Richards et al., 2018).

Benefits of the Interventions

Most people perceived the mobile interventions to be helpful in monitoring their treatment-related symptoms; and patients reported improved symptom management. There was evidence to suggest that this led to the reduced use of unnecessary health care services. In addition, symptom monitoring empowered patients to participate in their care and to better manage their condition due to increased knowledge and symptom management strategies provided by mobile interventions. Furthermore, many patients believed that communication with clinicians had improved or that their relationship with clinicians had strengthened as a result of the interventions. However, there were mixed findings regarding patients' reported anxiety and health-related quality of life. Some studies indicated that mHealth delivery reduced patient anxiety levels, thereby improving health-related quality of life. Other studies suggested that increasing patients' knowledge of treatment could potentially increase their anxiety (Richards et al., 2018).

Though the evidence demonstrates that mHealth delivery is effective in supporting self-management of cancer for patients undergoing treatment, many cancer survivors suffer from long-term symptoms that are sustained for years after treatment. Therefore, strategies for promoting self-care are integral to the adaptive functioning of patients at all stages of the cancer continuum (Dawlow & Wen, 2016). The incorporation of mHealth may be an effective approach to improve cancer survivorship by supporting self-management. Cancer survivors suffer from a variety of symptoms that can impact their quality of life when uncontrolled or poorly managed. Cancer-related fatigue is the most prevalent symptom, affecting between 58% and 90% of cancer survivors. An analysis of the National Health Survey revealed pain, psychological distress, and insomnia in 34%, 26%, and 30% of cancer survivors, respectively—figures which are significantly higher than in the general population. These symptoms are often considered as being part of a symptom cluster as they frequently occur together and may be linked by common biological processes. They are often under-diagnosed, difficult to treat, and can persist for years after the completion of treatment—thereby adversely impacting survivors' engagement in work, social, and personal activity. Therefore, a systematic review was done by Silva et al. (2019) in order to assess the effectiveness of mHealth apps for self-management in improving pain, psychological distress, fatigue, and sleep outcomes in cancer survivors.

The review included seven studies that used six different mHealth apps. Two interventions aimed to improve the management of patients' symptoms, two interventions aimed to promote health behavior change, and two interventions aimed to achieve both symptom management and behavior change. The apps displayed a range of features, including self-administered symptom questionnaires and progress tracking (being most common), survivorship care plans, and mind-body exercises. To engage users, interventions included features such as video demonstrations of prescribed exercises, and alerts sent via the app. All interventions were designed to be used daily. The duration of the recommended period of use ranged from 5 to 24 weeks. All outcomes were assessed using validated tools, most commonly the European Organization for Research and Treatment (EORTC) Quality of Life Tool (QLQ-C30). Four of the seven included studies reported data relating to fatigue, which improved across all studies. The improvement reached statistical significance in two weeks. Of the four studies that measured pain as an outcome, three (75%) reported that the mHealth intervention group had lower mean pain scores than the control group; however, only 2 of these studies reached statistical significance. Sleep outcomes also improved in both studies that reported data relating to insomnia or sleep disturbance; but only one of these studies was statistically significant. Two studies examined mHealth effects on psychological distress. One reported a mean reduction in psychological distress over 4-week and 8-week periods; however, the other study reported no difference in this outcome when comparing the intervention and control groups. Therefore, this review reveals that there is emerging evidence that mHealth interventions that support self-management can improve pain and fatigue symptoms in cancer survivors—with some data promising that psychological distress and sleep outcomes can also be improved. Despite differences in app delivery, they were designed specifically for the target population; delivered an evidenced-based intervention that was customizable; and could be accessed at any time to deliver real-time benefits to the user. These findings highlight a potential mechanism by which mHealth can be used to improve outcomes for cancer survivors, regardless of specific app design features (Silva et al., 2019).

Limitations, Conclusions, and Implications for mHealth in Cancer

The literature demonstrates that mHealth interventions are effective in supporting the self-management of cancer along the cancer care continuum. mHealth delivery has the ability to detect early stages of cancer through routine screening, as well as to improve outcomes in both patients undergoing treatment and long-term cancer survivors. The studies included in the systematic reviews report improvements in patients from various countries with different cancer types, indicating that mHealth has a large potential reach that can transcend diverse population characteristics. Interestingly, participant age was relatively high across studies, with the mean varying from 50 to 69 years, which challenges perceptions that mHealth is less likely to be used by older populations (Silva et al., 2019).

The review by Richards et al. (2018) established that diverse patients with cancer perceived mHealth to be an acceptable intervention that has various benefits for patients and clinicians. Symptom-monitoring interventions that provide treatment-related information have the potential to improve patients' self-management of their condition and provide clinicians with a better understanding of patients' symptom experiences, while improving the patient-clinician relationship. This may lead to the earlier detection of treatment-related side effects and timely intervention, which could reduce costs for the health care system. mHealth tools may also sustain or improve patients' well-being while receiving treatment. Although there was a high rate of acceptability and usability among patients, the review did not report on the perceptions and experiences of the healthcare professionals who participated in some studies. In order for these interventions to be fully realized in standard practice, it important to assess whether clinician engagement creates a significant burden beyond the normal workload.

The evidence synthesized by Silva et al. (2018) revealed that mHealth interventions that support self-management can improve pain, fatigue, psychological distress, and sleep disturbances in cancer survivors. However, few studies reported on outcomes related to sleep and psychological distress, and inconsistent results suggest the need for further research. Additionally, the review did not investigate specific characteristics of the population that may have influenced outcomes, nor did it report on mHealth adoption, usage, adherence, or user satisfaction. It is essential for further studies to assess such data in order to identify barriers, make improvements, and translate this research into routine cancer care.

Despite the potential utility of mHealth applications for supporting cancer self-management, research is still in its early stages; and there are relatively few interventions in this area as compared to emerging mHealth innovations for other chronic diseases. The mHealth interventions included in these reviews are reflective of the available literature, which focus solely on symptom monitoring and management (Silva et al., 2019). More comprehensive interventions are required for patients currently receiving treatment in order for them to meet their full range of illness-related informational needs. There are many cancer-related information resources and services that are developed and run by reputable cancer charities and health organizations. mHealth developers can incorporate these existing services within the interventions in order to provide patients with the necessary tools to obtain their desired information (Richards et al., 2018). It is also important to support patients at other points on the cancer care continuum. For example, researchers need to address the role of mHealth in the context of routine emergency care treatment as well as to incorporate methods for tertiary prevention and behavioral change—including smoking cessation, reducing alcohol intake, and increasing levels of physical activity (Darlow & Wen, 2016). Nonetheless, the included reviews are among the first to identify the positive impact of mHealth interventions in the self-management of cancer and recommend areas of further research needed for mHealth implementation into routine cancer care. Additional development of novel mHealth strategies in order to support self-management should be encouraged while incorporating evidenced-based models of care into app design and addressing the gaps in the needs of cancer patients whose needs are currently unmet (Silva et al., 2019).

CURRENT ISSUES IN MHEALTH: SAFETY, SECURITY, AND PRIVACY

In 2013, the U.S. Food and Drug Administration (FDA) issued a guidance document to a draw distinction between "mobile apps" and "mobile medical apps" that are subject to overt FDA approval. The FDA regulates "mobile medical apps," which are medical devices that could threaten patient safety in the event the app fails to function as intended and which include apps that engage in diagnostic testing and treatment. However, "mobile apps" are not overseen by the FDA and do not require its approval. These include apps that can assist patients with self-management of their disease, provide health-related information, and provide tools to track health information and communicate conditions with their health care providers— but which do not provide specific treatment suggestions. Therefore, the vast majority of apps intended for disease self-management are not regulated by the FDA; and although more than a third of physicians recommend mHealth apps to patients, many indicated they would not prescribe them because of a perceived lack of oversight. Furthermore, the FDA focuses on technical issues related to patient safety and the security and integrity of information—but not patient privacy (FDA, 2018).

Many healthcare providers are unaware of information privacy concerns relating to mHealth apps and mobile devices in general. In the United States, digitized health information is covered by legislation, including the Health Insurance Portability and Accountability Act (HIPAA) and the Computer Fraud and Abuse Act (CFAA). HIPAA was created to protect privacy of digital health information, but only applies to mHealth interventions within a medical institution and may apply to information transmitted between an app and a healthcare professional, such as an asthma app that monitors inhaler use and relays this information back to the patient's medical record. The CFAA functions to prevent access of a computer to retrieve patient information. Although the CFAA applies to the access of hospital databases, its extension to private usage of health apps remains unclear. Despite government legislation and guidance for app developers, poor handling of health information is prevalent among mHealth apps (Cook, Ellis, & Hildebrand, 2016). Even when a privacy policy exists, most fail to provide nontechnical information regarding precisely what data are shared and with whom; and placement of the privacy policy agreement as a barrier to app access discourages careful review by users. A recent assessment of 79 apps deemed clinically safe and trustworthy discovered gaps in compliance with data protection principles. None of the 79 apps encrypted personal information stored locally, 66% of apps sending identifying information over the Internet did not use encryption, and 20% did not have a privacy policy. Although mHealth apps may claim to meet privacy policy requirements, many do not actually comply due to the absence of enforcement. Therefore, there are limited guarantees regarding privacy protection (Prochaska, Coughlin, & Lyons, 2017).

Medical content within an mHealth app is also of concern. Medical expert consultation is currently poorly represented in app development, with evidence reporting involvement in less than 33% of mHealth apps. Additionally, a systematic review evaluated adherence to medical evidence in 6,520 mHealth apps within 52 studies and found that almost none of the apps adhered fully to evidenced-based practice. Therefore, the medical content within the mHealth app may be of poor quality. This could perpetuate misinformation, cloud the ability to study mHealth's effect on health outcomes, and may place the patients at potential risk for harm. Thus, healthcare providers have a responsibility to critically appraise mHealth apps before recommending their use by patients. It is also of essential importance for such providers to assess whether patients are already using mHealth apps for disease self-management; and to assess deficiencies, such as inaccurate content or privacy concerns (Cook et al., 2016).

CONCLUSION

Due to the exponential growth of the prevalence of chronic disease amidst the ubiquitously concurrent adoption of mobile devices, the utilization of mHealth interventions for chronic disease self-management is an emerging area of interest and research. mHealth technologies currently serve as a complementary treatment approach to disease self-management and have demonstrated an array of real and potential benefits for both patients and practitioners by increasing accessibility and efficiency of healthcare while reducing associated costs. Although the literature supports the effectiveness of mHealth technologies for chronic disease self-management, there is still insufficient evidence for healthcare providers and policy makers to integrate these tools into standardized practice. It should be noted, however, that small improvements discovered in clinical trials could translate into large benefits among the general population due to the broad reach of mHealth delivery (Silva et al., 2019).

The literature shows that mHealth interventions have unique effects on each chronic condition while also revealing that there are distinct gaps in research related to each disease process. However, such interventions all require rigorous clinical trials in order to determine adherence, sustainability, and long-term outcomes. The integration of theoretical frameworks for health-related behavioral change and the involvement of patients and clinicians are essential factors that warrant consideration during mHealth development. The expansion and utilization of mHealth technology is outpacing our ability to evaluate the privacy, safety, and efficacy of the interventions. Therefore, healthcare providers must assess patients' eligibility for mHealth use; lobby for safe and appropriate data handling of patient information; research and develop mHealth apps using evidenced-based medical content; and advocate for funding devoted to comprehensive evaluation of mHealth effects on patient outcomes (Cook et al., 2016).

REFERENCES

Al-Durra, M., Torio, M. B., & Cafazzo, J. A. (2015). The Use of Behavior Change Theory in Internet-Based Asthma Self-Management Interventions: A Systematic Review. *Journal of Medical Internet Research, 17*(4), e89. doi:10.2196/jmir.4110 PMID:25835564

Alessa, T., Abdi, S., Hawley, M., & de Witte, L. (2018). Mobile Apps to Support the Self-Management of Hypertension: Systematic Review of Effectiveness, Usability, and User Satisfaction. *JMIR mHealth and uHealth, 6*(7), e10723. doi:10.2196/10723 PMID:30037787

Alwashmi, M., Hawboldt, J., Davis, E., Marra, C., Gamble, J. M., & Ashour, W. A. (2016). The Effect of Smartphone Interventions on Patients with Chronic Obstructive Pulmonary Disease Exacerbations: A Systematic Review and Meta-Analysis. *JMIR mHealth and uHealth, 4*(3), e105. doi:10.2196/mhealth.5921 PMID:27589898

American College of Cardiology. (2017). *New ACC/AHA High Blood Pressure Guidelines Lower Definition of Hypertension.* Retrieved March 3, 2019 from https://www.acc.org/latest-in-cardiology/ articles/2017/11/08/11/47/mon-5pm-bp-guideline-aha-2017

American Heart Association. (2018). *More than 100 million Americans have high blood pressure, AHA says.* Retrieved February 9, 2019 from 2018/05/01/more-than-100-million-americans-have-high-blood-pressure-aha-says

Bonoto, B. C., de Aruajo, V. E., Godoi, I. P., Pires de Lemos, L. L., Godman, B., Bennie, M., ... Guerra, A. A. (2017). Efficacy of Mobile Apps to Support the Care of Patients with Diabetes Mellitus: A Systematic Review and Meta-Analysis of Randomized Controlled Trials. *JMIR mHealth and uHealth, 5*(3), e4. doi:10.2196/ mhealth.6309 PMID:28249834

Centers for Disease Control and Prevention. (2018). *Cancer Data and Statistics.* Retrieved February 22, 2019 from https://www.cdc.gov/cancer/dcpc/data/index.htm

Centers for Disease Control and Prevention. (2019). *About Chronic Diseases.* Retrieved February 8, 2019 from https://www.cdc.gov/chronicdisease/about/index.htm

Cook, V. E., Ellis, A. K., & Hildebrand, K. J. (2016). Mobile Health Applications in Clinical Practice: Pearls, Pitfalls, and Key Considerations. *Annals of Allergy, Asthma & Immunology*, *117*(2), 143–149. doi:10.1016/j.anai.2016.01.012 PMID:27499542

Darlow, S., & Wen, K. Y. (2016). Development Testing of Mobile Health Interventions for Cancer Patient Self-Management. *Health Informatics Journal*, *22*(3), 633–650. doi:10.1177/1460458215577994 PMID:25916831

Farzandipour, M., Nabovati, E., Sharif, R., Arani, M. H., & Anvari, S. (2017). Patient Self-Management of Asthma Using Mobile Health Applications: A Systematic Review of the Functionalities and Effects. *Applied Clinical Informatics*, *8*(4), 1068–1081. doi:10.4338/ACI-2017-07-R-0116 PMID:29241254

Food and Drug Administration. (n.d.). *Mobile Medical Applications.* Retrieved February 9, 2019 from https://www.fda.gov/medicaldevices/digitalhealth/mobilemedicalapplications/default.htm

Garabedian, L. F., Ross-Degnan, D., & Wharam, J. F. (2015). Mobile Phone and Smartphone Technologies for Diabetes Care and Self-Management. *Current Diabetes Reports*, *15*(109). PMID:26458380

Greenwood, D. A., Gee, P. M., Fatkin, K. J., & Peeples, M. (2017). A Systematic Review of Reviews Evaluating Technology-Enabled Diabetes Self-Management Education and Support. *Journal of Diabetes Science and Technology*, *11*(5), 1015–1027. doi:10.1177/1932296817713506 PMID:28560898

Hui, C. Y., Walton, R., McKinstry, B., Jackson, T., Parker, R., & Pinnock, H. (2016). The Use of Mobile Applications to Support Self-Management for People with Asthma: A Systematic Review of Controlled Studies to Identify Features Associated with Clinical Effectiveness and Adherence. *Journal of the American Medical Informatics Association*, *24*(3), 619–632. PMID:27694279

Kitsiou, S., Pare, G., Janna, M., & Gerber, B. (2017). Effectiveness of mHealth Interventions for Patients with diabetes: An Overview of Systematic Reviews. *PLoS One*, *12*(3), e0173160. doi:10.1371/journal.pone.0173160 PMID:28249025

Korpershoek, Y. J. G., Vervoort, S. C. J. M., Trappenburg, J. C. A., & Schuurmans, M. J. (2018). Perceptions of Patients with Chronic Obstructive Pulmonary Disease and Their Health Care Providers Towards Using mHealth for Self-Management of Exacerbations: A Qualitative Study. *BMC Health Services Research*, *18*(757). PMID:30286761

McCabe, C., McCann, M., & Brady, A. M. (2017). Computer and Mobile Technology Interventions for Self-Management in Chronic Obstructive Pulmonary Disease [Review]. *Cochrane Database of Systematic Reviews*, *5*(CD-11425). PMID:28535331

McClean, G., Murray, E., Band, R., Moffatt, K. R., Hanlon, P., Bruton, A., ... Mair, F. S. (2016b). Interactive Digital Interventions to Promote Self-Management in Adults with Asthma: Systematic Review and Meta-Analysis. *BMC Pulmonary Medicine*, *16*(83). PMID:27215329

McLean, G., Band, R., Saunderson, K., Hanlon, P., Murray, E., Little, P., ... Mair, F. (2016a). Digital interventions to promote self-management in adults with hypertension systematic review and meta-analysis. *Journal of Hypertension*, *34*(4), 600–612. doi:10.1097/HJH.0000000000000859 PMID:26845284

Mentor, D. (2018). Micro to macro social connectedness through mobile phone engagement. In *Encyclopedia of Information Science and Technology* (4th ed.; pp. 6184–6194). IGI Global. doi:10.4018/978-1-5225-2255-3.ch537

Mentor, D. J. (2011). *Exploring social connectedness via mobile phone texting* (Doctoral dissertation). Teachers College, Columbia University.

Morrison, D., Mair, F. S., Yardley, L., Kirby, S., & Thomas, M. (2016). Living with Asthma and Chronic Obstructive Airway Disease: Using Technology to Support Self-Management—An Overview. *Chronic Respiratory Disease*, *14*(4), 407–419. doi:10.1177/1479972316660977 PMID:27512084

National Cancer Institute. (2015). *What is Cancer?* Retrieved February 22, 2019 from https://www.cancer.gov/about-cancer/understanding/what-is-cancer

Nelson, L. A., Coston, T. D., Cherrington, A. L., & Osborn, C. Y. (2016). Patterns of User Engagement with Mobile- and Web-Delivered Self-Care Interventions for Adults with T2DM: A Review of the Literature. *Current Diabetes Reports*, *16*(66). PMID:27255269

Parati, G., Torlasco, C., Omboni, S., & Pellegrini, D. (2017). Smartphone Applications for Hypertension Management: A Potential Game-Changer That Needs More Control. *Current Hypertension Reports*, *19*(48). PMID:28501982

Pew Research Center. (2017). *Tech Adoption Climbs Among Older Adults.* Retrieved February 9, 2019 from http://www.pewinternet.org/2017/05/17/technology-use-among-seniors/

Pew Research Center. (2018). *Mobile Fact Sheet.* Retrieved February 8, 2019 from http://www.pewinternet.org/fact-sheet/mobile/

Prochaska, J., Coughlin, S. S., & Lyons, E. (2017). Social Media and Mobile Technology for Cancer Prevention and Treatment. *American Society of Clinical Oncology Educational Book*, *37*, 128–137. doi:10.14694/EDBK_173841 PMID:28561647

Raghupathi, W., & Raghupathi, V. (2018). An Empirical Study of Chronic Diseases in the United States: A Visual Analytics Approach to Public Health. *International Journal of Environmental Research and Public Health*, *15*(431).

Richards, R., Kinnersley, P., Brain, K., McCutchan, G., Staffurth, J., & Wood, F. (2018). Use of Mobile Devices to Help Cancer Patients Meet Their Information Needs in Non-Inpatient Settings: Systematic Review. *JMIR mHealth and uHealth*, *6*(12), e10026. doi:10.2196/10026 PMID:30552082

Shah, S. A., Velardo, C., Farmer, A., & Tarassenko, L. (2017). Exacerbations in Chronic Obstructive Pulmonary Disease: Identification and Production Using a Digital Health System. *Journal of Medical Internet Research*, *19*(3), e69. doi:10.2196/jmir.7207 PMID:28270380

Silva, E. H., Lawler, S., & Langbecker, D. (2019). The Effectiveness of mHealth for Self-Management in Improving Pain, Psychological Distress, Fatigue, and Sleep in Cancer Survivors: A Systematic Review. *Journal of Cancer Survivorship: Research and Practice*, *13*(1), 97–107. doi:10.100711764-018-0730-8 PMID:30635865

Simpson, A. J., Honkoop, P. J., Kennington, E., Snoeck-Strobrand, J. B., Smith, I., East, J., ... Fowler, S. J. (2017). Perspectives of Patients and Healthcare Professionals of mHealth for Asthma Self-Management. *The European Respiratory Journal*, *49*(5), 1601966. doi:10.1183/13993003.01966-2016 PMID:28461291

Sobnath, D. D., Philip, N., Kayyali, R., Nabhani-Gebara, S., Pierscionek, B., Vaes, A. W., ... Kaimakamis, E. (2017). Features of a Mobile Support App for Patients With Chronic Obstructive Pulmonary Disease: Literature Review and Current Applications. *JMIR mHealth and uHealth*, *5*(2), e17. doi:10.2196/mhealth.4951 PMID:28219878

Takahashi, D. (2018). *Smartphone users will top 3 billion in 2018, hit 3.8 billion by 2021.* Retrieved Feb, 2019 from https://venturebeat.com/2018/09/11/newzoo-smartphone-users-will-top-3-billion-in-2018- hit-3-8-billion-by-2021/

The Statistics Portal. (2019). *Number of Smartphone Users Worldwide from 2014 to 2020 (in billions).* Retrieved February 8, 2019 from number-of-smartphone-users-worldwide/

The Statistics Portal. (2019). *Smartphone and feature phone penetration in the United States in 2009 and 2011.* Retrieved February 8, 2019 https://www.statista.com/statistics/216340/us-smartphone-and- feature-phone-penetration-since-2009/

Tinschert, P., Jakob, R., Barata, F., Kramer, J. N., & Kowatsch, T. (2017). The Potential of Mobile Apps for Improving Asthma Self-Management: A Review of Publicly Available and Well-Adopted Asthma Apps. *JMIR mHealth and uHealth*, *5*(8), e113. doi:10.2196/mhealth.7177 PMID:28768606

Uy, C., Lopez, J., Trinh-Shevrin, C., Kwon, S. C., Sherman, S., & Liang, P. S. (2017). Text Messaging Interventions on Cancer Screening Rates: A Systematic Review. *Journal of Medical Internet Research*, *19*(8), e296. doi:10.2196/jmir.7893 PMID:28838885

Veazie, S., Winchell, K., Paynter, R., Ivlev, I., Eden, K., Nussbaum, K., . . . Hefland, M. (2018). Mobile Applications for Self-Management of Diabetes. Technical Brief No. 31. Agency for Healthcare Research and Quality. No. 18-EHC010-EF. doi:10.23970/AHRQEPCTB31

Wu, Y., Yao, X., & Li, S. (2018). Mobile App-Based Interventions to Support Diabetes Self- Management: A Systematic Review of Randomized Control Trials to Identify Functions Associated with Glycemic Efficacy. *JMIR mHealth and uHealth*, *6*(1), e20. doi:10.2196/mhealth.8789 PMID:29334479

Xiong, S., Berkhouse, H., Schooler, M., Pu, W., Sun, A., Gong, E., & Yan, L. (2018). Effectiveness of mHealth Interventions in Improving Medication Adherence Among People with Hypertension: A Systematic Review. *Current Hypertension Reports*, *20*(86). PMID:30088110

Yang, F, Wang, Y., Yang, C. Hu, H., & Xiong, Z. (2018). Mobile Health Applications in Self-Management of Patients with Chronic Obstructive Pulmonary Disease: A Systematic Review and Meta-Analysis of Their Efficacy. *BMC Pulmonary Disease*, *18*(147).

Chapter 5
Mobile Phones for Assessment:
Anywhere, Anytime, by Anyone

Füsun Şahin
American Institutes for Research, USA

Dominic Mentor
Columbia University, USA

ABSTRACT

This chapter focuses on using mobile assessment (m-assessment) for teaching and learning in formal and informal education settings. M-assessment has been helpful in making traditional classrooms more interactive and support the individual and class cognitive wealth. Unique affordances of m-assessment can also extend its benefits to digital classrooms, informal and formal learning settings, professional development settings, and other places where learning happens. M-assessment has increased accessibility to assessment, making it available anywhere, anytime, and to anyone.

INTRODUCTION

Assessment via mobile means, mobile phones in particular (referred to as mobile assessment or m-assessment in this chapter), has increased opportunities for formative and summative assessments. Mobile assessments can be used to assess various learning goals, ranging from basic goals, such as checking for students' knowledge of facts, to more complex learning goals, such as building critical thinking skills.

DOI: 10.4018/978-1-5225-9351-5.ch005

Mobile assessment has also increased accessibility to feedback— any user can receive feedback anywhere, at any time. With the help of mobile assessment, teachers and students can generate quicker and more immediate feedback, which promotes the link between learning and tracking learning and increases student engagement. Moreover, the anywhere, anytime, and anyone availability of m-assessment aligns well with the saturated technology environments that students traverse in and out of school. As such, m-assessment can facilitate the measurement of performance and success in many civic, professional, and entrepreneurial contexts and serve as an integral – and perhaps seamless – part of these contexts.

This chapter will offer a variety of mobile assessment examples from different contexts and will also highlight their transferability to learning and education. While some examples will be offered from settings outside of learning and educational—such as sports, physical and psychological health, museums, and civic activism—such examples will provide a progressive overview of how m-assessment continues to grow and evolve into innovative everyday practices. The main focus of the chapter will be using mobile assessment for in teaching and learning for individuals' personal development as well as for societal change in formal and informal education contexts. As m-assessment becomes an integral and almost a seamless part of education, before delving into mobile assessment, we first need to understand mobile learning (mLearning).

Mobile learning is the use of mobile or handheld technology to enable learning anytime and anywhere. Learning can take place in a variety of ways: users can use mobile devices to access educational resources, connect with others, create content both inside and outside of classrooms, and generate evidence of learning by taking mobile assessments. Even though this chapter will focus on using mobile assessments within the mobile learning context, we will also discuss how to leverage mobile devices in various other teaching and learning contexts. We believe that such discussions on using mobile assessments in contemporary classrooms can be applied to other cultural, personal, corporate and nonprofit spaces. Since mobile devices are highly accessible, and in some cases the only available technological tools, mobile learning becomes a natural and critical part of any learning infrastructure.

As mobile-learning can happen anywhere and anytime, the concept of a classroom has become more fluid. Today's contemporary classrooms are not bounded by physical space, but rather can be imagined and realized as any place where students can engage in student-centered, active learning using various technological means. Some examples of contemporary classrooms are kitchens, museums, one's commute, online massive open online courses (MOOCs), subscription-based online courses (e.g., Udemy, Khan Academy), and app-based skill building (e.g., Scratch, Duolingo). Unlike traditional classrooms—held in physical spaces created for formal teaching and learning—contemporary classrooms are not restricted by physical environment or fixed times. For example, every learner in a MOOC environment can progress through the courses at their own pace and return to content as needed. MOOCs and other similar services generally operate outside of traditional educational time frames and deadlines. There can be staggered start dates to set deadlines for assignments and grade submissions, or a course may not have suggested deadlines if students' progression is based on accomplishing milestones rather than attendance or time spent. Contemporary classrooms offer more temporal and physical freedom to the learners than traditional classrooms.

In contemporary classrooms, learners engage in both collaborative and individual knowledge building, and mutual respect to all members of this learning community is valued. The "students" in contemporary classrooms is expanded from a certain age-group registered to an institution to people from all walks of life who have access to technological and mobile devices and become more engaged, abled, and empowered to learn with these devices. Learners are expected to be more self-directed, self-paced, engaged, and active, rather than passively receiving information. With the concept of contemporary classrooms, learning becomes available everywhere, including both low-income and high-income areas and countries. As such, diverse users can be learners of the same content or skills at a given time. The varying needs of learners can be addressed more easily in contemporary classrooms than in traditional classrooms. With the flexibility of mobile phones, learners can be assessed according to how they could produce a variety of learning artifacts by enabling learners to show their comprehension as an app like Explain Everything. With the ubiquity of mobile phone ownership, there is relatively reduced cost in having students' or institutional mobile devices used as either producing learning evidence or for assessment of learning activities. The concept of "teachers" has also changed in contemporary classrooms, as teachers, trainers, or facilitators, as well as peers can take the role of more experienced or knowledgeable "mentors" in the learning process. In contemporary classrooms, learners are a community. In this community, there is typically a guide or mentor who facilitates the learning by providing resources and behaviors to model. In addition to this mentor, peers also build knowledge together and enhance each others' understanding and products by sharing and providing feedback.

Despite the differences between contemporary and traditional classrooms, the ultimate similarity between them is that the aim of both contexts is to provide meaningful learning. M-assessment plays an instrumental role in promoting meaningful learning in both contexts. The distinction between contemporary and traditional classrooms may not be as solid. These two contexts can be thought of as areas on a continuum, where traditional and contemporary classrooms are two ends of a continuum. While mobile assessment can be a practice common to both settings, using mobile assessment in a relatively traditional classroom can make shift the classroom environment shift towards the contemporary end of the spectrum. Incorporating mobile assessment can facilitate to further student-centered, active learning and fulfilling the promise of diverse learning approaches. To extrapolate the use of mobile assessment in contemporary classrooms, we make three main arguments: a) mobile assessment is a practice that fits well in contemporary classrooms, b) mobile assessment can aid the educational experience of both students and mentors (e.g., teachers, facilitators, trainers), and c) mobile assessment can monitor and capture progress in working toward learning goals and desired learning outcomes. To support the first argument, we start by narrating in detail how m-assessment increases accessibility of assessment anywhere, anytime, and by anyone. For the second, we offer examples of multimodal engagement, which makes use of all features of mobile devices (i.e., observing, interacting with, or producing images, audio, video). Research shows that students retain and comprehend information better if various senses are engaged simultaneously or in an integrative teaching and learning manner (Boyle, 1997; Clark & Mayer, 2011; Kondratova, 2009; Zheng, 2009). For the third argument, we will examine the effects of extended learning engagement time with m-assessment by comparing outcomes before and after mobile assessment sessions. In this chapter, we will also be cultivating and extrapolating richness from personal learning environments, expanding opportunities for formative and summative assessments, and discussing experiences of mentors and learners who used mobile assessment that made their life easier.

This chapter starts by explaining what mobile assessment entails and providing early examples of mobile assessment and use of mobile phones in data collection in general. The chapter continues by highlighting that mobile phones have increased accessibility to assessment and provided an extensive collection of examples of various mobile assessment tools and software programs. The chapter then provides in-depth examples of how mobile assessment increased accessibility by, making it available anywhere, at any time, and to anyone. In these sections, readers will find example studies and cases that used mobile assessment to combat challenges of having a fixed location, time, and number of users. Readers will then find detailed summaries of the studies that use mobile assessment in terms of the desired learner outcomes. In this chapter, learners' achievement on the topics assessed, their engagement, and their cognitive load were collectively summarized to provide an empirical account of the outcomes. Then, from a realistic perspective, the chapter mentions the challenges of using mobile assessment and provides recommendations on smart use of mobile assessment opportunities. Finally, the chapter ends with future directions and a summary. Figure 1 provides an overview of the major components of the chapter.

Figure 1. Visual summary of the major contents of the chapter.

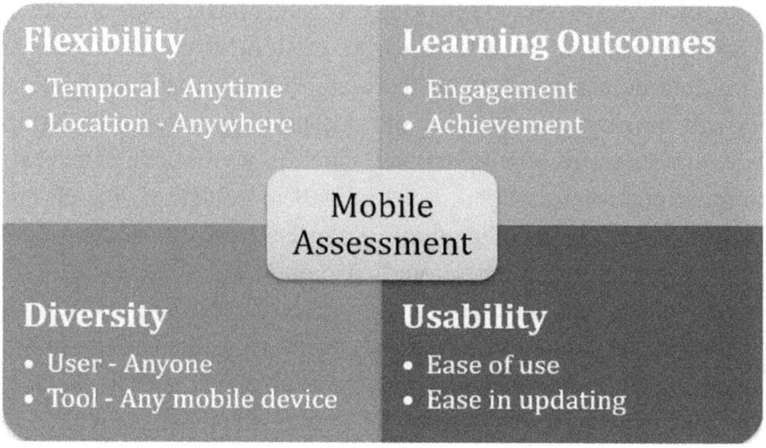

Background

Mobile assessment (m-assessment) can be defined as offering assessment of learning using mobile devices. Educational assessment activities include but are not limited to checking and testing for prior knowledge, engagement, and understanding of learners' mastery of the content and skills through formative or summative assessments. A wide range of skills can be assessed by m-assessment, including basic skills, such as literacy development, and advanced skills, such as scientific reasoning. It can be used as a formative tool to shape teaching and learning as well as a summative tool to measure mastery. Devices that can be used for m-assessment include mobile phones, tablets (e.g., iPads), personal digital assistants (PDAs), iPods, and other handheld devices. All of these devices have various communication capabilities, but mobile phones, which are have calling and texting functions, have a unique place due to their ubiquity, affordability, and affordances. Even though m-assessment includes all types of mobile devices, this chapter focuses on m-assessment using mobile phones.

To fully understand the contributions of mobile phones for m-assessment, the background of m-assessment using early mobile technology, such as PDAs and pocket PCs, is important. To provide more background of mobile assessments, one can go as far back as PDAs, which even predated Pocket PCs. PalmPilots were introduced in the 1990s by Palm Inc., who sued Microsoft for using the name as part of their Palm PC line, which was then changed to Pocket PCs (Niccolai & Gohring, 2010). Initially, PalmPilots were popular because they offered portable electronic organizing capabilities. As their computing power increased, their functionality expanded. One example of how the PDA was exploited for mobile assessment was for literacy assessment, which was pioneered by Greg Gunn, one of the founders of Wireless Generation (New Schools, http://www.newschools.org/venture/wireless-generation). The company took a compulsory paper-based reading assessment that every lower elementary teacher had to conduct on each student and streamlined the process with a PDA. Making the process more efficient for teachers was but one of the major benefits. Other benefits included the collection of valuable reading literacy analytics, which could be shared with the school, fellow teachers, parents, and district and regional superintendents. The reading analytics offered tracking of progress, issues, and challenges, which could be targeted by teacher and parent interventions. The mobile assessment involved elementary students reading from a prescribed text at their level and the teacher electronically capturing correct and incorrect letters, sounds, pronunciations, and words. The electronic capturing of this information via a handheld device allowed for tracking of patterns and progress for individual students as well as data collection within and across classes, schools, districts, and regions. Educators were able to see the baseline reading literacy levels after the data was uploaded via a docking cradle. Although Wireless Generation pioneered the mobile reading assessment, it was not the only entity working in this direction.

Vahey and Crawford (2002) reported a study of 100 Palm-equipped classrooms, where 90 percent of the teachers reported that these handheld devices were effective instructional tools that could affect student learning positively and do so across curricular topics and instructional activities (Vahey, Tatar, & Roschelle, 2006). Using PalmPilots for mobile assessment allowed for electronic note taking, voice recording, and sharing capabilities. As with Wireless Generation's later literacy assessments, students could also take polls, surveys, or tests using PalmPilots. While these devices were pretty robust for their time, there were other low-tech mobile devices used as part of learning and assessments.

Low-tech regular calculators in Southern Africa and Asian countries were used for simple yet creative English as a second language (ESL) learning purposes (Mentor, 2014). Turning calculators upside down and using the numbers as letters was a rudimentary but creative way to build ESL students' vocabulary. Assessment of vocabulary in this way was done both individually and collaboratively as a competition and followed up with contextual use of the words. With the advent of scientific calculators, more word permutations became available for such vocabulary building ESL learning. In the United States, science teachers used graphing calculators as handheld devices in several ways for capturing learning and assessment of learning (Roschelle, 2006). Another mobile device that predated and ran concurrently with smart phone use was the clicker (Mentor, 2011). Although clickers could check or assess understanding before, during, and after classroom learning, they have often been replaced with smart phone applications (i.e., apps, or small computer programs, which can be installed on smart phones).

Outside of the formal education system, various assessments were conducted in the health arena. It has been argued that mobile health (mHealth) was ahead of mobile learning in formal education spaces and that health care led the way in adopting mobile devices (Ally, 2009; Dobkin et al., 2007; Levine, McCright, Dobkin, Woodruff, & Klausner, 2008). In the health arena, there are many underlying reasons for assessing knowledge or soliciting feedback/opinions via text messaging. Apart from convenience, there is also a level of privacy, which is particularly useful in the areas of psychological and sexual health. Much of the successes reported by the aforementioned references were attributed to quantitative results and analysis. There is massive socio-emotional or social-connected value inherent in dealing with health issues through a relatively private device (Mentor, 2011; Mentor, 2018). These studies often showed significant savings in cost and time—benefits for both hospitals and patients (Dobkin et al., 2007; Kay, Santos, & Takane, 2011; Kelsey & Cavendish, 2014; Mahmud, Rodriguez, & Nesbit, 2010; Mao, Zhang, & Zhai, 2008; Milne, Horne, & Torsney, 2006). It should be noted that older mobile health studies are a far cry from what is possible with more recent smart phones but still show major benefits. The New Zealand government initiated a project around assessing the ownership of and ability to use mobile phones. This project attributed some differences in people's connectedness and quality of life to various levels of ownership and usage intensity of mobile phones (Jamieson, 2007). Another New Zealand project involving 1,705 smokers in a smoking cessation program allocated supportive text messages to control and participative groups. The group receiving the text messages was determined to have higher quit rates than other groups (Rodgers et al., 2005).

Amidst all the improvements in using mobile phones to assess different aspects of health, there is still a need for a multifaceted analysis that assesses the success of using mobile devices in mobile health initiatives. Such multifaceted analyses dealing with quantitative reports of socio-emotional values would highlight the reasons behind the current mobile health initiatives. Although supplemental interview inquiries might have revealed reasons for successes, failures, or areas of improvement of the campaigns, there are some drawbacks of using angled surveys. For example, while collecting data about sensitive and commonly socially awkward or taboo topics, the studies could have been suspended or held in abeyance for protecting confidentiality and rights of the participants. While interviews might have revealed deeper social factors around such private and personal topics, the data collection would still be time consuming during collecting, coding, and interpreting phases. With recent advances in mobile technology, simpler staggered methods could also deliver additional layered results. In fact, such practice is highly common and beneficial. Mosa, Yoo, and Sheets (2012) reviewed health care applications with mobile phones and provided a portrayal of various medical purposes, such as disease diagnosis, drug reference, medical training, and chronic disease management.

In addition to mobile health, m-assessment facilitated near synchronous surveying of people for evaluation purposes. For instance, Empowered Professional Internships with College Credits (EPICC)—a national nonprofit workforce development training institution that offers intensive internship preparation while allowing trainees to earn higher education college credits—delivered mobile surveys to alumni to evaluate their participation in the program and performance after completion of the program. Previous efforts (2000 - 2011) to garner alumni assessment and feedback took months, exhausted staff time, and resulted in very low completion rates (Mentor, 2018). A mobile text messaging survey system was thus implemented (2012), and alumni were engaged via their cell phones with one survey at a time via text message (Mentor, 2018). Upon each response, participants received another short question. This method of sending a short survey via text message proved highly successful, with a response rate of more than 90 percent. As todays' mobile device surveys are pretty powerful and can capture processes, pre- and post-photos, and videos, which can also be revisited by the student for reflection.

Şahin and Yan (2013) published a more general review examining 171 publications that used mobile phones for data collection in various fields, including health, education, and communication. According to the synthesis of the data collection methods, Şahin and Yan reported that mobile phones enabled data collection on human subjects and nonhuman subjects (such as insects) to provide large generic samples of all users of a carrier company as well as specific human samples, such as participants in a local study. The review also pointed out that researchers could use mobile phones for collecting data while users were engaged in a specific activity relevant to the study topic, engaged in their routine activities, or not engaged in any activity at all. For example, one study asked the participants to keep mobile phone diaries and used the content in the diaries for analysis. The mobile phone diary method could enable data collection at various times of the day as participants are intentionally involved with the activities requested by the study protocol. In addition to the intentional activities of participants, this type of data collection could also be used as an unobtrusive data collection method, which would not require participants to engage in specific activities. For example, relevant data can be passively collected as users are engaged in their routine activities, such as communicating. As an example of more unobtrusive data collection, some research studies use cellular network data, which consists of the logs of user activities such as calling, texting, and connecting to the Internet using cellular data. These very large datasets allow researchers to examine various patterns and relationships regarding people's mobile phone use. Another interesting opportunity that cellular network data provides researchers is being able to collect data even when users are not using their mobile phones. For instance, users' locations are tallied in cellular network data, which can be used anonymously for research that requires specific location information. For example, Caceres, Wideberg, and Benitez (2007) monitored the dynamic location of mobile phone communications in a network using location information in cellular network data. In addition, many apps nowadays include activity trackers that make use of various built-in sensors in mobile phones and can even give individualized feedback to users about their activity levels. Such capabilities of mobile phones provide opportunities for collecting various types of data. As Şahin and Yan reported, visual and text-based data can be collected as well metadata of activities (e.g., log data), which can be considered as ubiquitous data. For example, through increased mobile phone accessibility, data collected through mobile phone sensors. Such data can be collected ubiquitously as for collecting metadata, log data, app use, trackers, location tracker), or by data collected and sent by the users, such as pictures participants took using their mobile phone cameras.

INCREASING ACCESSIBILITY TO ASSESSMENT WITH MOBILE PHONES

Increased access to mobile phones has increased access to m-assessment. M-assessment introduced a plethora of opportunities for learning and assessment. One breakthrough was the smart phone, which enables the installation of specialized programs called applications (or apps). Many of these apps are available for free. Some campuses have equipped their classrooms with mobile phone compatible receivers enabling students to use their mobile devices as clickers. Similarly, some campuses that subscribe to learning management systems (LMS), which are web-based platforms that help instructors and students organize learning materials and activities, also subscribe to mobile apps of the same LMS systems that allow students to access the school's LMS from their phones. M-assessment has overcome some restrictions on the location and time of assessment and now reaches a wider range of users. Therefore, m-assessments in all forms has made assessment more accessible.

Mobile polls and surveys have also seen a shift and offer various types of assessments, ranging from assessing information in collated fashion or gathering understanding or opinion. Many mobile survey and poll tools, such as PollEverywhere, InFuseLearning, Socrative, and Mentimeter, are used for formative assessment purposes. There is certainly no shortage of how those tools can be used to check learners' understanding, both in and outside of the classroom. One such example is Answergarden, which is both a web and native app, that can be used for individual or group brainstorming purposes. Google Forms is another tool that has been repurposed in many different ways as a formative assessment tool. Animoto offers students an opportunity to record 30-second videos to check for and show their understanding, and AudioNote offers students the ability to take notes by recording audio. Another online web tool that offers multimodal engagement to assess student understanding is VoiceThread, which is a mobile-accessible online tool that offers video, texting, and audio engagement opportunities. Another powerful native app is ExplainEverything, which allows students to record audio and video and use tools to animate or highlight various aspects they wish to showcase to communicate their understanding of a subject or concept.

In addition to this range of mobile formative assessment tools, there are many mobile summative assessment tools. Various quiz and practice tools have become more mobile friendly and mobile responsive, such as cue cards. And summative assessments like A+ Spelling with crowdsourcing ability where students can create individual or group tests, and audio versions of the words are available. Many of these examples of formative and summative assessments function as independent assessment tools in the sense that they operate outside of an LMS. While an LMS can offer learning analytics opportunities by analyzing large datasets accumulated in LMS databases, m-assessment tools that work independent of an LMS can make learning analytics slightly challenging. Learning analytics tools that are integrated into an LMS or through an application protocol interface (API) can generate external or internal data so that in-house data can be used to track learners' activities to understand and facilitate the learning progress. Many apps that are not part of an LMS can also provide learner analytics and even back-end log data of learner actions for app developers and researchers to analyze further. For example, Duolingo, a widely used foreign language learning app, offers users a dashboard that shows their learning progress in various topics and milestones they've reached in the course. Duolingo also offers researchers the back-end log data to investigate learners' interactions in the app. Bolsinova et al. (2018) examined Duolingo users' scores on the course questions as well as the amount of time they had until the time limit for that question and whether they used any hints for the question using the "hint" button on the user interface. This way, Bolsinova et al. judged whether the learners were fast or slow, provided a correct or incorrect response to the question, and whether they used hints. They concluded that using hints was dependent not only on learners' abilities on the topic but also on their personal characteristics, such as gender. Bolsinova et al. suggested learners' tendencies to use or not use hints could be used to provide additional feedback to learners and understand their learning paths better.

The abundant availability and the growing creation of educational technology tools, such as LMSs, necessitated communication and transferability between various systems, which is known as *interoperability of learning tools*. Significant progress has been made to offer interoperability between learning tools, commonly known as learning tools interoperability (LTI), which was developed by the nonprofit Instructional Management System (IMS). One of the goals of IMS is enabling the integration of internet-based learning applications and their associated online platforms in a seamless LTI fashion (Abel, 2015). Within the LTI framework, IMS suggests educational technologies have some desired properties, such as online plug-and-play design and suitable architecture and ecosystem of the tools, which can enable other educational technology products to work and be deployed together.

A great example that fulfills the promise of LTI and increases accessibility of mobile phones comes from EPICC, which has been in existence for more than 15 years but only adopted e-learning and mobile learning as part of its academic learning and development in 2013. A well-positioned institution such as EPICC introducing e-learning and mobile learning into its long-standing system is an exemplary case study for analyzing and imagining interoperability. When EPICC started with the mission of integrating e-learning and mobile learning, it also recognized the need to revamp, digitize, and integrate its entire program—from admissions to alumni. To that end, a strategic approach was envisioned and put in place to grow digital learning in an organic manner, which suggested a bottom-up approach for generating e-learning and mobile learning parts. According to this holistic strategy, every dependent department would be able to generate digitized learning materials and assessments, and the global project would oversee the compatibility, integration, and sustainability of these digitized materials. To fulfill such efforts, a taxonomy of criteria was formulated that suggested specifications for having a baseline quality for user interfaces and user experiences within digital learning. In addition, considerations for long-term sustainability and evolutionary needs for both the front end and back end were established to make user experiences sustainable and reduce the number of issues, such as system glitches. The considerations in reformation of the system also included LTI and API for data migration and data analytics (Mentor, 2014; Mentor, 2018). The criteria matrix was suggested to define stakeholder and organizational needs and served the organization well in selecting a 21st-century LMS needs with mobile responsiveness. A mobile-accessible LMS extended teaching and learning engagement and offered a variety of assessments, including discussions, quizzes, essays, projects, and presentations that could be assigned and responded to via mobile devices. Significant successes were achieved in showcasing the mobile responsive assessments that saved mentors' and learners' time and effort, as well as providing real-time feedback on the assessment results. Adoption of a mobile compatible LMS along with learning and assessment materials was reported at a 100 percent adoption rate 6 to 12 months after the project first started.

While the mobile blended e-learning approach worked well within specific academic units, the academic vision encompassed the need for other departments—starting with admissions and alumni communications—to also adopt the ease, effectiveness, and efficiencies of the mobile responsive tools. To better serve learners and mentors, 100 percent of the organization's admissions departments across the country have offered potential learners digital assessments as part of their admissions processes since 2016. Much of the functionality of the processes and assessment tools have not been adopted, mainly due to the variation of admission practices as well as differentiation of criteria for hiring admissions staff. A possible scenario with varying hiring criteria is that staff at one or a few sites may swap departmental positions without necessarily checking for required qualifications to lead or work in admissions departments. Furthermore, seeing admissions tests and criteria changing at a local site in isolation from nationally set methods or standards is not uncommon. EPICC is certainly not alone in these ad hoc processes and practices, and they are also not alone in having varying admissions processes and practices even though they compromise data fidelity and integrity.

Although the constantly changing practices highlight the risks of data pollution, adopting mobile technologies continues to provide ease in delivering necessary trainings. Despite the benefits, adoption of mobile learning is still challenging. Some of the other challenges of mobile adoption are lack of understanding of the benefits of utilizing mobile phones as learning tools. Some of the benefits that are overlooked are self-directed learning, multimodal learning, ease of submitting assignments, and ease of delivering and taking assessments. Multimodal learning resonates highly with the 21st-century world experiences of many learners. These multimodal experiences build on learners' existing skills in navigating resources to satisfy their information and communication needs that they typically practice on a daily basis.

In the case study of EPICC, there is still hope that more mobile engagement and mobile assessments can be implemented with more academic oversight, influence, and collaboration. Oversight over the academic admissions assessment component would be beneficial. Furthermore, improving delivery of procedures that measure student performance throughout the program, prior to and after the training period, would add fidelity to the entire program. Thus far, contrary to the 6-month training component, only a few of the nontraining assessments have been converted to a mobile friendly version. Despite the observed success of mobile assessments, as seen in the EPICC example, staff that have access to data coming from mobile assessments were not fully prepared for and appreciative of mobile assessments that would motivate them to further build, investigate, and enhance baseline admissions assessments that would allow fruitful real-time performance analytics. Examples of such fruitful information that can come from mobile-friendly assessments are cohort-to-cohort, city-to-city, and divergent channel comparisons; student enrollment; students' learning styles, strengths, and learning needs. The EPICC example shows that e-learning and mobile learning can become an integral facet of the learning systems offered at institutions. This integration and remodeling typically requires proper training and a strategic vision. As the digital and mobile learning capacities of institutions take root and grow, they offer the promise to further aid the learning at institutions while saving organizations time, money, and effort.

Taking Mobile Assessment Anywhere: The Multi-Context Availability

A main advantage of m-assessment has been expanding accessibility to educational assessment from traditional classrooms to anywhere else assessment is needed. Both settings have benefited from m-assessment by overcoming challenges. For example, efficiently assessing a fairly large number of students by administering a quiz or an exam and offering them feedback quickly is a challenge in traditional classrooms.

Using mobile phones as clickers has been widely reported as a way to overcome the challenges of administering classroom assessments in large traditional classes. Mobile phones have replaced clickers used for checking students' prior knowledge before lectures, giving them feedback, and checking student mastery during or at the end of a class session or semester (Arnesen et al., 2013; Lin & Rivera-Sánchez, 2012; Stav et al., 2010; Susono & Shimomura, 2006). Arnesen et al. (2013) narrated in detail how mobile phones were used as clickers. First, a quiz consisting of multiple-choice questions was presented at the beginning of the class, and students were allowed to work individually or in groups to complete the quiz. Then, students were asked to provide their answers using mobile phones. After students provided responses, frequencies of each selected option were presented to the whole class. Answers from the whole class were used to provide immediate and anonymous feedback and facilitate a class discussion. It should be noted that clickers are limited in terms of the types of questions for which they can be used (Mentor, 2011). Replacing clickers with mobile phones not only makes the transition of clicker use via mobile phones more affordable, but also provides additional opportunities. One added value of mobile phones is the type of the questions they can assess. Most teachers use clickers to ask multiple-choice and true or false questions due to lack of training. However, such questions, which usually have one single answer, hit lower levels of cognitive stages in Bloom's taxonomy (Bloom, 1956). With the growth of mobile technology and the use of mobile phones, m-assessments can now also include an array of matching and ordering questions that demand and cultivate higher order cognitive levels, including critical and creative thinking (Anderson, Krathwohl, & Bloom, 2001). Moreover, porting clickers over to smart phones, which allow users to use a smart phone app to control clickers, also increases the capabilities of clickers. Another added value of mobile phones used as clickers is user authentication. Students rent clickers from the university and register the clickers under their name at the beginning of the semester (and do not need to register them again). When a class uses clicker quizzes to take attendance, however, students who do not want to show up to class may take advantage of having separate clickers. Some students give their clickers to one of their classmates and ask the classmate to answer the quiz question using the clicker so that the student who did not show up would not be seen as absent from the class. If mobile phones were used to answer the quiz question, however, students would be much less likely to leave their phone with a classmate, as students are much more attached to their phones and consider them private devices. Thus, mobile phones would provide better user authentication.

Mobile phones have been also beneficial for overcoming challenges of assessing learners in location agnostic ways. Making assessment independent of learners' location and embedding assessment into their learning process while extending their learning engagement are challenges of providing such innovative assessment practices. Flexible tools, which can be adapted to various learning environments, are needed to overcome these challenges. Mobile phones have provided enormous flexibility in many environments, such as digital classrooms held on online and mobile platforms, practicum sites, and school trips. For instance, m-assessment can be integrated with online or face-to-face courses that utilize LMSs for organizing course materials. Bogdanović et al. (2014) refers to the use of a mobile quiz app for testing student mastery on the content delivered by the LMS. Similarly, students submit long assignments via cell phones through LTI connections with Google Drive or Office365 or by uploading assignment documents. This is particularly helpful for submitting just-in-time deliverables. Students in EPICC also participated in formal and informal discussions and completed exit ticket quizzes or tests in or outside of class. For exit ticket quizzes and tests, students can receive immediate results and, if required, redo or retake the assessment. Likewise, Kizito (2011) integrated a mobile quiz before students enrolled in an online class for testing students' readiness. M-assessment was not only suitable for supporting online classes but also suitable for video classes such as workplace webinars. Stav et al. (2010) integrated m-assessment into a lesson held via video conference. Students in Hungary could simultaneously answer the questions posed by the instructor located in Norway using their mobile phones.

In addition to digital classrooms, practicum sites such as hospitals are places where m-assessment can be particularly handy. As one of the first examples of this, Taylor et al. (2010) introduced an m-assessment app for testing and storing performances of health care students on their practice in hospitals. This app consisted of tests and practice assignments developed by faculty members for checking student mastery as well as the feedback given on such assignments. The app also stored student performances for helping them build an e-portfolio, which was used to observe a student's overall performance and growth as a summative evaluation. Later, Dearnley et al. (2013) improved the m-assessment app introduced by Taylor et al. (2010) by enabling peer feedback, predictive texting, spell checking, and audio response recording functions. The improved m-assessment tool also encouraged students to reflect on their performances and compile feedback from various sources. Similarly, Ferenchick et al. (2013) used an m-assessment app for providing feedback to medical students on their patient evaluations. Expert doctors (i.e., residents and faculty members) who supervised medical students reviewed the students' evaluation of their patients, assessed the students' skills—such as communicating with patient, taking patient history, and physically examining patients—and provided feedback using the app. Sixty-eight percent of the expert doctors reported that giving feedback to students took less than 10 minutes.

M-assessment has been valuable in overcoming challenges in professional development, such as training and evaluating military personnel for field tasks. Heric and Carter (2011) used m-assessment to evaluate operational decisions of military personnel. Questions in this context consisted of time-critical scenarios designed for measuring cognitive skills such as perception, memory, and problem solving. The trainee was given a scenario and asked to choose the best course of action within that scenario. Then, new questions were assigned based on the trainee's performance in earlier questions. If the trainee's answer to the question was wrong, an easier scenario was sent to the trainee. After the testing session, the m-assessment app reported rose diagrams depicting relative strengths and weaknesses of the individual trainee or a group of trainees and recommended further training in the areas of weaknesses.

Training teachers for actual classroom situations is another challenging professional development context due to limited time for observing teacher trainees in classrooms and fidelity of observations in the existence of an observer. Gibson (2013) used an m-assessment platform that simulated a virtual classroom. The virtual classroom was a simulation of a real classroom filled with artificial students, representing real classroom challenges. The m-assessment app then recorded trainee's reactions in this virtual classroom and, after the session, allowed the trainee to reflect and try again.

Mobile assessment can support learning anywhere that learning happens as long as users have access to their mobile phones. Some examples of using mobile assessment everywhere are participating in online class discussions, receiving and uploading assignments, and taking tests on mobile devices. For instance, class assignments consisting of multiple-choice and free-response questions can be sent via text message, responses can be collected, and the instructor can send feedback on these assignments via text (Brett, 2011). In addition, students can take practice tests wherever they wanted thanks to mobile testing apps. Virvou and Alepis (2005) developed the first mobile testing tool capable of sending and scoring multiple-choice and fill-in-the-blank questions. The automatic scoring capability of mobile testing apps provides an opportunity for mobile assessment to grow further and enables the delivery of computer adaptive testing (CAT). Basically, CAT adjusts the difficulty of a test to the learner's level by automatically scoring the learner's responses to initial questions in a test and selecting further questions based on the learner's performances on those questions (Van der Linden & Pashley, 2010). Therefore, CAT requires fewer questions and less time than nonadaptive tests, making testing more efficient. Studies reported using mobile assessment apps for delivering CAT anywhere the learners desired (Huang, Lin, & Cheng, 2009; Romero, Ventura, & de Bra, 2009; Triantafillou, Georgiadou, & Economides, 2008). Moreover, multimedia stimuli can be added to CAT because of the functions of smart phones (Huang, Lin, & Cheng, 2009).

Informal learning settings, such as museums, also benefit from mobile assessment because of the mobility of mobile phones. In particular, mobile assessment apps that are capable of capturing users' locations in real-time could be used to triangulate information on students' location, the artifacts in that location, and the learning activity for assessing students. As opposed to a museum map or a static worksheet, such capabilities can make learning and assessment more authentic by using real, tactile stimuli and responses. Recent studies reported using context-aware m-assessment apps in museum trips for tracking students' activities in the museum, assessing their learning of the materials, providing instant feedback, and even providing hints and supplementary materials if needed (Hwang & Chang, 2011; Santos et al., 2011).

Additionally, location-based learning in museums offers further multimodal experiences via mobile devices. Examples of such learning in museums today show how museums are at the vanguard of utilizing context-aware mobile learning (Mentor, 2014; Mentor, 2018; Turchinsky, 2011). By providing additional information about an exhibit in an interactive, multimodal manner, the museum becomes an augmented experience. Such electronically enhanced experiences add to visitor's engagement and can enhance their experience and learning from the visit.

There are plenty examples from museums in big cities around the world, such as New York City, Boston, and Chicago. The Asia Society museum in New York City exhibited a mobile set of separate audio and visual engagements in 2008. In 2011, the Metropolitan Museum of Art built a mobile app to go with its guitar exhibit and created an interactive scavenger hunt leading up to an exhibit of Alexander McQueen's designs. At the time, with the advent of native apps and the technology at hand, these mobile phone engagement methods situated learning in informal contexts, with the possibility of converting these into formal educational opportunities with assessments. Other examples include the Grant Museum of Zoology in London providing a fully interactive experience—visitors could respond to questions posed by the curators, add to the discussion, and post comments about the exhibits (Giles, 2010). Visitors' comments then were collated and synchronized with the QRator website (University College London Centre for Digital Humanities, Centre for

Advanced Spatial Analysis, & University College London Museums, 2011, http://www.qrator.org), which allowed them to contribute to the continuing discussion asynchronously and away from the museum setting (Ross et al., 2012). The Chicago Historical Society also created an interactive scavenger hunt visitors could access via a handheld computer, which inspired other museums to create similar games (e.g., "The Mystery at the Museum" at the Boston Museum of Science). More recently, the Met established other interactive mobile assessment initiatives, creating a murder mystery app, which delivers an assessment activity that requires users exercise deduction technique. Although most of these interactive experiences are designed for individual enjoyment, collaborative group assessments can be added, and formative assessments as well as summative assessments can be offered.

Taking Mobile Assessment Anytime: Temporal Flexibility

In addition to providing flexibility in location, mobile assessment provides flexibility in the scheduling and duration of an assessment, which are particularly challenging in traditional classrooms. For example, scheduling a time slot for delivering a traditional, paper-pencil exam to a large classroom can be a hassle. Likewise, distributing and collecting papers during class time in a large classroom can be time consuming and interrupt the flow of the lecture. Such obstacles to efficiency can discourage educators from administering tests. Mobile assessment can provide ease to both educators and students by making assessment accessible anytime.

Mobile assessment provides tremendous ease in scheduling assessments, which is particularly helpful in test taking and instruction. First, as a stand-alone testing tool, mobile assessment can be accessible anytime. As exemplified before, mobile assessment is suitable receiving assignments without being constrained by fixed, predetermined times, made possible to use it on the go. Another example can be students' receiving class quizzes on their mobile phones after school, as exemplified in a study conducted by Bogdanović et al. (2014). A lot of assessment is also embedded in learning. For example, learners using the DuoLingo app progress through the levels of proficiency only by correctly answering the questions tailored specifically to their proficiency level. Learners can benefit from this functionality anytime they have access to their phone.

Mobile assessment is also used as a self- and peer-assessment tool. Samaie, Nejad, and Qaracholloo (2018) utilized a widely used mobile app called WhatsApp for self- and peer-assessment of English-speaking proficiency of 12- to 18-year-old Iranian students learning English. WhatsApp allows both one-on-one and group text and voice recording messages. The researchers formed a WhatsApp group consisting of their participants, provided a discussion topic to them, and asked them to talk about the topic and send a voice recording of their speech to the group for evaluation. The recordings could be only created while the users were on the WhatsApp interface, and everyone in the group could access the recordings simultaneously and replay them if needed. Learners in this WhatsApp group evaluated both their own voice recordings for self-assessment and the voice recordings of others in the group for peer-assessment. Samaie et al. also noted that learners would have some more preparation time for their speech in WhatsApp than they would have in a face-to-face environment. While this could lead to less stress on the user to produce impromptu speech, they noted that it might decrease the spontaneity of language production and affect the data; synchronous visual connections can be used to combat this threat in the quality of the data.

Chao et al. (2016) used mobile augmented reality (AR) in performance assessment. Performance assessment, similar to assessment of oral English language proficiency, is a skill that is difficult and expensive to assess with fidelity. AR provides human-computer-context interactions by enabling users to visualize environments in a real world setting with the digital information overlaid on real environments. As performance assessment is a labor-intensive task both for the learner and the evaluator, AR provides an efficient solution to assess all the learners simultaneously in real-time. Chao et al. used real-time peer-assessment of 50 culinary students on a training course about Western cuisine. The students formed groups of five and performed a recipe together and evaluated other students. Students first chose their role in the AR system as either the evaluator or the performer; then performers scanned the sample of the outcome dish they were assigned to cook. With the help of the context-aware AR functionality, the system visually recognized the assigned dish and provided learners the instructions. While performers were cooking the assigned dish, they were also asked to record their steps by taking videos using their mobile devices. The context-aware AR recognized the dishes the students were cooking and the ingredients they were using from the videos. Then, the system uploaded the videos from the students to the cloud and immediately sent them to the evaluator. The evaluator was provided with the assessment criteria on the screen so that he or she could provide immediate and consistent evaluation and provide feedback to the performer.

In addition to apps, mobile assessment facilitates delivering timely assessments aligned with instruction. Mobile assessment can be delivered before, during, and after instruction. Arnesen et al. (2013) used mobile phones as clickers in a large lecture hall to administer class quizzes to awaken the students' attention before the classes, check student understanding during class, and check student mastery of the content at the end of the class. Similarly, Kizito (2011) pretested students' readiness for a course offered fully online even before students enrolled in the course. Assessing learners as they are engaged in the lectures or learning activities allows assessment to be incorporated into the learning processes. For instance, the mobile assessment app in the study conducted by Santos et al. (2011) captured students' understanding of the learning material as they were visiting a museum. Likewise, the mobile assessment platform in Gibson's (2013) study captured preservice teachers' performances as they were interacting with artificial students in a virtual classroom.

Mobile Assessment Taken by Anyone: Mobile Assessment for All

As mentioned before, accessibility of mobile assessment has increased and the profile of mobile assessment users has diversified. Mobile assessment is available to anyone with a mobile phone and can be done as part of self-directed learning, formative assessment, or formal summative assessments. One reason to suggest that mobile assessment is available to a wider range of learners is that mobile phones are more ubiquitously available than other mobile technological devices such as laptops and tablets. In fact, a study conducted by the Pew Research Center in 2018 suggested that 95 percent of American adults own a mobile phone of some kind, and 77 percent own smartphones (Pew, 2018, http://www.pewinternet.org/fact-sheet/mobile/). Another study done by Pew highlighted that smartphone ownership is on the rise in emerging economies: at least 75 percent of adults reported owning a mobile phone in every country surveyed (Pew, 2018).

Given the device ownership percentages, particularly in developing countries, m-assessment can be more accessible than other computer-based assessments given affordability of mobile phones over tablets or laptop computers. For example, in South Africa, where the majority of students could not afford desktops or laptops nor the internet via those computer devices to access online learning tools, mobile assessment was used as a viable option (Kizito, 2011; Mentor, 2014). Limited access to online education in South Africa motivated institutions like Kizito's to offer a diagnostic test for testing student readiness before letting students enroll in online courses. Kizito required students to take a diagnostic test on basic calculus topics before joining an advanced calculus course. After taking the tests, students were recommended to review areas that they were not proficient in, as outlined by the test results, and decide to either fully prepare for the course or not enroll in the course at all. These tests were intended to help students manage and get the most out of their limited resources by informing them about their readiness and guiding them to study materials.

Limited access to devices is not a problem specific to the developing world. Even in developed nations, some classrooms may have limited computer equipment, class sizes or schedules may not be arranged in accordance with the number of computers in a classroom, or computer maintenance may not be timely. Such cases will disrupt interactive teaching activities or practice sessions that require students to work individually at an assigned computer. For instance, students who tried mobile assessment in a study conducted by Romero et al. (2009) indicated that mobile devices could be especially useful when there are more students in a classroom than available computers, which would help when they need to split the students into several groups for assessment activities. Moreover, using mobile phones instead of clickers can also increase users. Taking the cost of buying clickers and the probability of forgetting clickers at home into account, using mobile phones in classrooms can be a more feasible alternative.

Another way to increase accessibility with m-assessment can be by mobile surveys or live polling (i.e., mobile surveys, game shows, or sporting event voting). For instance, Poll Everywhere enables users to gather live responses in contemporary classes like presentation venues. Other examples include mobile adaptations of well-known online survey websites such as Survey Monkey and Qualtrics. Both Poll Everywhere and Qualtrics adopt offline surveys by collecting participant responses via text messages. Other live polling apps, including Mentimeter, SMSPoll, and DirectPoll, have been incorporated into social media platforms with immediate collation of results, as well as live sporting events. Depending on the participants' phones and carrier plans, some participants may find texting more accessible and cheaper than using cellular data and therefore find the cost of participating in mobile surveys via texts cheaper than online surveys. However, service provider carriers and some commercial companies also charge additional costs to use their live polling. Where there is low or no cost to the participants, there is generally an increase in people's willingness to participate in mobile polling and surveys. Therefore, sampling of such surveys can be more inclusive, and it can be concluded that m-assessment increases participation and accessibility to assessment to a wider range of people. Furthermore, just like commercial entities, surveys and live polling can be used to continuously engage a target audience to keep one's finger on the pulse of the participants and to offer learners a channel to have their voices heard. This continuity between live polling and surveys was visualized in Figure 2.

EFFECTS ON STUDENT OUTCOMES

The ultimate goal of educational assessment is to improve positive student outcomes. Although mobile assessment was suggested as a tool for catalyzing student achievement, one concern with using mobile assessment may be that, due to the interactivity and abundance of stimuli in mobile assessment activities, learners may not show their true potential. To understand the effectiveness of mobile assessment, major learning outcomes, achievement, and engagement were examined along with cognitive load requirements of mobile assessment.

Figure 2. A continuous loop of mobile feedback as an additional communication channel

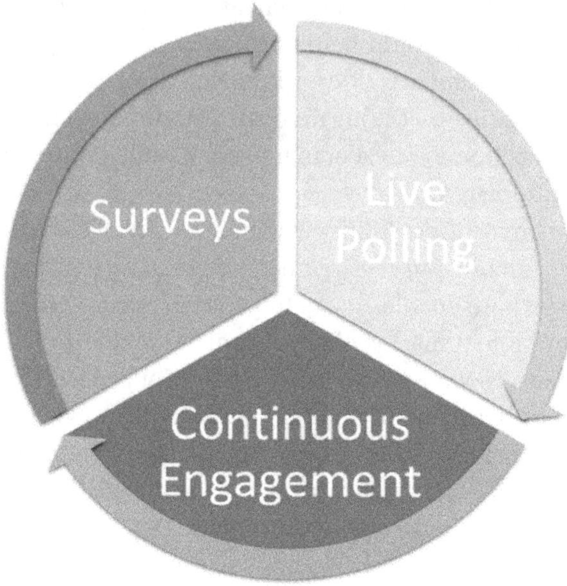

Achievement

Researchers have paid considerable attention to comparing achievement between groups receiving assessment activities enhanced with mobile assessment and control groups not receiving mobile assessment. Particularly, if mobile assessment was used as part of a learning session or course, it is hypothesized that mobile assessment facilitated learning and thereby increased achievement. Although some evidences suggested that learners' achievement increased by using mobile assessment (Hwang & Chang, 2011), some results are inconclusive (Arnesen et al., 2013; De-Marcos et al., 2010). Hwang and Chang compared achievement of two groups of students visiting a museum for a learning activity where only the experimental group used mobile assessment and received hints when students failed to respond to questions correctly. The results suggested that the experimental group significantly outperformed the control group, while controlling for initial differences between the two groups. Similarly, De-Marcos et al. compared experimental and control group pairs at the middle school, high school, and college levels, where experimental groups received mobile assessment and control groups did not. Mobile assessment was used to deliver class quizzes to facilitate students' self-assessment. Comparisons were based on the percentage of the course objectives for chosen courses that were aligned with mobile assessment that students mastered. Although significant improvement was found in the total scores at middle school level, no significant difference was observed at high school and college level courses. Similarly, Arnesen et al. also compared students using mobile assessment in the classroom to those who did not based on the percentage of correct responses given on classroom quizzes and final grades taken from the course. Evaluation of the quizzes suggested that the percentage of correct responses in the mobile assessment group was almost twice that of the control group. However, the final grades of the control group were higher. Arnesen et al. linked the difference in the final grades to not accounting for initial differences between groups.

Engagement

Engaging learners with content has been a goal of mobile assessment not only for improving learners' affective states, but also for increasing learning outcomes. Mobile assessment makes content available anytime and anywhere, which may increase learner engagement and thereby learning outcomes. With this motivation, studies investigated learner engagement in various learning experiences enhanced with mobile assessment. Studies consistently reported that learner engagement increased with mobile assessment (Arnesen et al., 2013; Santos et al., 2011; Stav et al., 2010). Learner comments and survey responses were used for identifying the effects of mobile assessment on learner engagement. For instance, more than 70 percent of students participating in the study conducted by Arnesen et al. indicated that mobile assessment increased both engagement and learning. In the study conducted by Stav et al., students responded positively to survey questions on whether mobile assessment encouraged and activated students. A possible reason for students feeling more encouraged could be that majority of students agreed that being in lectures with mobile assessment fun was fun. Santos et al. provided open-ended survey questions to students and the results were insightful for explaining why students perceived mobile assessment as more engaging. Santos et al. summarized students' comments and indicated that students found the mobile assessment enhanced learning activity to be innovative and fun, enjoyed the structure of the activity, which required being active in the museum rather than being static in front of the teacher taking notes, and enjoyed being able to perform the mobile assessment activities in groups even without the presence of a mentor.

Cognitive Load

One of the major challenges in designing mobile assessment activities is maintaining a balance between making assessments interesting, context rich, interactive, and dynamic while not burdening learners with excessive cognitive activities. This can be especially challenging on mobile devices with smaller screen real estate (Mentor, 2018). If learners are overwhelmed, the learning activities may not be effective and assessment activities may not represent their true abilities. Sweller (1988) suggested that overlapping cognitive processes use up much of the cognitive processing capacity of a person and make problem solving ineffective as a learning tool. Hung et al. (2012) designed a series of mobile assessment activities that aim to provide an assessment experience that provides a manageable cognitive load. In their design, mobile assessments were embedded into the mobile learning activities in three layers. These three layers consisted of guided observation, autonomous observation, and scientific inquiry tasks. Similarly, Hung et al. (2013) developed digital worksheets arranged according to this three-layer design, which were also used in further studies.

Various other designs were also used to accommodate an optimal cognitive load. Hwang and Chang (2011) measured the cognitive load of learners who were engaged in a mobile assessment enhanced learning task by adapting the cognitive load survey from Sweller et al. (1998). Two dimensions from the original survey (mental load and mental effort) were included in this adapted version. Hwang and Chang also compared the cognitive load of learners in experimental and control groups and found no significant difference. They also compared the cognitive load of learners with different achievement levels from experimental and control groups and found no consistent difference signaling a link between cognitive load promoting or inhibiting achievement in the assessment activities. Hwang and Chang concluded that the cognitive load requirements used in the mobile assessment enhanced learning tasks were appropriate.

CHALLENGES

Despite the abundance of benefits outlined here, many challenges persist with mobile assessment. While mobile assessment provides major advantages and extended accessibility with a plethora of devices and apps, it also brings up compatibility issues. Compatibility issues can arise between various operating systems of smart phones such as iOS, Android, and Windows and because of the variability of apps. Since each operating system works differently, apps should be specifically designed to work under a specific operating system. Because iOS and Android currently make up the biggest share of the market (at the time of publication), many well-known and widely used apps do have versions for both operating systems. However, compatibility issues may arise for custom apps, where the resources might be limited for producing a version compatible with other operating systems. Such compatibility issues are highlighted when learners bring their own mobile devices. Mentors and institutions should be aware of the availability of the custom apps for students using different operating systems. In addition, if instructions will be presented, learners should be provided with screenshots of the apps using different operating systems.

With smart phones getting larger screens, reading from the screen and inputting information using the phone becomes easier. The recently introduced foldable screen technology seems to allow users to customize and enlarge their screen sizes for complex tasks. Extending availability of m-assessment beyond classes can add to the challenges. For instance, Brett (2011) reported that students who received class assignments as text messages complained about classwork invading their personal time. Providers of mobile assessment should also consider the applicable charge rates for mobile services such as sending or receiving text messages or using cellular data.

Another challenge relevant to user experience is user readiness. It may not be realistic or useful to assume that all mentors and students are all familiar with mobile learning just because they have a mobile device. With the switch from using mobile phones as communication devices to using them as assessment devices, users may need to utilize functions of mobile phones they have not used before. Previous recreational uses of mobile phones and functions for mobile assessment may need to be intentionally highlighted as a new realization or utilization of the mobile device as a teaching, learning, or assessment tool. These initiatives can also not operate in a vacuum; they require a holistic approach that includes mentors, students, and administrators who are familiar with the technology and its use for educational learning purposes.

RECOMMENDATIONS

Suggestions for using mobile assessment effectively can be organized under two topics: (a) smart use and (b) smart design. First, mobile assessment should be used in a way that complements learning and increases effectiveness of learning. Practices that make the most out of what mobile assessment offers make the assessment more aligned with the learning experience. For instance, in museums or on field trips, mobile assessment can be offered as an integral part of the learning activities. In classes that use mobile phones as clickers, mobile assessment can be integrated at the end of each learning unit. In the online learning context, mobile assessment can be used as part of a holistic e-learning strategy, which caters to cross-device access that varies between desktop computer, laptop, tablet, mobile phone, or other handheld devices. Such practices can provide continuity in learning experiences. An all-inclusive approach will highlight not only the cost savings that come with integrating mobile devices into curriculum delivery or assessment but also the fact that the target audience will have 24/7 access to content from any location. With this always-available access, mobile learning and assessment can aid the development of personal learning environments as well as self-paced personal learning journeys.

Another suggestion for smart use of mobile assessment is utilizing mobile assessment for formative assessment purposes. As understood from the anywhere, anytime, anyone argument of mobile assessment, mobile assessment lends itself to providing and creating formative assessment opportunities. With mobile assessment, formative assessment is not limited to administering a content test; rather, learners can show mastery of concepts via recordings, images, presentations, and much more.

Second, smart designs of m-assessment help mentors learn about how students are learning. For instance, more authentic (i.e., natural, not standardized, integrated into real-life practice) assessments provide opportunities to evaluate learners during their interactions with peers or materials. Moreover, such assessments can also help to measure constructs that are typically hard to measure, such as cognitive readiness. Since high cognitive load is a concern, mobile assessment tasks should be designed so that learners can focus on a manageable number of tasks. In addition, avoiding m-assessments that are offered at a preset time can also help ensure that mobile assessment is available anytime (Pappas, 2015).

Making mobile assessment more engaging and usable requires smart design. Mobile assessments require a different approach that takes into consideration users' mobile phone habits and knowledge of phone functionality. For instance, if an assessment requires a long response, it may not be well suited for mobile delivery. If users use voice recording functions or social media sharing functions more frequently than text entry, using these functions or platform could increase the assessment's usability and. Moreover, the outputs of mobile assessments should also be easy to digest and use. For instance, input and output should be compatible with various devices such as tablets and laptops. Usability of mobile assessment via mobile phones should be similar to usability on other devices so that similar user engagement can occur. As a side note, cognitive load of the user should be also considered.

Another recommendation for smart design should be allowing hypertext and hypermedia functionality. Internet-connected mobile devices allow users to access hypermedia, which aids the development of clustered and connected thinking. Although allowing the user to explore various connections related to a topic can be a source of distraction, the mobile assessment could be set up in manner that would ask the learner to track its centrality of concepts or divergence from the central concept. These ideas can also be used for formative and summative assessments. Students and mentors would both need education or professional development to adopt and exploit such affordances and methods.

FUTURE RESEARCH DIRECTIONS

Collectively, the results of the studies on mobile assessment suggest that mobile assessment can be useful in various environments and indicate areas that needed to be explored. Because one of the biggest strengths of mobile assessment is representing real-life learning scenarios, more studies using mobile assessment in realistic learning scenarios are needed. Mobile assessment practices that represent virtual practice areas or real-life scenarios should continue to interrogate and explore effectiveness of mobile assessment. Outcomes of mobile assessment can also be diversified. In addition to measuring achievement, other indicators of meaningful learning, such as formative learning or retention, can be also studied. More research is also needed to explore the long-term affective outcomes of mobile assessment. Such research should investigate whether short-term affective outcomes, such as enjoyment and engagement, can continue long term, especially after the honeymoon effect of mobile assessment wears off.

Finally, future studies can make use of communication functions of mobile phones and develop mobile assessment practices that foster communication and collaboration. For instance, by connecting to various social media tools and cloud-based collaboration platforms, mobile assessments can facilitate skills such as collaborative problem solving and leadership.

CONCLUSION

Mobile assessment is an important tool to make learning accessible to anywhere, anytime, and to anyone. Looking at current practices and challenges of mobile assessment provided three lessons:

1. Incorporating various features of mobile assessment practices can allow assessment to be used in more innovative ways, even in traditional classrooms.
2. Mobile assessment facilitates desired learner outcomes such as engagement.

3. Mobile assessment cultivates rich learning environments and experiences while providing ease to teachers, mentors, and students. Practitioners and researchers who are interested in using mobile assessment should consider smart use (i.e., how to best use m-assessment for overcoming the challenges of their specific contexts) and smart design (i.e., how to best design m-assessment to measure learner outcomes with minimal error and ensure the optimal cognitive load). Future research areas indicate that mobile assessment is a promising and fruitful practice that can be applied to a variety of contexts and foster multiple learner outcomes.

REFERENCES

Abel, R. (2015, May 12). *Find out more about IMS*. Retrieved November 30, 2015, from https://www.imsglobal.org/aboutims.html

Anderson, L. W., Krathwohl, D. R., & Bloom, B. S. (2001). *A taxonomy for learning, teaching, and assessing: A revision of Bloom's taxonomy of educational objectives*. Allyn & Bacon.

Arnesen, K., Sivertsen Korpås, G., Hennissen, J. E., & Birger Stav, J. (2013). Experiences with use of various pedagogical methods utilizing a student response system – Motivation and learning outcome. *Electronic Journal of E-Learning*, *11*, 169–181.

Bloom, B. S. (1956). *Taxonomy of educational objectives: The classification of educational goals. Handbook 1-2*. Longmans: McKay.

Bogdanović, Z., Barać, D., Jovanić, B., Popović, S., & Radenković, B. (2014). Evaluation of mobile assessment in a learning management system (LMS). *British Journal of Educational Technology*, *45*(2), 231–244. doi:10.1111/bjet.12015

Bolsinova, M., Deonovic, B., Attali, M., Settles, B., Hagiwara, M., von Davier, A., & Maris, G. (2018). *Modeling hint use and response accuracy in learning environments*. Paper presented at Maryland Assessment Conference. Retrieved from https://education.umd.edu/file/11330/download?token=MYO8oBKU

Boyle, T. (1997). *Design for multimedia learning*. London: Prentice Hall.

Brett, P. (2011). Students' experiences and engagement with SMS for learning in higher education. *Innovations in Education and Teaching International*, *48*(2), 137–147. doi:10.1080/14703297.2011.564008

Caceres, N., Wideberg, J. P., & Benitez, F. G. (2007). Deriving origin–destination data from a mobile phone network. *IET Intelligent Transport Systems*, *1*(1), 15–26. doi:10.1049/iet-its:20060020

Chao, K. H., Chang, K. E., & Lan, C. H., Kinshuk, & Sung, Y. T. (2016). Integration of Mobile AR Technology in Performance Assessment. *Journal of Educational Technology & Society*, *19*(4), 239–251.

Clark, R., & Mayer, R. (2011). *E-learning and the science of instruction proven guidelines for consumers and designers of multimedia learning* (3rd ed.). San Francisco, CA: Pfeiffer. doi:10.1002/9781118255971

de-Marcos, L., Hilera, J. R., Barchino, R., Jimenez, L., Martinez, J. J., Gutierrez, J. A., ... Oton, S. (2010). An experiment for improving students' performance in secondary and tertiary education by means of m-learning auto-assessment. *Computers & Education*, *55*(3), 1069–1079. doi:10.1016/j.compedu.2010.05.003

Dearnley, C. A., Taylor, J. D., Laxton, J. C., Rinomhota, S., & Nkosana-Nyawata, I. (2013). The student experience of piloting multi-modal performance feedback tools in health and social care practice (work)-based settings. *Assessment & Evaluation in Higher Education*, *38*(4), 436–450. doi:10.1080/02602938.2011.645014

Ferenchick, G. S., Solomon, D., Foreback, J., Towfiq, B., Kavanaugh, K., Warbasse, L., ... Mohmand, A. (2013). Mobile technology for the facilitation of direct observation and assessment of student performance. *Teaching and Learning in Medicine : An International Journal*, *25*(4), 291–299. doi:10.1080/10401334.2013.827972 PMID:24112197

Gibson, D. (2013). Assessing teaching skills with a mobile simulation. *Journal of Digital Learning in Teacher Education*, *30*(1), 4–10. doi:10.1080/21532974.2013.10784720

Giles, J. (2010, April 13). Barcodes help objects tell their stories. *New Scientist.* Retrieved from http://www.newscientist.com/article/dn18766-barcodes-help-objects-tell-their-stories.html

Heric, M., & Carter, J. (2011). Cognitive readiness assessment and reporting: An open source mobile framework for operational decision support and performance improvement. *Performance Improvement*, *50*(7), 5–17. doi:10.1002/pfi.20227

Huang, Y. M., Lin, Y. T., & Cheng, S. C. (2009). An adaptive testing system for supporting versatile educational assessment. *Computers & Education*, *52*(1), 53–67. doi:10.1016/j.compedu.2008.06.007

Hung, P. H., Hwang, G. J., Lin, Y. F., Wu, T. H., & Su, I. H. (2013). Seamless connection between learning and assessment – applying progressive learning tasks in mobile ecology inquiry. *Journal of Educational Technology & Society*, *16*, 194–205.

Hung, P. H., Hwang, G. J., Su, I., & Lin, I. H. (2012). A concept-map integrated dynamic assessment system for improving ecology observation competences in mobile learning activities. *Turkish Online Journal of Educational Technology-TOJET*, *11*, 10–19.

Hwang, G.-J., & Chang, H.-F. (2011). A formative assessment-based mobile learning approach to improving the learning attitudes and achievements of students. *Computers & Education*, *56*(4), 1023–1031. doi:10.1016/j.compedu.2010.12.002

Kay, M., Santos, J., & Takane, M. (2011). Barriers to mHealth implementation. In mHealth: New horizons for health through mobile technologies (Vol. 3, p. 112). Geneva: World Health Organization.

Kelsey, T., & Cavendish, W. (2014, November 13). *Personalised health and care 2020: A framework for action.* Retrieved December 14, 2015, from https://www.gov.uk/government/publications/personalised-health-and-care-2020/using-data-and-

Kizito, R. (2011). Pretesting mathematical concepts with the mobile phone: Implications for curriculum design. *International Review of Research in Open and Distance Learning*, *13*(1), 38–55. doi:10.19173/irrodl.v13i1.1065

Lin, J., & Rivera-Sánchez, M. (2012). Testing the information technology continuance model on a mandatory SMS-based student response system. *Communication Education*, *61*(2), 89–110. doi:10.1080/03634523.2011.654231

Mahmud, N., Rodriguez, J., & Nesbit, J. (2010). A text message-based intervention to bridge the healthcare communication gap in the rural developing world. *Technology and Health Care, 18,* 137–144. doi:10.323 d3/THC-2010-0576

Martínez, M., & Hilera, J. R. (2010). A new m-learning application designed for self-assessment. *Pulso: Revista de Educación*, *33*, 207–215.

Mentor, D. (2011b). *Supporting students' wellbeing via a texting service*. Invited guest speaker iClicker Conference, Houston, TX.

Mentor, D. (2014a). *Showcasing a Criteria Taxonomy for selecting a 21st Century Learning Management System to support blended teaching and mLearning*. Poster Presentation. The 21st Century Academic Forum Conference at Harvard, Cambridge, MA. Retrieved from http://www.21caf.org/uploads/1/3/5/2/13527682/21st_century_academic_forum_conference_book.pdf

Mentor, D. (2014b). From app attack to goal oriented tablet use. In H. An, S. Alon, & D. Fuentes (Eds.), *Tablets in K-12 education: Integrated experiences and implications*. Hershey, PA: IGI Global. doi:10.4018/978-1-4666-6300-8

Mentor, D. (2018a). Teachers College, Columbia University, Department of Biobehavioral Science. *BBS 5199 002: mHealth - Mobile Health App Development CRN 32100*. Retrieved from tc.edu

Mentor, D. (2018b). TLC for MOOCS: Teaching and Learning Communities for Computer Programming. In Computer-Mediated Learning for Workforce Development (pp. 93-110). IGI Global.

Mentor, D. (2018c). mClass Planet of the Apps: The Rise of Mobile Learning. In Computer-Mediated Learning for Workforce Development (pp. 196-215). IGI Global.

Mentor, D. J. (2011a). *Exploring social connectedness via mobile phone texting* (Ph.D. thesis). Teachers College, Columbia University. Retrieved November 25, 2018 from https://www.learntechlib.org/p/11

Mosa, A. S. M., Yoo, I., & Sheets, L. (2012). A systematic review of healthcare applications for smartphones. *BMC Medical Informatics and Decision Making*, *12*(1), 67. doi:10.1186/1472-6947-12-67 PMID:22781312

New Schools. (n.d.). *Wireless Generation*. Retrieved from http://www.newschools.org/venture/wireless-generation

Niccolai, J., & Gohring, N. (2010, April). A brief history of palm II. *Network World*.

Pappas, C. (2015, April 21). How to develop mobile learning assessments [Web Blog Post]. Retrieved from http://elearningindustry.com/5-tips-develop-mobile-learning-assessments

Pew Research Center. (2014, February 13). *Emerging nations embrace Internet, mobile technology - Cell Phones Nearly Ubiquitous in Many Countries*. Retrieved from http://www.pewglobal.org/2014/02/13/emerging-nations-embrace-internet-mobile-technology/

Pew Research Center. (2015). *Device ownership over time*. Retrieved from http://www.pewinternet.org/data-trend/mobile/device-ownership/

Pew Research Center. (2015). Retrieved from http://www.pewglobal.org/2018/06/19/2-smartphone-ownership-on-the-rise-in-emerging-economies/#smartphone-ownership-growing

Poll Everywhere. (n.d.). *Live audience participation*. Retrieved from http://www.polleverywhere.com

Qualtrics. (2015). *Qualtrics Mobile*. Retrieved from http://www.qualtrics.com/research-suite/mobile-surveys

Romero, C., Ventura, S., & de Bra, P. (2009). Using mobile and web-based computerized tests to evaluate university students. *Computer Applications in Engineering Education*, *17*(4), 435–447. doi:10.1002/cae.20242

Ross, C., Gray, S., Warwick, C., Hudson-Smith, A., & Terras, M. (2012, July). *Engaging the museum space: Mobilising visitor engagement with digital content creation*. Paper presented at Twenty-third Digital Humanities Conference, Hamburg, Germany. Retrieved from http://www.dh2012.uni-hamburg.de/conference/programme/abstracts/engaging-the-museum-space-mobilising-visitor-engagement-with-digital-content-creation.1.html

Şahin, F., & Yan, Z. (2013). Mobile phones in data collection: A systematic review. *International Journal of Cyber Behavior, Psychology and Learning*, *3*(3), 67–87. doi:10.4018/ijcbpl.2013070106

Samaie, M., Mansouri Nejad, A., & Qaracholloo, M. (2018). An inquiry into the efficiency of WhatsApp for self-and peer-assessments of oral language proficiency. *British Journal of Educational Technology*, *49*(1), 111–126. doi:10.1111/bjet.12519

Santos, P., Pérez-Sanagustín, M., Hernández-Leo, D., & Blat, J. (2011). QuesTInSitu: From tests to routes for assessment in situ activities. *Computers & Education*, *57*(4), 2517–2534. doi:10.1016/j.compedu.2011.06.020

Stav, J., Nielsen, K., Hansen-nygård, G., & Thorseth, T. (2010). Experiences obtained with integration of student response systems for iPod touch and iPhone into e- learning environments. *Electronic Journal of E-Learning*, *8*, 179–190.

SurveyMonkey. (n.d.). *Mobile Surveys*. Retrieved from http://www.surveymonkey.com/mp/mobile-surveys

Sweller, J. (1988). Cognitive load during problem solving: Effects on learning. *Cognitive Science*, *12*(2), 257–285. doi:10.120715516709cog1202_4

Sweller, J., Van Merriënboer, J. J. G., & Paas, F. G. W. C. (1998). Cognitive architecture and instructional design. *Educational Psychology Review*, *10*(3), 251–297. doi:10.1023/A:1022193728205

Taylor, J. D., Dearnley, C. A., Laxton, J. C., Coates, C. A., Treasure-Jones, T., Campell, R., & Hall, I. (2010). Developing a mobile learning solution for health and social care practice. *Distance Education, 31*(2), 175–192. doi:10.1080/01587919.2010.503343

Triantafillou, E., Georgiadou, E., & Economides, A. A. (2008). The design and evaluation of a computerized adaptive test on mobile devices. *Computers & Education, 50*(4), 1319–1330. doi:10.1016/j.compedu.2006.12.005

Turchinsky, M. (2011). Smartphone Apps: The Hype, the Reality and Everything in Between. In *Proceedings of The American Alliance of Museums*. Houston, TX: PROLibraries.

University College London Centre for Digital Humanities, Centre for Advanced Spatial Analysis, & University College London Museums. (2011). *QRator*. Retrieved from http://www.qrator.org

van der Linden, W. J., & Pashley, P. J. (2010). Item selection and ability estimation in adaptive testing. In Elements of adaptive testing. New York: Springer.

Zheng, R. (2009). *Cognitive effects of multimedia learning*. Hershey, PA: Information Science Reference. doi:10.4018/978-1-60566-158-2

Chapter 6
Visualizing the Evolution of Mobile Learning Research

Hüseyin Özçınar
Pamukkale University, Turkey

ABSTRACT

The aim of this chapter is to investigate the scope and change of m-learning literature over a period of thirteen years. The chapter takes an author co-citation approach where patterns are revealed in mlearning publications between the years 2002 and 2015. The author co-citation analysis is a kind of bibliometrics, and Power Flow Network Library (PFNET) analysis method to obtain the intellectual structure of the m-learning field. A PFNET is a psychometric scaling method where a network of associative nodes is generated using specific values. The method is based on graph theory and has been used to explore knowledge connections. The method is referred to as a pathfinder network for representing segments of knowledge in a branched manner. The method can also be used for developing consensus building or recognizing expertise in a specific area. In this study, it was found that the m-learning research focus on design and effectiveness of m-learning environments, developing a pedagogic framework for m-learning, adoption and proliferation of m-learning-assisted language learning, collaborative m-learning, and augmented reality. Assessing the results of factor analysis together with the PFNET graphics, the subfields show the fundamental orientations of the field, and the main themes will be the focus of future research.

DOI: 10.4018/978-1-5225-9351-5.ch006

INTRODUCTION

Rapid developments in information and communication technologies (ICT) has had an impact on education and learning just as in other areas of life. Today mobile technologies are widely used in transforming education and learning. In a similar vein, conducting research on mobile technologies is one of the main research trends in educational technology journals (potential citation needed or rephrase). Considering there was limited research conducted on mlearning (or mobile learning) ten years ago (Hung and Zhang, 2011), with the rapid growth and adoption of mobile devices, one can clearly understand the rapid change in this research area. As Naismith et. al. (2004), Sharples, Taylor and Vavoula (2005) stated, there is a co-evolution of learning and mobile technology.

There were previous studies that reviewed mobile learning research area. For example, Naismith et. al. (2004) focused on theories supporting mlearning and mlearning examples. Mentor (2016) contends that the examples supplied in Naismith et.al. had unintended social connectedness embedded in its approach leveraging the main facets of mobile communication to build portable communities. Hung and Zhang (2011) studied the research trends in mlearning research area using text mining technique. Data for Hung and Zhang's (2011)study consisted of 119 articles and conference proceedings that were published between the years 2003 and 2008 and indexed by SCI/SSCI database. Hung and Zhang (2011) investigated the publication date, publication journal, country and institution of origin and research themes. A rapid increase in mlearning publications found Taiwan and the USA are the most prolific countries in some areas of mobile learning. Main mobile learning research themes in their study were effectiveness, evaluation, and personalized systems.

To analyze mlearning literature, Wu et al. (2012) conducted a meta-analysis study on 164 articles which had been published between the years 2003 and 2010. Research results showed that mobile phones and Personnel Digital Assistants (PDAs) at that time, were the most widely used devices for mlearning, while effectiveness and system designs were the most prolific research themes. These studies shed light on the structure of the field between the years 2003 and 2010.

In a more recent study, Hwang and Wu (2014) determined 214 publications published between 2008 and2012 on mlearning indexed 7 journals publishing research on technology-enhanced learning. They investigated these publications in terms of learning domain, context of mobile learning, and the type of mobile devices adopted. At the end of the research, they reported that mlearning holds significant promise for the future in terms of increasing academic success, raising interest and motivation towards learning on condition that satisfactory support is granted for the learners and suitable strategies are used during the learning process. Other findings obtained in the research reveal that smartphones and tablet PCs are the most commonly used tools in the recent period and mlearning is used both in formal and informal education. It has also been revealed that many studies using the mlearning method mostly in language learning fields do not focus on a certain subject .

The ever-changing environment of mobile devices and mlearning research makes it important to map the structure of the research field and trends. Providing an up-to-date and in-depth map of the field could be very informative for policy makers in educational regulations, funding agencies and researchers in the field. In the literature review, studies conducted on mlearning especially before the year 2010 were analyzed. This period is named "*embryonic stage*" of mlearning field (Motivalla, 2007, p.582). After the mlearning keywords were searched in the web of science database, it was found that nearly 60 percent of mobile learning research published after 2010 end this sentence with what was found. Besides, mobile device usage increasing daily, research on mlearning went through changes concurrent with technological developments (Motivalla, 2007). This case indicates that the literature may have substantially evolved after 2010. Therefore, it is believed that a research embracing this period will be useful in terms of depicting the development and current picture of the literature. One of the common characteristics of the studies on mlearning research is the focus on method, mobile device and learning domain rather than the formation of sub-fields of the main field in these studies. Motvilla's study, unlike similar studies, tries to determine sub-fields and orientations of the field through bibliometric method. The aims of this chapter are (1) to determine the subfields and reference disciplines of the mlearning field, (2) to determine the changes in intellectual structure of the mlearning field and (3) to determine the emerging mobile learning research themes.

METHOD

Bibliometrics

This study uses the author co-citation analysis which is a type of bibliometrics approach. Bibliometrics is the application of statistics to the publications and written professional communications (Diodato, 1994). The main assumption behind bibliometrics is that "*bibliographic citations are an acceptable surrogate for the actual influence of various information sources on a research project*" (Culnan, 1987, p. 342). Citations, not singularly but when aggregated over time, reflect the perceptions of citing authors about the resemblance between citing and cited documents (White and McCain, 1998). Studies which use citation models to examine development of scientific fields acknowledge that a shift in the intellectual state of the field is presented in the citation patterns of documents (publications) in the field (Small, 1977).

Co-Citation Analysis

In a co-citation analysis, how many times two authors are included in the references of different documents is considered as a measure of the similarity between these two authors in terms of their research perspectives or theoretical perspectives. Author co-citation analysis ensures the compilation of co-citation counts in matrix form; and to analyze this matrix through statistical techniques which yields a snapshot of the field in a distinct point in time. Moreover, one can get insight into changing and evolving structure of the research field by comparing these images (Small, 1993). Author co-citation analysis is commonly used in different fields to interpret their subfields, the relationships between these subfields, and thus possible research trends in the future.

Data

Data for this study was obtained from the Web of Science Database which is the world's leading citation database, and has broad coverage which provides systematic ways to analyze bibliometric information efficiently (Hsiao and Yang, 2011). In order to identify papers related to mlearning, a search was designed by using "m-learning", "mobile learning", "ubiquitous learning", "wireless learning" and "wireless network learning" as keywords. Search for these keywords in title, abstract and keyword sections produced 2,339 source documents published between January, 2002 and June 2015. Search results were filtered, and book reviews, editorials and meeting abstracts were not included in the data set. The remaining 2294 articles and conference proceedings were included in the study. In pre-2010 period mlearning studies were discussed in field-dedicated congresses rather than academic journals (Traxler, 2007). For that reason, this study involves not only the articles published in academic journals but also congress proceedings.

In this study, the most cited authors in the field were selected as source authors (Özçınar, 2015). The top-cited authors, according to citation count, were ranked in descending order, and the most-cited authors were included in the study. For the early period (2002-2010), 96 authors who were cited at least 15 times in the mlearning literature were included in the study. For the later period (2011-2015) 98 authors who were cited 20 times or more in the mobile learning literature were included in the study.

In ACA research in the literature, there is no consensus on the size of data set. This number was determined as 47 in the study of Culnan (1986), as 25 by Estabrooks et al. (2008) and as 100 by White and McCain (1998). In the present study, data set was determined as 97 authors cited for more than 15 times in mlearning field in the period up to the year 2010 which is called as early period and as 98 authors cited for more than 20 times in 2011-2015 period. An author who was not co-cited with any other author in the data set was omitted and the data set of the early period consisted of 96 authors. Afterward, the co-citation frequency matrix consisting of co-citation counts of authors was formed. One of the most commonly discussed topics about ACA research is how the diagonal data in the co-citation frequency matrix should be determined. The diagonal data (a_{ii}) are derived from the citations of varying works of an author in the same publication. In the co-citation studies, the data on the matrix (a_{ij}) show the number of co-cited publications of authors (the ith and jth authors) and it is used as the criterion of the similarity in conceptual perspective and research themes of two authors. As each author had the most similarity with oneself independently from the citation number, this study determines the data on the diagonal as the maximum of related column +1 (Ma, Da,Ni and Li, 2009).

Bibliographic data of these source documents were downloaded and prepared for the analysis via the Bibexcel program (Persson, Danell and Schneider, 2009). Ucinet program (Borgatti, everett and Freeman, 1999) was used to convert the data into matrix format and to prepare the input data, the Pajek program (Batagelj and Mrvar, 2005) was used for the pfnet drawing. Steps followed in the data analysis process are summarized in Figure 1.

Factor Analysis

In the author co-citation analysis, general factor analysis or cluster analysis are used to reduce the large number of original variables to a smaller number of factors or clusters which can be used to study the relationship between original variables. In the author co-citation analysis, principal component analysis and varimax rotation are commonly used as factor analysis procedures. In this process, authors who are co-cited in the references of source papers are clustered together. The variance explained by a factor can be considered as a contribution made by the factor or subfield to the field (McCain, 1990) and factor loadings of each author can be considered as strength of the relationship between factor and author.

Visualization of the mLearning Field

Garfield, Malin and Small (1979) claimed that "*mapping science is an attempt to arrive at a spatial representation of fields and disciplines-and, at a lower level, individual papers and scientists-in which the relative locations of entities is depicted*"(p. 192). The visualization of abstract concepts and relations may contribute to our understanding. In this regard, the visualization of the mobile learning field may contribute to our understanding of the structure of the field and the relationship between subfields. In this study, PFNET graphics obtained by dissimilarity matrix was used to obtain the spatial representation of the mlearning field.

One step of the factor analysis process is to create Pearson correlation matrix. The values of Pearson correlation matrix differ between -1 and +1. In this matrix, if the values indicating the correlation between two authors is close to -1, it is interpreted that their research fields are different. Values closer to +1 are considered as an indicator of the similarity between research perspectives of two authors. In PFNET analysis, dissimilarity matrix is used. Therefore, Pearson correlation matrix being a similarity matrix should be converted into dissimilarity matrix to conduct PFNET analysis using Pearson correlation matrix. To this end, each value in the matrix was subtracted from 1 and the dissimilarity matrix was obtained. The dissimilarity matrix consists of the values between 0 and 2. In this matrix, the values being close to 0 show that the research perspectives of two authors are highly similar with each other and the values being close to 2 show that the research perspectives are dissimilar. In PFNET graphics, similar authors (values being close to 0) are placed close to each other, the positions of two authors get further from each other on the map depending on the decreasing similarity level.

RESULTS AND DISCUSSION

The ACA studies dividing data according to publication dates can yield significant findings on how the field has evolved. Papers published before 2010 are less than half of the total number of publications in the mlearning field; furthermore, mobile learning research dramatically increased after 2010 (see figure 2). Thus, 2010 can be considered as a turning point in mlearning field. In order to ensure a better understanding of the evolution of the mlearning research, the data were divided into two time-frames (papers published before 2010 (2002-2010) and after (2011 to 2015), were examined separately and the results of first (2002-2010) and second time period (2011-2015) are compared in this section.

Figure 1. Author co-citation analysis steps (Özçınar, 2015)

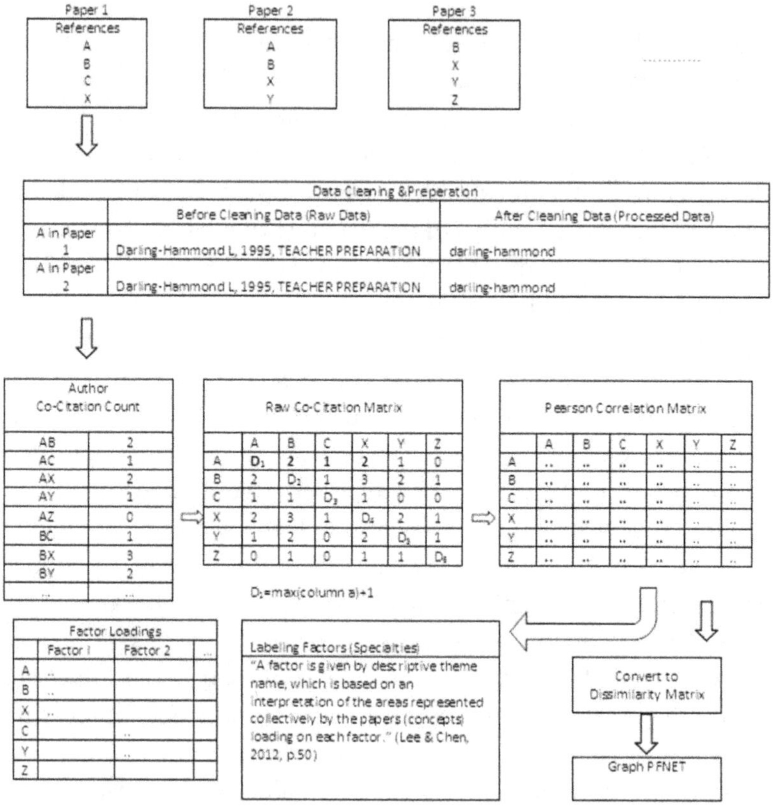

Figure 2. Publication counts by years

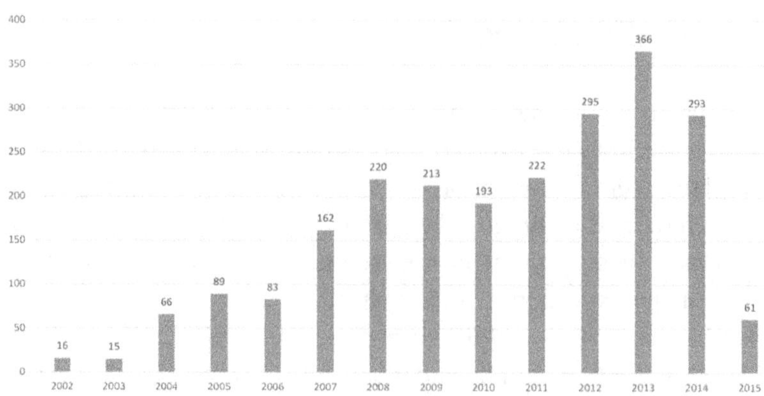

Articles and conference papers published between 2002 and 2010 in mlearning cited 8859 authors from 63 different countries. Taiwan was the most prolific researcher country in this period. Researchers in the Taiwan produced 18.32% of total mobile learning research. The Peoples Republic of China was the second prolific country (11.6%), and Japan was the 3rd one (6.4%) (see fig 3.). For the second period, 13703 authors cited between 2011-2015 in mlearning literature indexed by the web of science were analyzed. Considering the countries of the authors, it can be seen that the studies from 84 different countries were cited in the field of mlearning in this period. China, which ranked the 2nd in 2002-2010 period, went up to the first rank with 143 studies. On the other hand, Taiwan which was at the top fell back to the second rank. In the aforementioned time period the USA and Spain, which were ranked 5th and 6th in the previous period took 3rd and 4th place . Figure 1 shows the major contributing countries.

In the analysis of the distribution of studies by years (see Figure 2.), it can be seen clearly from the figure that mlearning research papers are increasing steadily each year. Perhaps the only exception, seen from the figure, is the years 2014 and 2015. The main reason of the decrease in research in the years 2014 and 2015 can be an indexing delay. The findings of the present study on the fact that more countries started to make contributions to mlearning research and the number of research increases are in parallel with the findings of Hwang and Tsai (2011) and Hung and Zhang (2012).

Figure 3. Publication counts by countries

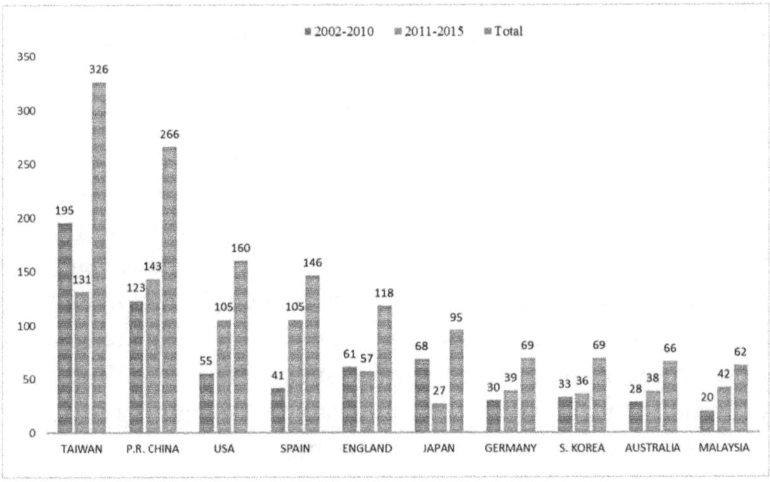

Papers included in this study were published in more than 500 different journals or conference proceedings being indexed by the web of science. Nearly 21% of papers were published in the five journals. Table 1 shows the top 10 source titles.

The Intellectual Structure of mLearning Field Between 2002 and 2010

Co-citation counts of 96 authors cited for 15 or more in this period were calculated via the Bibexcel program, and the obtained co-citation frequency matrix was analyzed via principal component analysis and varimax rotation method. At the end of the factor analysis, 22 factors having eigen value over 1 in 2002-2010 period were obtained, and these factors explained 85% of total variance. 11 factors consisting three or more authors, and explaining 68.50% of total variance (see Table 2) were named and analyzed. Each conceptual substructure of the factor represents a sub-field (specialty), which was made by the authors and the conceptual substructure of which was loaded on that factor.

In this period, the biggest factor in the field was the "*general issues and reviewing the progress in the field*" which explains 19% of total variance and includes 17 authors, 13 of whom had factor loadings above 0.6. In this specialty, four authors having a factor load above 0.8 (Kukulska-Hulme, Traxler, Naismith, and Cobcroft) had only loaded on this factor. White and Griffith (1982) call the authors who produce factor loading above .08 in a factor, yet do not produce factor loading above 0.4 in any other factor as "crystallized authors". According to White and Griffith, crystallized authors are considered as specialized people in a field by those who cite them. It is known that the authors who take place in this factor have publications in different subfields of mlearning; however, in the analysis of the most cited works in this time period (Kukulska-Hulme, 2009; Traxler, 2007; Naismith, 2004; Cobcroft, 2006), it can be seen that the common theme of this factor is to develop a pedagogic framework for the field, define the field and to determine the research trends.

The second biggest factor in this period was the "*effectiveness*" subfield that explains 13.4% of total variance and consisted of 12 authors having factor loading above .6. In this subfield, the focus is on the effect of the methods and strategies to be used in the design of mlearning environments and different mlearning designs on learning achievements and attitudes toward mlearning. The inclusion of authors related to constructivism and situated learning such as Jonassen (Year), Lave (Year), and Bruner (Year) in the factor can be an indicator that mlearning designs are based on situated learning and constructivist approaches.

The 3rd biggest factor called "*system design*" explains 6.6% of total variance and consists of 12 authors, 11 of whom have factor loading over .6. The publications in this factor focus on subjects like the presentation of innovative designs, fundamental design characteristics of mlearning systems, structuring human-computer interaction and mlearning systems according to the needs of users. Analyzing the studies cited in the subfields "effectiveness" and "system design", we can see that both factors are related to the design of mlearning environment. Besides, effectiveness factor mostly focuses on mlearning design principles, different learning designs and the students' success of these designs. On the other hand, system design factor focuses on the design of technically different tools and the interaction between human and computer.

The subfield "*collaborative mlearning*" represented by the crystallized authors; Liu, Roschelle, Norris and Zurrita is the 4th biggest subfield of the early-period mlearning studies. In this subfield, the effects created by each student having wireless hand-held devices and access to resources and classroom interactions as well as the collaborative learning and teaching means are discussed (Liu et. al., 2003; Norris and Soloway, 2004; Roschelle, 2003). In the factor "mlearning in higher education" which ranks 5th in terms of percentage of explained total variance, mlearning environment designs developed for higher education level (Sariola, 2003) and especially for teacher training (Seppala and Alamaki, 2003; Seppala and Sariola, 2002) have been discussed.

In the PFNET map used to visualize the structure of the field, the size of nodes representing authors is in proportion with the citation number of authors. Authors taking place in the same factor are colored the same and factor numbers are additionally depicted within parenthesis. It can be interpreted that the authors in the center of the PFNET graphics bring contributions to fundamental orientations about the field (Lee and Chen, 2010). Analyzing the PFNET graphic made for the period 2002-2010, it can be seen that the prominent subfields in terms of percentages of explained total variance such as "*effectiveness*" (2), "*general issues and reviewing the progress in field*" and "*collaborative mlearning*" are also on the main axis and could be considered as fundamental subfields for this period. Ranking 3rd in terms of the percentage of explained total variance, the subfield "*system design*" is positioned away from the center. The position of the subfield "*system design*" having high percentage of explained variance away from the center in PFNET graphics can be interpreted as the studies related to system design are significant for the field, yet they have "a supplemental role" in the field (Lee and Chen, 2010, p. 269). The subfields "*mlearning in higher education*", *virtual experiential learning with mobile technologies*","*adaptation in mlearning*" and "*usability in mlearning*" having relatively lower percentage of explained total variance seem to be peripheral fields in mlearning literature.

Taking place in the subfield "*acceptance of information technologies*" which is ranked relatively low, with 16th factor having low percentage of explained total variance mostly focus on foreign language teaching via mobile devices but take central position in the PFNET graphic. This case can be considered as the indicator that the subject of teaching English through mobile devices and acceptance of mobile technologies are going to be widely discussed in the post-2010 period and beyond.

Analyzing the PFNET graphics in terms of the relation between clusters, Keegan and Triffonova function as a bridge between 1st (general issues and reviewing the progress in the field) and 4th (collaborative mlearning) factors. And the factor "*acceptance of information technologies*" and "*collaborative mlearning*" subfields are attached to the "*effectiveness*" subfield through Chen. It can be seen that Lave J. on the PFNET graphic, functions as a bridge between both "*effectiveness*" and "*system design*" research through a focus on theories espoused by Jonassen (year), Vygotsky (year), Wenger (year) axis.

The Intellectual Structure of mLearning Field Between 2011 and 2015

The principal component analysis with varimax rotation method was used to analyze the symmetrical co-citation frequency matrix made upon the calculations of co-citation counts of 98 authors who were cited for 20 times or more in the period 2011-2015. At the end of the analysis, 98 authors are grouped into 16 factors in the period 2011-2015, and these 16 factors explain 80% of total variance. In this period, 48 of 98 authors have factor loading in first two factors. The first nine factors explaining 66.7% of total variance were named by the researcher after the analysis of most cited works of the authors in this period.

The 1st factor in which 10 out of 12 authors, having taken place in the "*effectiveness*" factor in previous period, three authors from "*system design*" factor and one author from the field "*collaborative mlearning*" take place together with 14 authors who

Figure 4. PFNET graph of authors from 2002 to 2010

had not taken place in data set in previous period explains 19% of total variance. This factor was named "*Design and Effectiveness*" after the analysis on most cited works of the authors taking place in this factor. Considering the factor loading of authors, apart from the "crystallized" authors of the "*effectiveness*" field in the period 2002-2010, like Hwang and Chu; Peng, Shih and Jonassen are also "crystallized" authors of the subfield. The fact that the "design and effectiveness" subfield is a big factor and 24 of 28 authors in this factor produce factor loading above .6 indicate that this subfield reflects fundamental orientation of the field.

The themes of the "*design and effectiveness*" subfield are to reveal general methods and principles for the design of ubiquitous learning activities and the effects of these methods and principles on academic success and interest, attitudes towards ubiquitous or mobile learning. Hwang and Tsai (2011) investigated the studies on mobile and ubiquitous learning fields between 2006 and 2010.They reported that these studies, instead of focusing on a specific learning field, generally focus on students' perception, motivation and attitudes towards mobile and ubiquitous learning. Similarly, Wu et al. (2012) asserted that the aim of most of the preliminary research is evaluating the effectiveness of mlearning. Researchers stated that this case emerges in parallel with other fields related to technology enhanced learning (citation or citations needed here. The finding of fundamental research orientation in both periods in this study and the matching of these findings with previous research show that the "effectiveness" subfield has a significant place for the mlearning field.

Being the biggest factor of the period 2002-2010, "*general issues and reviewing the progress in the field*" was divided into two factors in this period. The factor named as general issues which focus on principles of mlearning, developing a general framework for mlearning and building conceptual base for mlearning (Traxler and Kukulska-Hulme, 2005) explains 13% of total variance while the "*review*" factor consisting of the authors of review studies on the field explains 3.75% of total variance and ranks 8th. The crystallized authors of the "*general issues*" subfield are Cochrane, Traxler and Pachler. In review factor, five authors are included and no one of them has factor loading above .8. Out of 5 authors in the "*review*" factor, three authors have factor loadings in both "*review*" and "*general issues*" factors.

The main idea in the 2nd factor called "*general issues*" is to transform learning in accordance with the affordances of mlearning. Sharples, Taylor and Vavoula (2005) summarizes this case as follows:

There is a need to re-conceptualize learning for the mobile age, to recognize the essential role of mobility and communication in the process of learning, and also to indicate the importance of context in establishing meaning, and the transformative effect of digital networks in supporting virtual communities that transcend barriers of age and culture. (p. 1)

Considering the subfields "*design and effectiveness*" and "*general issues*" which have ranked the first and second in both periods, it can be seen that the "ubiquitous learning" concept is preferred in the subfield "design and effectiveness" while mlearning concept is used in "general issues" subfield. In addition, the analysis on the authors taking place in these subfields indicates that the majority of the authors taking place in the "*design and effectiveness*" subfield are Asian and especially Taiwanese. Similarly, most of the authors in the subfield "*general issues*" are European and especially from the United Kingdom. In the studies within the "*design and effectiveness*" subfield, technical details on the design of mlearning environments take more place. However, the "*general issues*" subfield focuses on the transformation of learning in accordance with the affordances of mlearning tools and the development of the pedagogical framework for mlearning. This case makes one think that themes of mlearning research and the concepts used in this research can differ by regions.

Ranking 6th in terms of the percentage of explained total variance in the early period, the "*acceptance of information technologies*" factor explains 7.79% of total variance in the period 2011-2015 and becomes the 3rd biggest factor. Mobile devices are considered among the most rapidly proliferated technologies in the history. It has been observed that mlearning research have rapidly developed to reveal effective and innovative methods of the use of mobile devices in learning field thanks to their widespread usage. However, a severe gap can be seen between the affordances offered by mobile technologies and current usages despite all these efforts (Liu, Han and Li, 2010). Researchers assert that mlearning practices are at "*embryonic stage*" yet (Motiwalla, 2007). It is thought that the understanding of learners' mlearning adoption behaviors will enable proliferation of mlearning and transform the educational landscape (Wankel and Blessinger, 2013). Questions about how the mobile technology affordances can be used better for educational purposes and how mlearning adoption of learners can be raised have created a significant research subfield. The crystallized authors of the factor including a total of 10 authors are Venkatesh, Ajzen, Davis, Wang and Rogers. Considering the most cited authors in this field and crystallized authors, it can be seen that they use technology acceptance model (TAM) and the developed versions of this model (TAM2, UTAUT) (Venkatesh and Davis, 2000; Venkatesh et al., 2003) and theory of planned behavior (Ajzen, 1985) in order to understand learners' mlearning adoptions.

The subfield "*collaborative mlearning*" was the 4th biggest factor in this period just as in the early period. In this factor, there are nine authors having no greater factor loading than .8. Although mobile devices are used for learning/teaching activities designed with different pedagogic point of view (Taylor and Sharpless, 2006), highly situated and personal and collaborative features of mlearning is one of the commonly agreed finding of different review studies (Cobcroft, Towers, Smith and Bruns,

2006; Naismith, Lonsdale, Vavoula and Sharples, 2004). Mobile technologies offer new means to ensure classroom discussions in learning activities and interaction, collaboration and communication opportunities during the process of creating content (Bryant, 2006, O'Reilly, 2005). Sharples (2000) asserts that changes in pedagogic perspective in recent period started to converge with technologic developments. In this scope, the increase in mobile technologies and means of internet access also increase the means of collaboration and communication and in parallel, collaborative learning activities have a fundamental importance in mlearning design. In this regard, findings of this study are in parallel with the findings of the research conducted by Hwang and Wu (2014). In their review study, researchers reported that in most of the learning activities designed with mlearning, learners learn in groups, and collaborative mobile learning activities are widespread in the field.

The "*mobile technologies in language learning*" factor revealed a very limited percentage of total variance and ranking 16th in the period 2002-2010, yet having a central position in PFNET graphic of the period, ranks 5th in this period. In this factor including 4 authors which do not have factor loading above .4 in any other factor, all the authors have very high factor loading. This case results from the fact that researchers studying on a certain learning field are frequently co-cited. One of the common findings obtained from the review studies on mlearning is that researchers generally do not focus on a certain learning domain and the most extensive learning domain is foreign language learning among the focused learning domains (Hwang and Tsai, 2011; Hwang and Wu, 2014; Thornton and Houser, 2005; Ogata and Uosaki, 2012). Hwang and Wu (2014) state that it is reasonable that foreign language learning stands forefront as a learning domain in mlearning field because students "need to be situated in an environment that enables 'any time' and 'anywhere' practicing to learn a language well"(p. 85).

One of the prominent subfields of mlearning field in the period 2011-2015 is augmented reality(AR). Rapid development of mobile devices in terms of software and hardware and increased internet access enabled proliferation of AR applications in mobile devices. In the same period, the widespread acception of the situated learning theory which asserts that learning is not the transfer of concepts free of the context but situated in a specific context and embedded within a particular social and physical environment (Lave and Wenger, 1991) ensured the technical and pedagogic basis to increase the use of mobile AR applications for the purpose of learning. FitzGerald et al. (2013) investigated the use of AR for the purpose of mlearning and reported that the use of AR for the purpose of mlearning is in the infancy period yet and constructivist and situated learning activities which encourage the collaboration and inquiry of learners can provide immersive and engaging experiences for students. In parallel with the assertions of FitzGerald et al, the findings of the present study show that AR applications started to take more place in mlearning literature after the year 2010, and it is possible for this case to go on in the following period as well. The "*augmented reality*" factor which did not take place among the fundamental factors of the period 2002-2010, ranked 6th in terms of percentage of explained total variance in the period 2011-2015. In this factor including a total of 6 authors, crystallized authors are Dunleavy M and Dede C.

The 7th factor including Mayer R and Churchill D who are known for their studies on multimedia design as crystallized authors explains 4% of total variance. The "*review*" factor being the 8th one explains 3.7% of total variance while the "*perceptions on mobile learning*" factor being the 9th one explains 3.5% of total variance.

Analyzing the PFNET graphic in Figure 3, the main axis of mobile research in the period 2011-2015 is made of the subfields "*general issues*", "*Collaborative mlearning*" and "*design and effectiveness*". Considering the percentage of explained total variance and positions in PFNET graphic, it can be suggested that these subfields will preserve their place in the following period. The branching within especially 1st and 2nd factors in PFNET graphics show that these fields are not homogenous and they form sub-clusters within their own factors. These sub-clusters can be subfields in the future. Analyzing the PFNET graphics, one of the obtained results is that Hwang who is among the most cited authors in the field is cited from different subfields of mlearning field and has a bridging role to merge different subfields. Similarly, Sharples, Traxler and Kukulska-Hulme are the mostly cited authors in the subfield "*general issues* (2)" and function as bridge.

Being one of the subfields at high ranking, "*augmented reality*" is attached to the PFNET graphics through the main axis and is attached to the "*collaborative mlearning*" subfield and thus to the main axis. Having a high percentage of explained the variance and taking place on the main axis, "*augmented reality*" studies seem to be among remarkable trends of the related studies. According to some researchers, educational technologies try to adapt pedagogic trends (Crombton, 2013) while other authors suggest that technologic development are useful on pedagogic

Figure 5. PFNET graph of authors from 2011 to 2015

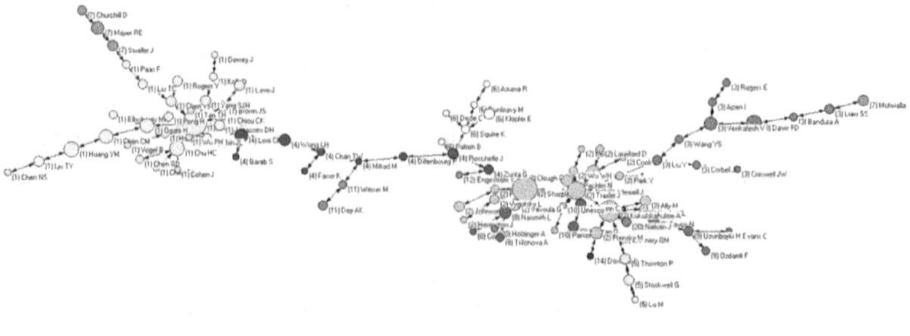

orientations (Sharples, 2005). Although the cause and effect relationship at this point is controversial, it is obvious that there is a convergence between technologic developments and pedagogic approaches. The affordances offered by mobile technologies led to a common discussion on socio-cultural constructivist pedagogies in the mlearning field (Crombton, 2013).

Being another factor at high ranking, the subfield "*acceptance of information technologies (3)*" is related with the subfield "*general issues* (1)", yet it is positioned in a distant place compared to other subfields. In other words, the authors studying on "*acceptance of information technologies* (3)" have limited number of co-citation with the authors of other subfields of the mlearning field and they could not be completely integrated with the field.

As expected, the subfield "*multimedia design*" is positioned close to the subfield "*design and effectiveness*", yet positioned distant from other subfields. It is seen that the 11th factor including AK. Dey and M. Weiser who have research on "*context-aware computing*" subfield and do not take place in the first 9 factors is attached to the "*collaborative mlearning*" factor on the main axis. This case can be considered as an indicator that the context-aware computing theme will take place in the field in the periods to come.

CONCLUSION

This research has investigated the scope and change of mlearning research literature taking a stand from the author co-citation patterns revealed in the publications between the years 2002-2015 in mlearning field. Conference proceedings and journal articles indexed by the web of science which is one of the broadest and most reliable databases for bibliometric data were used for this investigation. In this study, it has been found that the mlearning researches focus on certain subjects such as design and effectiveness of mlearning environments, development of pedagogic framework for mlearning, adoption and proliferation of mlearning assisted language learning, collaborative mlearning, and augmented reality. Assessing the results of factor analysis together with the PFNET graphics, it can be said that these subfields are the fundamental orientations of the field and the main themes of the field could be these subjects in following periods as well. Also, the findings of the research showed that studies on mlearning are constantly increasing, and more and more countries bring a contribution to this field.

This study has investigated the history of mlearning research with quantitative methods by means of the citations of mlearning research papers and a perspective point of view has been provided. It is believed that the findings of this study will shed light to educational policymakers, funding agencies and especially novice researchers in terms of a better understanding of the structure and orientations of the field. Apart from this, some limitations of the research should be taken into consideration.

As in many other academic databases, mostly English publications are indexed in the web of science as well. In this study, 97% of the analyzed publications are English. This case shows that the research shared in national journals and congresses in different languages are not satisfactorily presented. Due to the publication delay and indexing delay, the orientations in post-2013 period are not satisfactorily presented in the present study that can be considered as another limitation. The stylistic errors such as the mistakes in recording author names into the database (lack of middle name, long or missing writing etc.) and misspelling in citations were corrected both via the Bibexcel program and by the author manually.

REFERENCES

Abu-Al-Aish, A. (2014). *Toward mobile learning deployment in higher education* (Doctoral dissertation). Brunel University, School of Information Systems, Computing and Mathematics.

Acedo, F. J., & Casillas, J. C. (2005). Current paradigms in the international management field: An author co-citation analysis. *International Business Review, 14*(5), 619–639. doi:10.1016/j.ibusrev.2005.05.003

Ajzen, I. (1985). *From intentions to actions: A theory of planned behavior*. Springer Berlin Heidelberg.

Batagelj, V., & Mrvar, A. (1998). Pajek: A program for large network analysis. *Connections, 21*, 47–57.

Borgatti, S. P., Everett, M. G., & Freeman, L. C. (1999). *UCINET 6.0 Version 1.00*. Natick: Analytic Technologies.

Cobcroft, R. (2006). *Literature review into mobile learning in the university context* (Unpublished doctoral dissertation). Queensland University of Technology.

Cobcroft, R. S., Towers, S. J., Smith, J. E., & Bruns, A. (2006). Mobile learning in review: Opportunities and challenges for learners, teachers, and institutions. In *Proceedings Online Learning and Teaching (OLT) conference 2006* (pp. 21-30). Brisbane: Queensland University of Technology. Available at http://eprints.qut.edu.au/5399/1/5399.pdf

Crompton, H. (2013). A historical overview of mobile learning: Toward learner-centered education. In Z. L. Berge & L. Y. Muilenburg (Eds.), *Handbook of mobile learning* (pp. 3–14). Florence, KY: Routledge.

Davis, F. D. (1989). Perceived usefulness, perceived ease of use, and user acceptance of information technology. *Management Information Systems Quarterly*, *13*(3), 319–340. doi:10.2307/249008

Diodato, V. (1994). *Dictionary of Bibliometrics*. Binghamton, NY: Haworth Press.

Felder, R. M., & Silverman, L. K. (1988). Learning and teaching styles in engineering education. *Engineering education*, *78*(7), 674–681.

Garfield, E., Malin, M. V., & Small, H. (1975). A system for automatic classification of scientific literature. *Indian Institute of Science*, *57*(2), 61–74.

Glavinic, V., Ljubic, S., & Kukec, M. (2008, September). A Holistic approach to enhance universal usability in m-learning. In *Mobile Ubiquitous Computing, Systems, Services and Technologies, 2008. UBICOMM'08. The Second International Conference on* (pp. 305-310). IEEE. 10.1109/UBICOMM.2008.54

Hung, J. L., & Zhang, K. (2012). Examining mobile learning trends 2003–2008: A categorical meta-trend analysis using text mining techniques. *Journal of Computing in Higher Education*, *24*(1), 1–17. doi:10.100712528-011-9044-9

Hwang, G. J., & Tsai, C. C. (2011). Research trends in mobile and ubiquitous learning: A review of publications in selected journals from 2001 to 2010. *British Journal of Educational Technology*, *42*(4), E65–E70. doi:10.1111/j.1467-8535.2011.01183.x

Hwang, G. J., & Wu, P. H. (2014). Applications, impacts and trends of mobile technology-enhanced learning: A review of 2008–2012 publications in selected SSCI journals. *International Journal of Mobile Learning and Organisation*, *8*(2), 83–95. doi:10.1504/IJMLO.2014.062346

Kukulska-Hulme, A. (2009). Will mobile learning change language learning? *ReCALL*, *21*(02), 157–165. doi:10.1017/S0958344009000202

Kukulska-Hulme, A., Sharples, M., Milrad, M., Arnedillo-Sánchez, I., & Vavoula, G. (2009). Innovation in mobile learning: A European perspective. *International Journal of Mobile and Blended Learning*, *1*(1), 13–35. doi:10.4018/jmbl.2009010102

Lee, M. R., & Chen, T. T. (2010). Visualising intellectual structure of ubiquitous computing. In *Knowledge Management and Acquisition for Smart Systems and Services* (pp. 261–272). Springer Berlin Heidelberg. doi:10.1007/978-3-642-15037-1_22

Liu, T. C., Wang, H. Y., Liang, J. K., Chan, T. W., Ko, H. W., & Yang, J. C. (2003). Wireless and mobile technologies to enhance teaching and learning. *Journal of Computer Assisted Learning*, *19*(3), 371–382. doi:10.1046/j.0266-4909.2003.00038.x

Liu, Y., Han, S., & Li, H. (2010). Understanding the factors driving m-learning adoption: A literature review. *Campus-Wide Information Systems*, *27*(4), 210–226. doi:10.1108/10650741011073761

Ma, R., Dai, Q., Ni, C., & Li, X. (2009). An author co-citation analysis of information science in China with Chinese Google Scholar search engine, 2004–2006. *Scientometrics*, *81*(1), 33–46. doi:10.100711192-009-2063-x

Martín, E., Andueza, N., Carro, R. M., & Rodríguez, P. (2006). Recommending activities in collaborative m-learning. *Lecture Notes in Learning and Teaching*, 197-204.

McCain, K. W. (1990). Mapping authors in intellectual space: A technical overview. *Journal of the American Society for Information Science*, *41*(6), 433–443. doi:10.1002/(SICI)1097-4571(199009)41:6<433::AID-ASI11>3.0.CO;2-Q

Merton, R. K. (1973). *The sociology of science: Theoretical and empirical investigations*. University of Chicago Press.

Motiwalla, L. F. (2007). Mobile learning: A framework and evaluation. *Computers & Education*, *49*(3), 581–596. doi:10.1016/j.compedu.2005.10.011

Naismith, L., Lonsdale, P., Vavoula, G., & Sharples, M. (2004). Literature review in mobile technologies and learning. Futurelab. *Literature review in mobile technologies and learning: Futurelab*.

Norris, C., & Soloway, E. (2004). Envisioning the handheld-centric classroom. *Journal of Educational Computing Research*, *30*(4), 281–294. doi:10.2190/MBPJ-L35D-C4K6-AQB8

Ogata, H., & Uosaki, N. (2012). A new trend of mobile and ubiquitous learning research: Towards enhancing ubiquitous learning experiences. *International Journal of Mobile Learning and Organisation*, *6*(1), 64–78. doi:10.1504/IJMLO.2012.046883

Özçınar, H. (2015). Mapping teacher education domain: A document co-citation analysis from 1992 to 2012. *Teaching and Teacher Education*, *47*, 42–61. doi:10.1016/j.tate.2014.12.006

Persson, O. (2001). All author citations versus first author citations. *Scientometrics*, *50*(2), 339–344. doi:10.1023/A:1010534009428

Persson, O., Danell, R., & Schneider, J. W. (2009). How to use Bibexcel for various types of bibliometric analysis. *Celebrating scholarly communication studies: A Festschrift for Olle Persson at his 60th Birthday*, 9-24.

Roschelle, J. (2003). Keynote paper: Unlocking the learning value of wireless mobile devices. *Journal of Computer Assisted Learning*, *19*(3), 260–272. doi:10.1046/j.0266-4909.2003.00028.x

Sariola, J. (2003). The boundaries of university teaching: Mobile learning as a strategic choice for the virtual university. *Mobile Learning*, 71-78.

Schwabe, G., & Göth, C. (2005). Mobile learning with a mobile game: Design and motivational effects. *Journal of Computer Assisted Learning*, *21*(3), 204–216. doi:10.1111/j.1365-2729.2005.00128.x

Seppälä, P., & Alamäki, H. (2003). Mobile learning in teacher training. *Journal of Computer Assisted Learning*, *19*(3), 330–335. doi:10.1046/j.0266-4909.2003.00034.x

Seppälä, P., Sariola, J., & Kynäslahti, H. (2002). Mobile learning in personnel training of university teachers. In *Wireless and Mobile Technologies in Education, 2002. Proceedings. IEEE International Workshop on* (pp. 136-139). IEEE 10.1109/WMTE.2002.1039236

Sharples, M. (2000). The design of personal mobile technologies for lifelong learning. *Computers & Education*, *34*(3), 177–193. doi:10.1016/S0360-1315(99)00044-5

Sharples, M. (2005) *Learning As Conversation:Transforming Education in the Mobile Age*. Paper to presented at Conference on Seeing, Understanding, Learning in the Mobile Age, Budapest, Hungary.

Taylor, J., & Sharples, M. (2006). *The Pedagogical Perspectives of Mobile Learning*. Available at: http://www2.le.ac.uk/projects/impala/presentations/Berlin/The%20 Pedagogical%20Perspectives%20of%20Mobile %20Learning/view

Thornton, P., & Houser, C. (2005). Using mobile phones in English education in Japan. *Journal of Computer Assisted Learning*, *21*(3), 217–228. doi:10.1111/j.1365-2729.2005.00129.x

Traxler, J. (2007). Defining, Discussing and Evaluating Mobile Learning: The moving finger writes and having writ..... *The International Review of Research in Open and Distributed Learning*, *8*(2). doi:10.19173/irrodl.v8i2.346

Traxler, J. (2012). Mobile learning-The future already behind us. *Proceedings of 2012 International Conference on Interactive Mobile and Computer Aided Learning (IMCL)*. 10.1109/IMCL.2012.6396442

Traxler, J., & Kukulska-Hulme, A. (2005). *Evaluating mobile learning: Reflections on current practice*. Paper presented at the mLearn 2005: Mobile technology: The future of learning in your hands. Retrieved from http://oro.open.ac.uk/12819/1/

Venkatesh, V. (2000). Determinants of perceived ease of use: Integrating control, intrinsic motivation, and emotion into the technology acceptance model. *Information Systems Research*, *11*(4), 342–365. doi:10.1287/isre.11.4.342.11872

Venkatesh, V., & Davis, F. D. (2000). A theoretical extension of the technology acceptance model: Four longitudinal field studies. *Management Science*, *46*(2), 186–204. doi:10.1287/mnsc.46.2.186.11926

Venkatesh, V., Morris, M. G., Davis, G. B., & Davis, F. D. (2003). User acceptance of information technology: Toward a unified view. *Management Information Systems Quarterly*, *27*(3), 425–478. doi:10.2307/30036540

Wankel, L. A., & Blessinger, P. (2013). *Increasing Student Engagement and Retention Using Mobile Technologies: Smartphones*. Skype and Texting Technologies. doi:10.1108/S2044-9968(2013)6_Part_D

White, H. D., & McCain, K. W. (1998). Visualizing a discipline: An author co-citation analysis of information science, 1972-1995. *Journal of the American Society for Information Science*, *49*(4), 327–355.

Wu, W. H., Wu, Y. C. J., Chen, C. Y., Kao, H. Y., Lin, C. H., & Huang, S. H. (2012). Review of trends from mobile learning studies: A meta-analysis. *Computers & Education*, *59*(2), 817–827. doi:10.1016/j.compedu.2012.03.016

Chapter 7
Electronic and Mobile Learning for Workforce Development

Dominic Mentor
Columbia University, USA

ABSTRACT

This chapter covers the construction of a digital ecosystem grown organically to digitize a workforce development organization. A criteria matrix was developed from a listening tour to plan and adopt blended, electronic, as well as mobile learning (e&mLearning) as part of its practices. The chapter describes the organic approach to diffuse the technological innovation for cultivating supportive teaching and learning communities. The process started with the establishment and implementation of an academic vision and strategy; the building of a blended, electronic, and mobile learning criteria taxonomy; as well as a theoretically informed, technology integrated educational framework. The results and findings of the implementation of an LMS as the foundation of the e&mLearning vision is shared, followed by recommendations from successes and organizational needs.

INTRODUCTION

This chapter focuses on the vision and strategic implementation of electronic, blended and mobile learning into the educational processes of a national training organization's workforce development program. The perspectives are offered from a diffusion of educational technology approach, as well as from program design and leadership lenses. Through a six-month intensive training program followed by a six-month internship, Empowered Professional Internships with College Credits

DOI: 10.4018/978-1-5225-9351-5.ch007

(EPICC) provides urban young adults, between the ages of eighteen to twenty-four, with a combination of hands-on technical and professional skills, college credits, and corporate internships.

In order to initiate systems change and introduce mobile accessible eLearning in an integrated manner, a research framework was established. The framework consisted of field observations through site visits, semi-structured interviews and focus groups. Existing organization survey data was inspected and enhanced to create survey feedback loops to gather information from students and staff as customers. The results of the research site visits is reported as well as informed the construction of the academic vision and strategy. The chapter also covers the construction of an educational taxonomy, and the building of a learning criteria matrix to help select a mobile friendly Learning Management System (LMS). The results and findings of the LMS implementation is also shared followed by recommendations from current successes and organizational needs.

BACKGROUND

Seven years ago, EPICC, the then twelve-year-old organization, was running its six-month learning and development program primarily with paper-based models. In one week of an Information Technology course, students were printing five to ten assignments with screen grabs. The assignments would typically be 20 or more pages per student. Apart from the heavy printing cost and printer maintenance, which was not tracked separately as budget items, the "instructors" would be saddled with massive stacks of assignments to grade. Imagine the demotivation that would be induced when looking upon those heaps of assignments to grade. Additionally, instructors were only required to submit grades at the end of seven weeks, which is the time frame of one module. This impacted whether students and staff could see immediate or regular evidence of a student's learning and development progress. For the most part, aside from a few instructors at all sites, there was a heavy reliance on anecdotal evidence of a students' progress. A lack of criteria-based assessments and evidence of learning was overwhelmingly present within the organization. The issues of clarifying why criteria-based assessments were important, was further compounded by the lack of hiring educators that have undergone any type of teacher training. In a few cases, other program staff would be allowed to switch to teaching positions even though they had no training, no certifications or experience to educate methodically the young adults of the program. At times, the struggle to attract experienced or teacher trained staff was a result of program staff without any relevant academic background, industry or higher education work experience writing the job descriptions, running the hiring process and managing the training staff. Low paying instructor salaries further hampered the proper attraction of teaching talent. Furthermore, the short time frame in which to catch up students with less than desirable K- 12 educational experiences, and to deliver internship-ready young talent, requires well-trained and experienced trainers. Additionally, if EPICC is to justly serve the trainees, programming staff with a depth of industry and higher education experience would go far to support the training operation.

Additionally, EPICC students often face a unique set of preparation and professional challenges on their path to careers and post-secondary education. These challenges include systemic educational inequalities in their K – 12 learning experiences, no ownership or limited access to computers or the internet, and limited access to professional networks. Another somewhat universal aspect that affects EPICC's students is the predominantly massive gap between the operations of high schools, colleges, and the work place. Furthermore, many EPICC students combine their participation in EPICC with additional jobs and family responsibilities. The young adults being served either just completed high school, earned a GED, or just started or dropped out of college. The latter of which, on average, makes up 40% of the cohort each cycle. Despite having a high school diploma or GED, many EPICC students were registered for remedial classes when they did attend college. For its first twelve years, the EPICC organization addressed all of the above with paper based models, with sparse and piece-meal electronic engagement.

Notably, while the organization was predominantly paper based, there were two or three instructors out of eighty that employed the use of wikis and blogs. One site's students had access to their college partners' Moodle learning management platform, but not its instructors. The latter scenario was obviously not conducive to teaching and learning, as the instructors were deemed not qualified by their college partners, and these instructors could not see students' submitted work, nor grade the students' work and identify teaching and learning gaps. The instructors who were using wikis and blogs are to be commended for their efforts. However, neither the wikis nor the blogs were scalable or replicable across the eight sites, given the high learning curve of each, as well as the different technology proficiency levels of all academic staff (Dohn, 2010; Holtzblatt, Damianos, and Weiss, 2010; Parker and Chao, 2007). The blogs and wikis were problematic from FERPA and PIP regulatory perspectives, and had low adoption amongst staff. Other enterprise issues (e.g. scalability, support) abounded with blogs and wikis as online tools to transition the entire, multi-city national training organization to blended learning. While the blogs and wikis were media rich and stimulating, they could not, back then, capture feedback, comments and grades on a site wide or national scale, which would offer the students and the organization, valuable learning analytics.

Amidst the minimal use of wikis and blogs by a few creative instructors, EPICC still effectively deployed an in-person, class time specific, instructor-led training (ILT) model in its academic and program curriculum for twelve years. These practices persisted while students would enter 21st century workplaces as trainees in a digital age that offered mobile and blended eLearning in abundance. With a wealth of eLearning systems available to them, up until 2012 the organization was mainly focusing on capturing only students' grades at seven-week cycles. The grade repository was researched and selected by operations staff. The selected tool was categorized and described as a Learning Management System, but from an eLearning perspective, the grade repository tool did not qualify as a modern, interactive Learning Management system that allowed for student discussions, communications, video, or immediate feedback on submissions or tests. This grade capture method also permeated the educational philosophy of using one data point to assess and describe a student. The lack of educational or educational technology expertise from Operations and Programming staff is also highlighted here as not understanding the need for formative assessments, but given oversight over academic learning and development processes. Moreover, the long lapse on offering students' grades and feedback on their work added to the notion that students were but mere passive receivers of knowledge, and not, as Greene and Crespi (2012) or Robertson (2009) would describe, the modern student as twenty-first century netizens.

The grade entry process deserves elaboration, as it was a major source of frustration for instructors and academic directors. The seven week grading cycle was highly dependent on an outdated online grade entry system. While there were no malintent in selecting the grade entry online tool, the user interface of the system was extremely awkward to navigate, and at times, it took more than twelve clicks to get to the point of entering grades. The tool was also not mobile friendly. While at some sites, a few instructors' grades came in after the seven-week mark, it should be noted that many students were not getting grades on their assignments within a daily, weekly or two week cycle. Thus, students, the academic program staff, and the organization had to rely on anecdotal and infrequent reports from their instructors to know how students were doing.

While the stand-alone grade book was known as Grade Book Wizard, instructors' spoke of the tool as a major drain of their energy. They were tentative with each grade entry session, citing extreme trepidation and complained about the terrible user interface and experience. Program operation staff who selected the online tool and was coordinating the capture of grades, took an approach of "they (teaching staff) must just use the system and comply with this directive". This observation is included here, only to serve as a means for other administrators and educators who are required to follow instructions and use ill-designed online interface, and then impact future attitudes around migrations to online academic services. The afore-mentioned directive certainly decreased the levels of motivation even more to input grades and left the academic staff feeling that they were not being heard or that there was no understanding of how much time and effort it takes to prepare, organize, teach and offer feedback with grades. The seven week grading cycle also made it hard to identify and immediately act on students' academic challenges. Apart from the grading system, there were needs for more pedagogy and andragogy to be considered as added models.

The paper based models also meant that there were unclear and inconsistent places and methods for students and teachers to access course materials. Furthermore, with the short six-month period to prepare students for internships, the paper-based model also limited engagement with instructors outside of classrooms, restricting the educational engagement only to classroom time. For veteran and new teachers, curricula were hard to find and share which led to significant duplication of efforts across sites. Furthermore, national corporate partners would receive interns from disparate curricula.

Overall, a more holistic academic vision and strategy was needed. EPICC had to change its culture, getting away from seven-week grading cycles and the mono-directional teaching styles, to more facilitating methods of teaching. The e-&-m blended learning initiative had to address all the issues above. To this end, an organic diffusion of technology was envisioned to aid adoption of new multimodal methods and engaging educational technology. The first step to introduce mobile learning (mLearning) was to start with a simple, user friendly LMS that had mobile learning capabilities, which could also be used to aid curricula and quality alignment across city sites, as well as introduce the concept of adult learning in preparing these young adults for their internships.

THEORETICAL FRAMEWORK

To enact the digitization process, systemic change that was informed by sound educational theories, as well as andragogical and pedagogical practicalities were essential. Malcolm Knowles In *The Modern Practice of Adult Education* (1970), defined andragogy as "an emerging technology for adult learning." Ronkowitz (2013) summarizes Knowles' andragogical assumptions into four main categories showing the need for a shift in teaching and learning methods for adults. Those four categories are:

- Adult learners move from dependency to self-directedness;
- Draw upon their reservoir of experience for learning;

- Adult learners are ready to learn when they assume new roles; and
- Want to solve problems and apply new knowledge immediately.

Andragogy is an acknowledgement that alternative teaching and learning practices are required, different from what is often referred to as pedagogy, which is commonly used to describe teaching and learning methodologies for K-12 students. Over the years, as adult learning has grown and crystalized its different position and much needed approach, adult teaching and learning practitioners have commonly acknowledged a learner-focused centeredness for adult learners, but also for people of all ages (Abedi, Badragheh, & Varamin; Holton, Swanson, & Naquin, 2001). These tenets were to form both the philosophical and practical pedagogy in implementing the blended e-& -mLearning strategies. There was an intentionality behind the pedagogy and andragogy. These approaches were informed by theories that envisioned the academic digitization process, but also envisioned the foundations for other programmatic departments from admissions, student services, internships and alumni. The array of well-established educational theories was used to inform and aid planning, interactivity, delivery of curricula and cultivating transferable learning, informed by educational goals and objectives. All of which cater to multi-dimensional learners that engage with technology on a daily basis. For this multi-layered approach, other educational theories included in this blended e-& -mLearning initiative, ranged from blended and virtual collaborative learning in line with Vygotsky's Zone of Proximal Development (Afflerbach, 2016; Chaiklin, 2003; Wartella, Beaudoin-Ryan, Blackwell, Cingel, Hurwitz, & Lauricella, 2016) and Social connectedness in terms of creating community and a sense of belonging (Chayko,2012; Koole, 2009; Mentor, 2018), which formed a part of catering for social learning fundamentals. While Vygotsky's socio-cultural theories rang true here, behaviorism, and behaviorist aspects were also fully recognized as crucial elements that could help ensure positive reinforcement systems (Andrade, Ares, García, Rodríguez, Seoane, and Suárez, 2008; Mentor, 2015; Shih, 2007). While valuable in face-to-face (F2F) learning, the behaviorist theories would be extremely important for the blended engagements so that personal and group motivation are constantly cultivated and practiced through various pieces of the online processes. The latter could range from immediate automated system feedback on tests and quizzes, be it formative or summative assessments, or the accrual of badges for intrinsic or extrinsic acknowledgement or reward. This hybrid theoretical method within a blended e-&-mLearning context with multiple devices and operating platforms, would allow for the cultivation and maintenance of motivation through behaviorist angles, while also enabling cognitive constructive methods to thrive (Koole, 2009; Looi, Lim, Pang, Koh, Seow, Sun, & Soloway, 2016; Uden, 2006; Zurita and Nussbaum, 2004) as well as scaffolded knowledge construction opportunities and social collaborative learning methodologies. The collaborative social learning could be in the form of simple crowdsourcing from mobile phone voting, native or web apps, real time or near synchronous web surveys to capture a snapshot of students' prior knowledge, or even low-level Delphi method consensus building and understanding. A production of clustered data or connected thinking through crowdsourcing could also offer insight into individual, group or class views and nurture the ability to host multiple perspectives or be used for the recognition and construction of patterns from complex and collectively generated information and data inputs. Active engagement would also be developed through the researching, sifting, sorting and entry of data, thereby promoting digital data extraction and collation literacy. Additionally, increasing the chances of developing critical thinking from linear and non-linear research opportunities gathered from e-&-mLearning tools. The hybrid theories formed a part and parcel of the vision, strategies and methods that would be needed to evolve the organization from its paper-based models.

METHODOLOGY

Surveying the Landscape: Listening Tours and Field Observations

Before any blended, electronic or mobile teaching and learning initiatives were to be pioneered, the national organization's academic landscape needed to be surveyed and plotted. Capturing the current state would greatly inform the needs, starting points, and pace of attempting to change the organization's ecology and grow the technological adoption and expertise of the educational and non-academic program staff.

From semi-structured interviews with students and staff, as well as from student and program survey data, it was clear that apart from the grading practices, there were other facets of the learning and development program that needed attention. For example, at the time of the listening inquiry tour, it was reported that different admissions policies and practices were exercised in a fluctuating manner, impacting the student pool that the academic department would receive. With the academic department itself, an initial curricula landscaping and mapping also revealed that the eight city site locations, were not aligned around syllabi and syllabi templates, lesson templates, curricula or methodologies. While an observation form was instituted across the board by the previous National Director of Academics, the form was deemed laborious and time consuming. It certainly attempted to be thorough and tried to hit some of the growth areas previously identified and mentioned in this chapter. Apart from this uniform instrument, it was clear from the findings that

Figure 1. Theory in practice framework

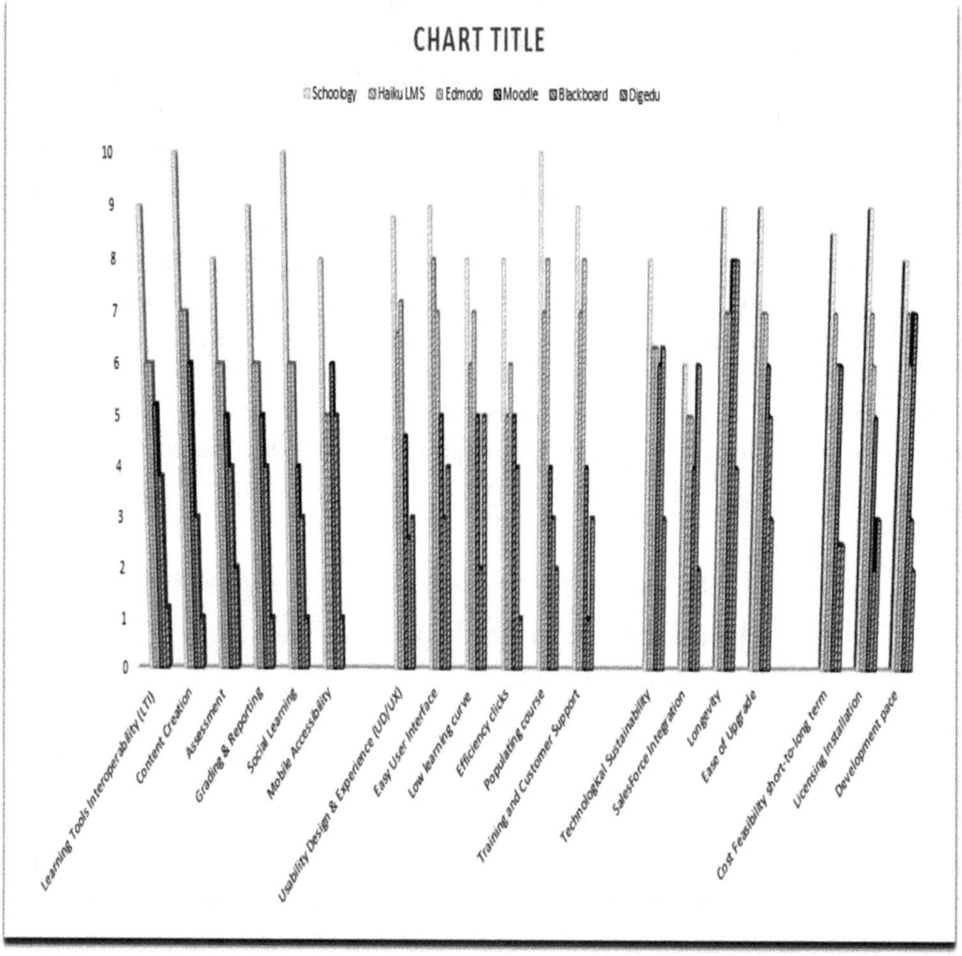

**For a more accurate representation see the electronic version.*

the national organization was offering eight different systems to navigate and to try and improve. To further compound challenges, the eight city sites, were also all partnered with different college partners and paid large amounts of money to have their students "dual enrolled" in the non-profit's technical training program and in the college. At seven of the sites, the students were offered instruction by the non-profit hired staff while one site needed an affiliate professor as part of its college partner contract. All of the courses were offered in intensive seven week time-framed courses while earning college credits through the paid for college partnerships. While earning these specific college credits, the students were prepared with knowledge, values, and professional skills for paid internships with major corporations ranging from American Express, Bank of America, Facebook, Google, JP Morgan Chase, Salesforce, Zynga, to a host of other city based, regional or national corporations. To summarize, the current state of the workforce development organization from 2011 to 2013 before implementing the e-&-m Blended learning was:

- Instructor led training
- Paper based submission of assignments
- No consistent place for students to access course materials
- Limited engagement with instructors outside of the classroom other than on site
- Seven-week grading cycle; hard to identify and immediately act on academic challenges
- Curriculum hard to find and share; led to significant differences in content and curricula delivery across sites and within the same city sites
- No alignment across subjects or cities teaching the same courses.

Constructing an Academic Vision and Strategy

From the above, it was important that a clear academic vision needed to be constructed to address the previously stated issues holistically as part of a blended e-&-mLearning strategy. To inform a vision and to build on the twelve year success of EPICC, necessitated that previous quantitative data had to come under heavy scrutiny as the then data gathering instruments were not aligned, nor necessarily informed by qualitative or quantitative specialists. Each site, like the curricula, could adjust surveys and processes as they saw fit making the then existing data difficult to use or to compare sites. Setting up research questions for a listening tour to get a proper national look at the landscape of the learning and development (L&D), was also problematic, as there were so many differences reported during semi-structured video and telephone interviews.

Apart from leveraging instructional and curricula quality, the listening tour revealed a need for student empowerment in the teaching and learning processes. Some students benefitted from differentiated instruction models, while others experienced daily lectures. Students were often assigned project based learning material, as well as collaborative tasks, but there were only a few rubrics to inform the students what was expected from them and what qualified as exceptional or met standards. In 2011 - 2012, some Business Communication (BC) classes did not maximize twenty-first century business communication tools. However, as noted previously, for many of the students, there was a need to make up for less than desirable K–12 educational experiences and some of the BC courses rather attempted to address those gaps. Some instructors were sharing material with one another, but without establishing what the unbiased baseline quality should be, nor was the sharing done in a structured, vetted or from an evidence-based manner that could be gauged through action reflection teaching methodologies.

While some ideas were being shared, the listening tour also included mapping the curricula of the different sites. The goal of the curriculum mapping was multifold:

1. to see for similarities,
2. to look for success stories that needed to be replicated across the country,
3. to provide a more than high level comparison to see which sites provided which career track curricula, and for what volume of internships and corporate partners,
4. to offer insights or help understand formulas and best practices of the academic program across sites
5. to identify sustainable internships career tracks (e.g. financial operations and computer programming), and lastly,
6. identifying partially developed tracks to grow and solidify.

All of which could be leveraged across sites where there were similar labor market demands in other cities. The curriculum chart was not just for the benefit of the National Academic Director, but also for all local site Academic Directors, Instructors, the Chief of Operations, as well as other National leaders and staff. The curriculum mapping could act as a general reference on site variability, to identify analysis in subject areas which we might need to be delved into more detail, or budget for curricula development, or require an overhaul in design. Once the curriculum mapping and listening tours were completed, a vision based on thematic needs identified across the country, steadfastly emerged. All of which was extremely valuable to document in pursuant of a blended e-&-mLearning strategy.

In the third quarter of 2012, an academic vision and strategy model was introduced at EPICC that encapsulated a blended, electronic and mobile learning approach. The vision and strategy established a focus on three foundational frameworks: instructor quality, student empowerment as well as curriculum mapping and alignment. All three pillars touched on important stakeholders and facets of the academic L&D program, while also building foundations for future program extensions and enhancements inside and outside of the academic part of the program. While the curriculum mapping looked at the foundational skills needed by students, corporate partner internship demands as well as college level and industry recognized courses and certificates, the first two pillars focused on attempting to replicate and enhance instructors' teaching impact while also empowering our students. There was also a need to holistically digitize the L & D program to hit all three pillars of the academic vision and strategy. In order to impact all three of the pillars and strengthen them, the academic team started by launching a set of eLearning and mobile learning initiatives, in a blended manner, to support those key pillars of our framework. As previously noted, a need existed for curricula development with access to online structured courses, which could be used to foster collegial interaction and virtual learning communities. The switch to mobile accessible and blended eLearning could also offer real time data and learning analytics and strengthen the visibility and transparency of the instructors' and students' work during L&D. The mobile and online access of course work extended individual and collaborative engagement opportunities for students. However, there were on average, at least 10 – 18% of students at all of the sites who did not own a computer, did not have access to a computer nor had access to the internet at their homes. For these students, the mobile access to the material was crucial for just-in-time work, post questions or respond to short discussion posts as 99% of students had smart phones with access to internet data.

For the student empowerment, and from a social theory and collaborative learning perspective, the vision of blended learning with mobile accessibility would provide the students a network of supportive relationships, and be used to cultivate a powerful sense of agency for students — encouraging them to be active participants in their internship, learning and development, and preparing them for setting a path and embarking on their career workforce trajectories. The vision also cultivated the notion that empowered students would also plant the seeds for instructors to shift from being the knowledge provider, but to rather act as facilitator inviting students to be fellow knowledge contributors. The latter was and still is quite a shift for many educators, and all of this amidst other progressive changes.

It must be noted that the twelve year old non-profit had just gone through a theory of change process, in part, to address the need to align staff's view of whether the organization was a college prep, or a workforce development organization. An important distinction to be made because, while students do earn college credits while studying at EPICC, they do so in preparation of and for an internship which would set them up for an entry level job above the labor wage limit and/or a career. Thus, the academic vision was anchored in EPICC's Theory of Change, and aligned to the growth goals described in EPICC's scaling efforts. Simply put, the academic vision was designed and written to address current and future scaling needs with e-&-m blended learning acting as a foundation on which to address current needs and accommodate its aggressive growth strategy.

E-and-mLearning

The national and local academic teams embarked on a three year initiative covering the integration of mobile accessible and blended eLearning into the national non-profit workforce development organization. The shift in moving the twelve-year-old organization from paper-based models to interactive multimodal engagement was to not only better engage young talent or provide open access to the workforce development curricula through mobile learning, but also to gather and act on valuable digitized learning analytics. The mobile learning facets were integrated into the eLearning drive and both were informed by learning goals and objectives, and well established educational theories. Before the e-&-mLearning drive could take root, the organization first needed to understand the academic vision and strategy to ensure that the technology did not become the main focus, or a new shiny toy to be dropped or added when the next hype or silver bullet mythical promise comes around, but rather remain a vehicle in support of its mission, and its learning outcomes (Dillon and Åhlberg, 2006). Data from the listening tour documented the current state to establish what would be needed to aid the diffusion of mobile and blended learning.

As a first phase, EPICC implemented e-&-mLearning in the Learning & Development phase to prepare students for the second phase of a six-month corporate internship. Given the intensive six-month time frame in which instructors achieve so much while also addressing the previously mentioned learning gaps from high school and some college work that students exhibited, extending the learning through technological opportunities was paramount. The hybrid tactic was to effectively extend the academic program from a brick and mortar restricted, class based program, to a program where the teaching and learning engagement could now continue after class hours. Effectively, students could learn on the go while commuting, waiting in line, during down time at work or at home. The purpose of this blended mobile and eLearning plan was to implement an operational foundation on which to integrate the academic vision: an enterprise-wide, mobile accessible, Learning Management System (LMS) to support EPICC's national academic goals, as well as help evolve the organization's program delivery. One key piece of the change management process, was to define mobile learning, eLearning, blended learning, and how the three could be integrated for better service delivery. Defining technologically enhanced teaching and learning terminology was important, because there were a myriad of definitions per different contexts. Deciding on the needs of our contexts and selecting definitions of the terms was important, as it would also help build common understanding and aid adoption.

The Delphi method was employed to build consensus on the new direction and terminology that encapsulated that direction. Delphi may be characterized as a method for structuring a group communication process so that the process is effective in allowing a group of individuals, as a whole, to deal with a complex problem (Linstone & Turoff, 2002). The Delphi method was also used during academic meetings to inform group decision making, research design and strategic planning for identifying and prioritizing managerial decision-making issues (Okoli and Pawlowski, 2004). Thus, as similarly indicated in Hsu and Sandford's (2007) paper, the Delphi system was also used to extract and build common understanding and consensus. While eLearning was deemed to refer to educational processes that utilize Information Computing Technologies (ICTs) to mediate blended a/synchronous learning and teaching activities, mLearning was spoken of as learning opportunities via portable technologies from interpersonal mobile communication devices. As with technology and society, the concept of e-learning is subject to constant change. It must be noted that different understandings of eLearning are conditioned by particular professional attitudes and interests (Dillon and Åhlberg, 2006; McPherson and Baptista Nunes, 2006; Sangrà, Vlachopoulos and Cabrera, 2012). While some can occur in or out of the classroom, it could still be self-paced, both a/synchronous or may or may not be instructor-led.

However, why was it meaningful for us to establish the difference? Apart from the fact that EPICC was embracing a real, robust and dynamic LMS as an engagement tool for the first time, and for many instructors, program and administrative staff, much of this technology-infused engagement would be a major paradigm modification that would require adopting new mindsets, processes and practices. There was also a hope that knowing the definitions and knowing the difference, would affect consensus building and how one will design and deliver curricula, as well as build formative and summative assessments.

Building the consensus up front with information from the listening tour and quantitative data, also helped to develop a criteria taxonomy as well as a decision support system hinging on mobile accessible blended eLearning.

Table 1. Criteria taxonomy for selecting an LMS for EPICC

Category	Criteria
Complete Feature Set	Assessment Tools (question formats, online assignment submission, quiz auto-grading, higher order question format capabilities) Course Authoring Tool (create assignments, link to resources, share with other instructors) Reports (online participation, individual, class and school level) Advisors can see students' academic progress
Usability	Easy to use, intuitive interface for teachers, students and administrators Short low learning curve required for full teacher and student participation Availability of training and customer support Mobile web responsive, Mobile Learning App. (features available via phone or mobile device)
Social Connectedness	Social Learning capabilities (e.g. document collaboration, threaded discussions, blogs, integration with web 2.0 online apps, Learning Tools Interoperability (LTI)
Technological Sustainability	Ability to easily integrate with EPICC's CRM/Salesforce system and data Ease in upgrading technology when necessary Anticipated longevity of product and company supporting product
Cost	Installation and training cost (e.g. per enterprise or per site) Development costs for data and Salesforce integration Licensing costs (e.g. one-time vs. annual, based on size / scale)

In this way the criteria taxonomy was also used to verify the best suited systems that would aid the three pillars of the academic vision listed above. Building a criteria matrix also catered for a mobile friendly LMS that would be widely adopted and offer value to the three main stakeholders and drivers of success; instructional quality, empowered students, and curriculum mapping and alignment. The relative weight of each evaluation category was measured for importance, and weighted against other categories and criteria. Additionally, long term sustainability was one of the main criterion and included not just a the LMS providers user interface, years of existence, their responsiveness, and rate of improvements, longevity, but also how well it subscribed to the ability to interact with other online services, which is commonly known as Learning Tools Interoperability (LTI) or external tools.

Examples of interoperable learning tools are OneNote, Office365, Google Drive or DropBox which can now seamlessly interact or be embedded within a learning portal. These mechanisms interact with the selected learning portal and allow LMS users to interact with LTI compliant resources from other websites and online resources when used in course(s), external LTI systems that support digital handshake and info pass-back to send log in or grading results and information back to the LMS gradebook. Other examples are Turn-it-in, Moodle, integrating quotes, Respondus quiz generator, Blackboard, and Brain Honey or Soft Chalks Score Center. Integrating these services into the LMS, allows for the staff and students to leverage commonplace digital resources that they are already using. Not only it saves time, but also allows for seamless cross-over connections and transferability of not only documents, but also knowledge to better serve students in their preparation for corporate internships.

The e/m-learning initiative was designed to better prepare EPICC students for success while in the program, and to solidify EPICC's high expectations and high support model leveraging technology to support our students, interns and alumni. EPICC is recognized as a best-in-class service provider, and in 2013, through this e-&m blended learning drive, embraced technology innovatively to have them better flex as a workforce development organization with an evolving labor market. The hope of further integration of educational technology into EPICC programming was not only to provide for a more effective teaching and learning environment, but also to create cost-savings, allow for improved student learning and outcomes, and enable the organization to scale dramatically and serve more young adults in need. Success of these innovative initiatives was envisioned so that it can also be replicated globally and share what we know in consultation with EPICC and its educational technology subject matter experts.

The e/m-learning initiative support EPICC's efforts to grow and strengthen its program model and curriculum while scaling to serve thousands more students, interns and alumni annually.

Figure 2. LMS E-&-mLearning criteria matrix (Mentor, 2014)

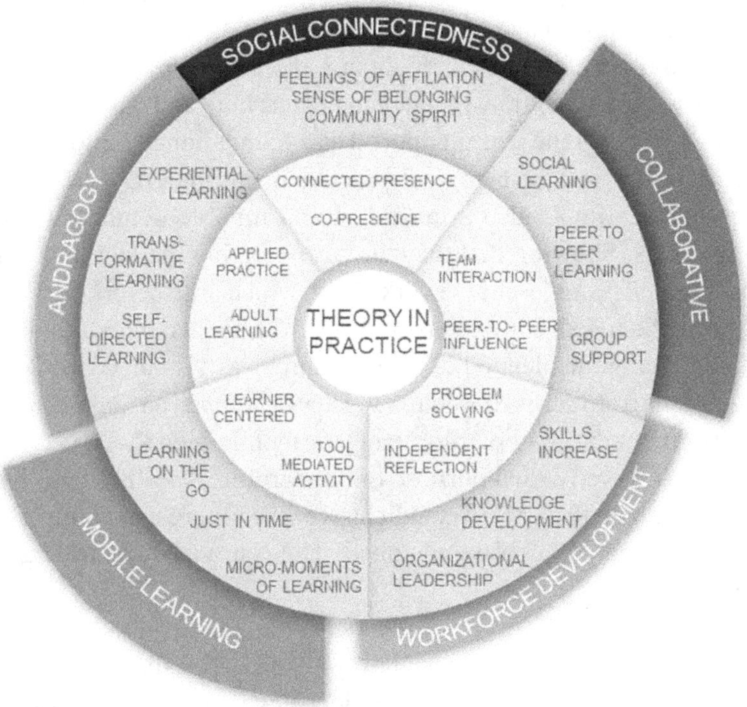

IMPLEMENTATION AND RESULTS

As previously indicated, while there was a need to digitize the then 12-year-old organization, there was also a clear need to extract and share instructional quality, align courses around outcomes and objectives, as well as leverage modern technological means. Furthermore, the need to empower students by pursuing a fine balance of aligning content with internships and higher education requirements were also envisioned. To test hypotheses in the form of a listening tour, an academic vision and set of strategies was formulated to enhance the educational culture and extract even more benefits from the highly dedicated staff. Based on the listening tour, the strategic roll-out envisioned training, identification of early adopter champions, as well as organic and intentional methods. Given the vision as well as the introduction of blended e-&m-Learning to build the foundation for further online learning, the academic directors decided as its first step to adopt a mobile friendly LMS. They did so in a holistic manner to address, in an overarching method all of the needs previously expressed. Within six months EPICC achieved 100% adoption of the e-&-m-Learning initiatives across eight (8) cities, geographically dispersed across the USA. With a further partial expansion to seven (7) other cities in limited academic and program aspects. EPICC was also able to standardize and harmonize their curricula development and delivery. With their e-&m-Learning initiatives they were also able to diversify their curricula at a more rapid pace while maintaining quality as well as clarity and transparency to the learning and development process.

Envisaging the use of the learning portal as a foundation to build on and putting fundamental theories into pedagogical practice as an integral part of the e-&m-Learning initiative, EPICC also arranged virtual mini conferences (webinars) to share best practices. Showing that they are also utilizing the learning portal as an online repository of sample coursework from which instructors can draw when addressing particular needs for various courses. Additional methods of diffusing the adoption and growing the blended e-&-mLearning was visualized in the formation of LMS champions from across the country. They developed an interactive instructor handbook for departmental onboarding using the mobile friendly LMS for this digital engagement. This served a dual purpose of onboarding as well as familiarizing the new staff member with the LMS. As indicated earlier, there were other overarching needs to be addressed and goals envisioned. The National Academic goals of the LMS project included:

1. Promote standards of instructor quality – an LMS will enable instructors to support EPICC's pedagogical style to learning and student outcomes through the use of interactive, customized, engaging and easily accessible technological features
2. Use academic transparency to empower students: through LMS-based exportable calendars, assignments, assessments, feedback, and grading, both students, instructors and advisors will be empowered to understand and manage the student's individual learning experience more effectively
3. Unify sites with a market-relevant curriculum – LMS tools will allow Academic Directors and instructors across sites to:
    a. Create a cross-site academic community within EPICC to inform best practices in delivering student outcomes
    b. Easily create and access a common library of course and curriculum materials that all sites and instructors can draw from, and that can serve as input into the identification and articulation of a set of common learning objectives for EPICC students and career tracks
    c. Reduce curriculum development and maintenance expense by efficiently re-using best in class materials across sites

EPICC's strategic goals in meeting with the organization's human computer digitization needs required checks for positive user interfaces and user experiences to promote and enhance learning engagement. The latter would be measured in terms of return on investment (ROI) from student feedback, staff time saved and results. The following goals and guidelines were outlined as part of the present and future needs of the digitization process, so as to serve the academic, program staff and the organization as a whole:

- All teachers and students (& program staff) learning to teach and learn online
- Reduced cost – using technology, where possible, to deliver high quality program to students
- Access to content across geographies – using technology to allow people from different geographies to access material
- Increased market responsiveness – using technology to better adapt to the skill needs of our internship clients
- Increased differentiated instruction – using technology to manage/support different paces or styles of learning
- Better formative and summative assessment through low and high tech methods to support student learning
- More developmental education – using technology to help students do more, on their own time, with respect to learning and development

The National Academic leader's approach to e-&-mLearning with its LMS as a unifying factor was also a creative response to several external trends. The program was in an ideal position to take advantage of established eLearning models and a growing trend toward mobile learning. In building its criteria matrix for its LMS at this late stage, EPICC was poised to learn from others' eLearning mistakes and pitfalls, and to build mobile learning into its e-&-mLearning and LMS criteria taxonomy. Adopting an LMS that would fit EPICC's twenty-first century needs and Customer Relations Manager (CRM) strategy, with e- and m-learning in mind, allowed EPICC to envision expanding its six (6) months of L&D to a full year, and engage with students while they are on their internships. This would effectively extend their learning opportunities and skill building as well as their connection to their learning communities. While the vision saw an extended use, the LMS also allowed EPICC to be more market responsive to labor market demands as well as raising the level of its ability to own its own courses.

Implementing an E-and-mLearning Management System

Underpinning nearly all of EPICC's work in transitioning to a program and curriculum that actively integrates e-learning and m-learning principles was the selection and implementation of a Learning Management System (LMS). This native and web mobile accessible LMs needed to support learning initiatives in every EPICC classroom and the vision was to tie together student data, not just from the academic phase and activities, but across each stage of its involvement with EPICC – from admissions through L&D, the corporate internship program, program completion and post-graduation job placement.

Installing a Learning Management System was not the goal in and of itself, but with accompanying goals of effecting changes in its pedagogical and cultural programs, EPICC's Learning Management System allowed instructors and students to:

- Understand and manage the learning of individual students in a more responsive and dynamic way, allowing instructors more discretion in where to spend their time in the classroom and with students one-on-one while customizing and providing students with greater accountability for their own learning
- Create a more complete profile and career plan for each student at each stage of their relationship with EPICC by drawing together student data from admissions through to post-graduation job placement. After achieving alignment around quality higher education standards, there is a plan for students to document and earn rewards in many different forms for learning where they can evidence their skills and knowledge gained during the entire program.
- Become comfortable with electronic performance management systems, and receive real-time feedback on students' performance on assignments and quizzes

In addition, the LMS rolled out across EPICC sites also created greater consistency in how program elements are delivered at each site, and create an important opportunity for identifying the most effective teaching methods and course content for distribution across sites. The latter was brought into fruition in the form of mini virtual conferences that advanced the organic alignment and harmonization of myriad different sets of site practices, which were the formation of online teaching and learning groups or communities amongst the teaching staff. Another one of the most significant pieces of the development and roll-out of the Learning Management System was the up-front investment in translating EPICC curriculum components from a traditional offline classroom setting into a useful and usable set of online resources for instructors, staff and students.

RESULTS

90% of instructional staff across the USA in eight (8) city centers adopted the LMS immediately. This soon grew to 100% with some late adopters bringing up the tail end of the implementation effort. Additionally, with expansion to admissions, student services and other parts of program, adoption now stands at 120%. Another expansion of the adoption beyond 100% of academic staff within the organization, was the addition of all staff to the LMS, not just the academic program staff. All staff are advisors and thus can see their advisee's progress. The need for advisors to see their students' progress was identified as a priority by the senior leaders' years before the introduction of the LMS and was a part of the criteria matrix and vision. However, due to the lack of staff resources and bandwidth on a local and national level, the advisors monitoring the online L&D progress of their advisees as a practice has been reported as sluggish in its uptake and difficult to fully train for, monitor and evaluate.

Nonetheless, students were fully utilizing the advantages of being able to connect with one another via the mobile accessible LMS. This helped them to getting to know one another online and forging further ties and bonds through the mobile electronic engagement. An example here, before the academic program commenced, and at the introduction of the LMS for their use before they started their program, students posted questions to the LMS wall. One student took a photo of his intended clothes for the next day with his phone and posted it to the front page bulletin board of the LMS to ask if it met the dress code. Apart from getting likes for being brave to ask that question, the student also received positive comments and affirmations for his choice of clothing. Students were also observed supporting and encouraging one another with assignments. Offering support, resources and words of encouragement, and even saying goodnight to one another via the LMS. Students were using the LMS during their commute to read over work and submit assignments, while also utilizing other features of the LMS.

Other features were the calendar feature that not only helped with student time management, but also with the ability to submit assignments and contribute to discussions via their phones. The ability to export and link the LMS calendar to their mobile phone calendar with reminders was a simple feature, but powerful for students' time management and empowerment as was the visibility of grading of assignments and results from quizzes. There are still reports from student surveys that there are at least one or two instructional staff who do not place every assignment in the LMS, nor do they use the assignment feature that automatically feeds into the calendar. Students comment on the inconsistent practice due to missing assignments, but also because they lose money from their earned stipend for not submitting assignments on time. The latter is part of the contract that the students sign when they join EPICC.

While there has been 100% adoption of the LMS, there are still some struggles from that small number of staff to adopt more of the helpful features of the LMS.

Staff use of the assignment feature and grading of assignments was sparsely adopted at first, but grew from 60% to 90% of all staff across the country. Adoption of the automatic quiz grading aspects was a little better and ran at 70% at initial adoption. However, 60% of the staff still struggled with separating open questions in quizzes with automatically graded questions. While the IT and Financial Operations instructors found the use of the quiz instrument very useful and easy, the BC instructors found the use of the feature slightly more difficult. Apart from offering some creative solutions, as well as a few LTI options to overcome these challenges, uptake of these ideas was slow. There were other aspects that were high to be adopted by individuals, but slow to be incorporated or continued by the organization.

The LMS offered various notifications that could be sent to teachers or student users for assignments and quizzes added, discussions posted, or changes to assignment deadlines. Both students and teachers could receive notifications via email or via mobile phone Short Message Service (SMS) texts. Initially, during the first year, only 20% of the sites used the notification system or the LMS front page bulletin wall to post site closures related to snow days or emergencies. Three years later, 90% of sites are using the LMS bulletin board landing page to communicate with students. While the LMS also has the ability to serve the students after the program, students' in the program unfortunately lose their associated student accounts, due to organizational policy, and budget cuts as successes without realizing the immediate, short, and long term impact on the organization and the phenomenal cadre of young talent that have developed. The text messaging feature has also grown, but only to 40% of use, and more so by students as opposed to staff and sites. Ideas to grow these adoption percentages ranged from updating and standardizing the job descriptions for instructors as well as including educational technology training or experience to the requirements and qualifications.

There was also the visualizing and sharing with the senior leaders of the need to change hiring practices and the need to employ national instructional supporting staff. The national academic department soon grew from 1 National Academic Director to 4 staff members to help implement, oversee, manage, and measure additional academic and technological innovations. Another suggestion was the need to hire instructional and program staff with blended eLearning experience to develop and deliver more responsive curriculum content utilizing e-learning and m-learning platforms. The organization should provide expanded opportunities to use ICTs to create intensive, specific training programs tailored for specific market driven career tracks and corporate partners. Just as EPICC graduates increasingly rely on their ability to use technology effectively in the workforce, the organization has a transformative opportunity to leverage blended e-&-mLearning tools and teaching methods to increase short-term student outcomes gains, and long-term graduate success with strategic academic vision and goals to promote enhanced teacher and student user experiences.

Improved Learning and Development

E-&-mLearning not only empowered EPICC's non-traditional student body to bridge specific gaps in their preparation for professional careers, but also allowed the organization to utilize interactive technologies and communication systems to improve the overall teaching and learning experience. E-&-mLearning started a slow but steady transformation of the pedagogy and curriculum delivery process, as well as how our students engage with the learning process. The implementation of the LMS as educational technology allowed the academic department to extend training beyond the classroom and allowed for access to knowledge and support. The hope and vision is to bring this capability to all phases of the program: from admissions to alumni, through continuous learning and engagement after internship graduation. From initial reports, the introduction of the LMS raised the teaching and learning standards for teachers and students, as both parties were enabled to track real-time data; student grades, quiz results, as well as the ability for teachers to act quickly on teaching and learning analytics, and deepen understanding and participation in lifelong learning. Additionally, the implementation of the LMS as educational technology enabled EPICC to support students and alumni throughout their journey in the program, as well as to empower staff.

Given the short time period that the organization has with their young students, efficiencies are required in order to maximize learning time and support our students' effective learning. E-&-mLearning enabled approach extended learning ideas and opportunities for every student, while implementing already-proven flip learning methods to immediately expand the amount of individualized learning and transform the in-classroom and out-of-classroom experience. The organization is now also in a position to pilot an efficient and repeatable market-responsive process for curriculum development and delivery that can be scaled and replicated across our national network with the implementation of e-&-m learning technology.

To summarize, the mobile accessible LMS enabled the organization to move beyond the constraints of their previous traditional, mono-modal, offline classroom teaching methods, by utilizing innovative e-learning and m-learning technology that will support highly interactive instructional methodologies in both the Learning & Development and Corporate Internship stage of the EPICC program.

EPICC has now shifted its approach to become a training program that is:

1. able to offer instructors, academic teams, and the organization, valuable teaching and learning analytics that would aid data-informed planning and professional development;
2. data-driven, measurable, and research-supported;
3. extended beyond the limitations of class time and space
4. accessible to all our staff advisors, as well as our interns and alumni;
5. personalized for students' needs, interests, and abilities; and
6. focused on the demands of the local market and our corporate partners' needs.

In order to become a proper human capital solutions provider for their existing national corporate partners and other major U.S. employers, the non-profit training program must invest in educational technology. There is a recognition that the former classroom-based program model and content delivery structure limited their ability to quickly design and implement standardized instruction based on national corporate partner demands. An investment in further educational technology professional development for staff would strengthen its ability to tailor training for specific corporate partners and career tracks, while better preparing their students for success in their professional careers. The blended e-&-mLearning model introduced, offered a transformational investment in the organization's ability to effectively serve all of their constituents; from students to staff, as well as its corporate partners, to receive well prepared interns.

When the initiative was piloted in 2012 and then implemented in 2013, it was introduced as a holistic e- and m-learning program to create a consistent experience for students and staff across the organization. There were two internal powers that inspired the fast adoption to enhance EPICC's 12-year old model — the students and their instructors. The students' eagerness and enthusiasm for the mobile eLearning opportunities was awe-inspiring, and the unfailing dedication of the instructional staff, who spend an amazing amount of energy with students in a very short, intensive six months to redress 18 -20 years of educational deficits is a marvel to behold. Instructors, who act as managers, can now spend more one-on-one time with their students, provide tailor made individual or group lessons, workshops and resources.

The LMS also afforded EPICC opportunities to create a new online curricula development and delivery framework. This allows them to not only improve programming quality and efficiency across several of our departments, but also to develop skills based courses at a much faster rate and deliver faster on labor market demand. Based on that success, their Emerging Leaders Program has also started utilizing the program for their focused staff training efforts. The intentionality behind the decision to select a 21st century interactive, nimble LMS portal with SCORM and LTI compliant packets, was to build toward the future with successful participatory collaborative teaching and learning community models. Multimodal teaching and learning aided one of the pillars of the academic vision and strategy which empowered students to be active participants, contributors, and owners of their learning. Students and all organizational staff could now monitor student progress, as well as identify and deal with academic challenges immediately. The e-&-mLearning also extended active off-site access to course materials, assessments, as well as to peers and teachers. Additionally the LMS also aided organic and strategic alignment of curricula across sites via a centrally located and shared resources space. Apart from alignment, new courses could also be supplanted to other sites far faster than during the paper-based approach that the organization was stuck in for 12 years.

The electronic and mobile learning initiative has been a great success as it has allowed EPICC to more efficiently roll out several new courses at various sites. From the present and future needs criteria taxonomy, the organization is also utilizing the learning portal as part of our virtual/distance learning efforts. They could also launch new curricula tracks ranging from Computer Scripting and Programming languages (Atlanta, Chicago, Miami, Providence SFO, San Jose, and Seattle), Project Management (NY), as well as Cyber Security (Chicago, San Francisco, and Washington DC,) courses. Through their e-&m-Learning models they have been able to expand their Quality Assurance/Software Testing track to more of a Software Development track. The e-&-m-Learning system has allowed for course work that is standardized around quality and hitting corporate partner needs, while still allowing for the flexibility necessary to meet the differing demands of each market. For example, in Miami they created front-end web development and computer programming courses with JavaScript, JQuery and Java courses in response to a need expressed by American Express, that they were then able to quickly and easily transplant to their Atlanta and Phoenix sites when similar corporate partner needs arose in that market (Mentor, 2018). Thus, e-&-mLearning technology has enabled them to be more nimble as an organization and respond more quickly to evolving market demands. Now they are in the early days of further deepening and expanding their front-end web/software development courses to Chicago, San Francisco, San Jose, New York, and Boston.

Over the last three years, the National Academic Director crafted a strategic vision for EPICC that incorporates an aggressive strategy to fully integrate technology into classrooms. This would allow for increased transparency in student progress, improvement in instruction and teaching techniques, and to align the curriculum nationally across sites. EPICC has successfully introduced blended learning into its educational system to offer transparency and accountability as it worked towards earning college credits that much easier without losing quality, but actually enhancing quality. It makes use of both eLearning and mobile engagement to increase student access to learning and development, while also cultivating and building virtual/ distance learning models.

CONCLUSION

The organization achieved 100% adoption in less than six months in over eight (8) cities and has been able to situate and connect the LMS to other parts of the program to benefit learning analytics, course offerings and develop tailored personal learning environments for trainees. The organization was also able to start cultivating curriculum alignment for all of its subjects across eight of its core sites. Likewise, the organization was able to transform their instructional program by moving from a Victorian instructor-led style to a 21st century 'flipped' learning experience extending the content delivery and engagement beyond the confines of time and physical space constraints.An LMS can serve as a foundation in creating a learning environment that makes use of the best elements of face-to-face experiences and online technologies while offering a blended learning experience, supported by an LMS that would continue to maximize EPICC's high-touch, high expectations attitude to learning. The latter includes developing students' soft skills for workplace contexts; develop a rapport and bond with students through its general processes and systems, which include appointing an advisor to a student and providing that advisor with access and insight into that student's L&D performance and progress as well as their social collaboration and professional networking.

Through the LMS the organization was also able to incorporate social collaboration and mLearning tools to cultivate and accommodate multimodal teaching and learning styles. Required features were selected to accommodate formal, LMS based and social learning opportunities. The new LMS served as a conduit in creating a conducive learning environment with virtual learning communities that also transfers and makes use of the best aspects of face-to-face experiences as well as on-the-go mobile technologies. The curriculum content delivery via e-&-mLearning offered flexibility to these non-traditional students who are either working one or two jobs. The hybrid eLearning approach also allowed the academic department to standardize the organizations' soft skills and hard skills curricula. From Professional Skills (ProSkills), IT (Hardware, Networking) curricula updates and new offerings to meet Cyber Security, as well as Scripting and Programming labor market demand.

The initial efforts and gains of building the e-&-mLearning foundations allowed the workforce development organization to reap huge benefits and return on their investment. Not just in terms of money, but also in terms of time and human resources. There have also been numerous reported successes by the corporate partners (American Express, Microsoft, Salesforce, JP Morgan Chase, etc.). With American Express, for example, the academic department successfully ran two distance-learning pilots using a Small Paid Online Community (SPOC) model with virtual teaching and tutorial support (Mentor, 2018). All of the participants in the

SPOC were converted to full time employees after completing the internship for which they were trained.

Apart from being recognized with additional grant money for its e-&-mLearning successes, as well as two awards for enacting a technology vision and blended learning, EPICC was also offered to participate in a virtual distance learning program by AT&T and a MOOC service provider. AT&T sponsored 20 Technology Learning and Development $2000 scholarships per student to participate in a MOOC. The academic department used its lessons learned from the LMS and converted a MOOC to a c3MOOC (Mentor, 2018). The c3 was for a focus on establishing a connected, collaborative community of high support while cultivating rapport through a remote virtual and on the ground supportive academic infrastructure. AT&T advised all other sponsored and non-sponsored participants to use EPICC's c3MOOC formula and increased their scholarships from 20 to 500 scholarships over three years. The organization was also offered access to additional courses that had high labor market demand. In this case, the organization piloted one of its many options and evaluated curriculum content delivery via eLearning and mLearning across a national network taking a successful model and enhancing it with further innovative methods to serve additional learning communities geographically distributed. Ultimately, they created a learning model that could scale to other departments and facets of the program. However, in order to do so, they will need to develop an Application Interface Protocol (API) to enable live, daily or weekly reporting of students' progress to adequately leverage the organizational structures and processes.

REFERENCES

Abedi, M., Badragheh, A., & Varamin, I. (2011). Review definitions and principles relating to Andragogy. *The Journal of American Science*, *7*(5).

Afflerbach, P. (2016). Reading Assessment. *The Reading Teacher*, *69*(4), 413–419. doi:10.1002/trtr.1430

Anderson, L. W., Krathwohl, D. R., & Bloom, B. S. (2001). *A taxonomy for learning, teaching, and assessing: A revision of Bloom's taxonomy of educational objectives*. Allyn & Bacon.

Andrade, J., Ares, J., García, R., Rodríguez, S., Seoane, M., & Suárez, S. (2008). Guidelines for the development of e-learning systems by means of proactive questions. *Computers & Education*, *51*(4), 1510–1522. doi:10.1016/j.compedu.2008.02.002

Brandon, B. (2006). *311 Tips for the Successful Management of an LMS or LCMS*. The eLearning Guild.

Chaiklin, S. (2003). The zone of proximal development in Vygotsky's analysis of learning and instruction. *Vygotsky's Educational Theory in Cultural Context, 1*, 39-64.

Chayko, M. (2012). *Connecting: How we form social bonds and communities in the Internet age.* Suny Press.

Dillon, P., & Åhlberg, M. (2006). Integrativism as a theoretical and organizational framework for e-learning and practitioner research. *Technology, Pedagogy and Education, 15*(1), 7–30. doi:10.1080/14759390500435747

Dohn, N. B. (2010). Teaching with wikis and blogs: Potentials and pitfalls. In *Proceedings of the 7th International conference on networked learning* (pp. 142-150). Lancaster University.

Ellis, R., & Goodyear, P. (2010). Students' Experiences of e-Learning in Higher Education: The Ecology of Sustainable Innovation (Open and Flexible Learning Series). Sydney, Australia: Routledge.

Greene, H., & Crespi, C. (2012). The value of student created videos in the college classroom–an exploratory study in marketing and accounting. *The International Journal of the Arts in Society, 5*(1), 273–283.

Holton, E. F., Swanson, R. A., & Naquin, S. S. (2001). Andragogy in practice: Clarifying the andragogical model of adult learning. *Performance Improvement Quarterly, 14*(1), 118–143. doi:10.1111/j.1937-8327.2001.tb00204.x

Holtzblatt, L. J., Damianos, L. E., & Weiss, D. (2010, April). Factors impeding Wiki use in the enterprise: a case study. In CHI'10 Extended Abstracts on Human Factors in Computing Systems (pp. 4661-4676). ACM. doi:10.1145/1753846.1754208

Hsu, C. C., & Sandford, B. A. (2007). The Delphi technique: Making sense of consensus. *Practical Assessment, Research & Evaluation, 12*(10), 1–8.

Knowles, M. S. (1970). *The modern practice of adult education* (Vol. 41). New York: New York Association Press.

Koole, M. (2010, May). The web of identity: Selfhood and belonging in online learning networks. *Seventh International Conference on Networked Learning 2010.*

Linstone, H. A., & Turoff, M. (2002). The Delphi Method. *Techniques and Applications, 53.*

Looi, C. K., Lim, K. F., Pang, J., Koh, A. L. H., Seow, P., Sun, D., & Soloway, E. (2016). Bridging Formal and Informal Learning with the Use of Mobile Technology. In *Future Learning in Primary Schools* (pp. 79–96). Springer Singapore. doi:10.1007/978-981-287-579-2_6

McPherson, M., & Baptista Nunes, M. (2006). Organizational issues for e-Learning: Critical success factors as identified by HE practitioners. *International Journal of Educational Management, 20*(7), 542–558. doi:10.1108/09513540610704645

Mentor, D. (2010). *Book Review of Students' experiences of E-learning in higher education; the ecology of sustainable innovation by Robert Ellis and Peter Goodyear*. New York: TC Record.

Mentor, D. (2011, October). Digit-all texting thumbs as a social driving force: A texting service study catering to the socio-emotional challenges facing the Higher Education student. In *World Conference on E-Learning in Corporate, Government, Healthcare, and Higher Education* (*Vol. 2011*, No. 1, pp. 1493-1498). Academic Press.

Mentor, D. (2014). From app attack to goal oriented tablet use. In H. An, S. Alon, & D. Fuentes (Eds.), *Tablets in K-12 education: Integrated experiences and implications*. Hershey, PA: IGI Global. doi:10.4018/978-1-4666-6300-8

Mentor, D. (March, 2014). *Showcasing a Criteria Taxonomy for selecting a 21st Century Learning Management System to support blended teaching and mLearning*. Poster Presentation delivered at the *21st Century Academic Forum Conference*, Harvard University, Cambridge, MA.

Mentor, D. (2018). Micro to macro social connectedness through mobile phone engagement. In *Encyclopedia of Information Science and Technology* (4th ed.; pp. 6184–6194). IGI Global. doi:10.4018/978-1-5225-2255-3.ch537

Mentor, D. (2018). TLC for MOOCS: Teaching and Learning Communities for Computer Programming. In Computer-Mediated Learning for Workforce Development (pp. 93-110). IGI Global.

Mentor, D. (2018, March). *TLC for MOOCs - Teaching and Learning Communities for MOOCs*. Paper presented at UNESCO's Mobile Learning Week Sympoisum at UNESCO headquarters, Paris, France.

Okoli, C., & Pawlowski, S. D. (2004). The Delphi method as a research tool: An example, design considerations and applications. *Information & Management, 42*(1), 15–29. doi:10.1016/j.im.2003.11.002

Parker, K. R., & Chao, J. T. (2007). Wiki as a teaching tool. *Interdisciplinary Journal of Knowledge and Learning Objects, 3*(1), 57–72.

Peña, J. J. S., Tello, J. C., & de la Cámara, S. P. (2011). *A comparative study between free e-learning platforms*. Academic Press.

Robertson, M. (2009). Young "netizens" Creating Public Citizenship In Cyberspace. *International Research in Geographical and Environmental Education, 18*(4), 287–293. doi:10.1080/10382040903251158

Ronkowitz, K. (2013, February 1). *Andragogy and pedagogy*. Retrieved December 20, 2015, from https://web.njit.edu/~ronkowit/teaching/andragogy.htm

Sangrà, A., Vlachopoulos, D., & Cabrera, N. (2012). Building an inclusive definition of e-learning: An approach to the conceptual framework. *The International Review of Research in Open and Distributed Learning, 13*(2), 145–159. doi:10.19173/irrodl.v13i2.1161

Shih, Y. E. (2007). Setting the new standard with mobile computing in online learning. *The International Review of Research in Open and Distributed Learning, 8*(2). doi:10.19173/irrodl.v8i2.361

Trilling, B., & Fadel, C. (2009a). *Career And Life Skills: Work Ready, Prepared for Life. In 21st century skills: Learning for life in our times*. San Francisco, CA: Jossey-Bass.

Trilling, B., & Fadel, C. (2009b). *21st Century Learning and Teaching. In 21st century skills: Learning for life in our times*. San Francisco, CA: Jossey-Bass.

Uden, L. (2006). Activity theory for designing mobile learning. *International Journal of Mobile Learning and Organization, 1*(1), 81–102. doi:10.1504/IJMLO.2007.011190

Wartella, E., Beaudoin-Ryan, L., Blackwell, C. K., Cingel, D. P., Hurwitz, L. B., & Lauricella, A. R. (2016). What kind of adults will our children become? The impact of growing up in a media-saturated world. *Journal of Children and Media, 10*(1), 13–20. doi:10.1080/17482798.2015.1124796

Williams van Rooij, S. (2011). Open-source learning management systems: A predictive model for higher education. *Journal of Computer Assisted Learning, 28*(2), 114–125. doi:10.1111/j.1365-2729.2011.00422.x

Zurita, G., & Nussbaum, M. (2004). A constructivist mobile learning environment supported by a wireless handheld network. *Journal of Computer Assisted Learning, 20*(4), 235–243. doi:10.1111/j.1365-2729.2004.00089.x

Chapter 8
Power up Learning With Mobile Games:
The Educational Arcade – An Innovative Approach to Learning

Kristin Villanueva
Columbia University, USA

Jeanny Vaidya
Columbia University, USA

ABSTRACT

The evolution of mobile games and the subsequent evolution of designing mobile gaming experiences to support learning is captured in this chapter. The authors will address the affordances and limitations of mobile gaming as well as the learning theories and frameworks that lay the foundation for learning through mobile games. Current perceptions of mobile gaming in the classroom and the implementation of mobile games in schools are discussed. In addition, this chapter explores mobile games for social change, their prevalence, and their role in facilitating learning. The chapter also discusses differing perceptions on the issue of mobile games in and out of the classroom as well as the potential for marketability.

INTRODUCTION TO MOBILE GAMING

Trends in the gaming industry and technological advances have spawned innovative ways of gaming with new technologies. In particular, the evolution of mobile gaming has penetrated and prevailed among populations of all ages and experience

DOI: 10.4018/978-1-5225-9351-5.ch008

levels. Market data shows that mobile gaming has transformed into a 4.6 billion dollar industry, with 202.8 million mobile gamers in the U.S. alone, it's penetration being 61.6% (Statista, 2019). This chapter will describe the evolution of mobile games and the subsequent evolution of designing learning experiences to support learning. The chapter will address the affordances and limitations of mobile gaming as well as the learning theories and frameworks that lay the foundation for learning through mobile games. In addition, this chapter explores mobile games for social change, their prevalence and their role in facilitating learning and discusses differing perceptions on the issue of mobile games in and out of the classroom and potential for marketability.

EVOLUTION OF MOBILE GAMES AND LEARNING EXPERIENCE DESIGN (MGAMES+LXD)

When considering mobile games in the context of learning, it is highly important to understand the nature and concurrent evolution of media, design of learning experiences and subsequent presentation of target instructional content. One cannot differentiate educational games from the learning objectives that it is trying to accomplish.

In the last 20 years, there have been major advances in technology. The evolution and changes in the medium of delivery for games, from console to desktop to mobile, are important factors to consider in order to understand the potential of each platform. The current conceptualization of instructional technology emphasizes the significance of both the systematic design process and the technology mediums. The medium and the instructional method work conjoinedly to facilitate the learning process. Kozma (1991) proposes, "Learners will benefit more from the use of a particular medium with certain capabilities if the capabilities are employed by the instructional method to provide certain representations or perform or model certain cognitive operations that are salient to the task and situation and that the learners cannot or do not perform or provide themselves." This view supports that learning results from the dynamic interaction between the capabilities of the media, the instruction (information), and the learner.

The most current conceptualization of instructional technology sees media as serving two primary functions: to represent information and to communicate information. Media in and of itself does not stimulate learning, but offers a way for learning to occur. As demonstrated by Collins, Neville and Bielaczyc (2002), because different types of media afford different ways of representing information, the type of media dictates how the information is delivered. This demonstrates how media functions to represent information. The authors consider this "educational technology". Bodily, Leary and West (2019) share that there is a large presence of computer-based technologies in published literature for instructional design and technology and that understanding research trends is vastly important to the future of educational technology. The content to be delivered and the way in which the material is prepared and structured for delivery influences learning outcomes; demonstrating how media functions to communicate information (Clark, 1983). The authors consider this learner-centered design of learning solutions from the perspective of the learner's actions and interactions with target content "learning experience design", or "LXD". Educational technology and learning experience design are interdependent and work in conjunction to structure learning environments to facilitate learning. The types of educational technology we will discuss in this chapter include mobile devices such as phones, tablets, and mobile compatible virtual reality (VR) technology. The evolution of mobile and VR technology has presented new and exciting challenges for learning designers. As the technology has evolved, subsequent design considerations have evolved as well. Later in this chapter, we will discuss the implications of mobile games and VR experiences on learning experience design, and how one might approach designing learning experiences within games informed by learning theories.

OVERVIEW OF LEARNING THEORIES

The portable nature of mobile devices inherently makes learning informal in the classical view of education and instruction, but many learning theories are relevant to mobile learning. For example, the Behaviorist Theory, originally conceived by Skinner (1968), promotes learning activities in which there are rewards and observable changes (Mentor, 2015). Rewards and punishments or drill and response systems for feedback in computer games are used to reinforce player behavior. Unlike such stimuli – response systems, the Constructivist Theory (Bruner, 1966) builds on new ideas and concepts based on previous knowledge. A historical example of this is the proliferation of personal home computer in the 1980s advanced constructivism in the technological arena. With the improvement terms of text, graphics, video and sound, information was no longer simply being provided, but users could actively manipulate what was shown on screen. Similarly, mobile devices have potential to provide such constructivist learning through the user's in-game experiences. Collaborative Theory promotes learning through social interaction and argues that learning more effective when learners converse either through technology or between people (Pask, 1976). Pask and Bruner's ideas are still relevant in a technologically rich era as presented by Ajegbomogun & Oduwole's (2017) study. Their research finds that social media and collaborative tools such as wikis, web blogs and instant messaging are important means for post-graduate students for scholarly research in a Nigerian University. For example, mobile – computer supported collaborative learning (MCSCL) facilitate and interaction between students without replacing face-to-face conversations as online forums do. However, the Experiential Learning, Constructivism, Situated Learning theory, and Csikszentmihályi's Flow theory (McCann, 2017), all set the foundation for learning with mobile games (Figure 1). These theories are further elaborated in the chapter.

Taking into consideration theories of learning, cognition, motivation, and multimedia theories, this chapter will suggest guidelines for the design of mobile games to promote learning.

Experiential Learning and Constructivism

Mobile learning provides the affordance of mobility as it allows for users to move freely into different spaces and contexts for learning. If we look at various technologies of the past such as Walkman's or stereos, these portable devices allowed us to carry audio content into different venues and experiences. We were able to listen to a song or podcast in whichever context suited or benefited us, be that for comfort or context of learning. Similarly, mobile devices, such as smartphones, allow for learning in varied contexts. For example, one can use a mobile application to learn how to tune their guitar in a music class or use an app to learn the language of the country they

Figure 1: Intersection of game elements and player behavior with learning theories.

	Experiential Learning & Constructivism	Situated Learning	Flow
Game Space	Context lays the foundation for player experience.	Context and the cognitive processes of the player are interdependent.	
Autonomy	Players freely move about and perform actions within the game space. Their autonomic interactivity determines their experience.	Player's actions influence their behavior and choices within the game, determining the progression of the game.	Freedom to move about the game, enabling player choice and agency precipitates flow.
Rules	Rules define the parameters for a player's experience.		
Goals	Defined goals influence a player's goal-seeking behavior within a game, aiding in constructing their play experience.		Clear goals paired with informational ambuguity and appropriate levels of challenge lead to curiosity and motivation, precursors to cognitive enjoyment.
Feedback	A player's actions are shaped by feedback from past actions. Construction is a result of feedback a player receives after performing actions within the game.		Feedback is a vital requirement for cognitive enjoyment and thus flow. The goal-feedback discrepancy triggers greater attention and motivation for the activity.
Attention			Deep focused attention and ability to narrow in on a task is characteristic of flow behavior.
Reflection	Construction is a product of reflection and building upon prior knowledge.		

are currently visiting. As we move more towards VR technology, computer generated experiences may genuinely replicate many authentic experiences, which may lead to more flexibility for learning growth that supports the principles of experiential learning and constructivism.

Constructivist theory is based off the idea that learners construct new ideas and concepts based on their previous knowledge and thus learning is an active process (Halupa, 2015). The idea that previous experiences affect new learning means that students are frequently creating meaning and self-reflecting at various stages of any learning process. Therefore, experiential learning, or learning by doing and reflecting, is already at play when considering constructivism. For Knowles (1978) "experience" plays a large role in adult learning. He says, "learning is facilitated by student's responsible participation in the learning process," (Knowles, 1978, p. 95). However, while adults are self-reflective and can relate to their new learnings in ways that help shape their process for future knowledge acquisition, mobile and VR designers must still give thought and provide scaffolded experiences. This will ensure learners are provided with the opportunity to reflect on their past experiences within a game or simulated environment.

To a certain extent, if VR mimics real life situations, we can even say that learners do and can bring their previous knowledge from the outside world into a VR experience. Like any other education program or software that is being built, we must consider the design and implementation questions and limitations. Hason and Shelton (2008) provide a framework for designing for a virtual reality experience. They suggest the challenge that educators and designers must consider, is making choices between creating a technologically rich and accurate environment versus modeling and generating a more attainable one due to resource limitations. They ask that designers carefully research the field of VR, articulate learning goals and consider budget. After doing thorough research, one can try to create an environment that not only mimics authentic experiences, but is structured in a way that provides optimal learning.

A basic feature that is commonly used in VR is haptics, often used as means to give learners feedback. Vibration is a common form of haptic feedback used in tablets and mobile phones to give feedback to the user. It has been used in games and virtual reality similarly (Li, Sareh, Xu, Ridzuan, Luo, Xie,... & Althoefer, 2016). Haptic technology has also been used in fields such as disabilities and education (Bargerhuff, Cowan, Oliveira, Quek, & Fang, 2010). Bargerhuff et al. tested a haptic glove which allowed visually impaired students to receive vibrations/haptic feedback when teachers moved their fingers on a diagram or chart. This aided instruction as someone else did not have to guide and support student learning in a classroom.

Haptics provides a way to give learners direct feedback so that they can make decisions moving forward in a game or a simulated environment. Therefore, haptics plays a crucial role in experiential learning and constructivism in VR. Haptics create a sense of touch and is important as it gives immediate feedback to users while interacting with technology. When designing for VR, one should consider questions such as "level of desired immersion" and "modes of sensory feedback" as proposed by Hason and Shelton (2008, p. 122). In addition to the visual and auditory effects, haptic feedback can be added to enhance the learning experience. It provides real-time feedback. However, as Hason and Shelton have pointed out, it can be costly, but there are ways to consider the type of haptic feedback technology one may build into their game environment.

Experiential learning and constructivism are important theories to consider in mobile learning and even virtual reality experiences. Designing experiences that allow for feedback are important aspects of game design which allow students to reflect and act on their learning in real-time.

Situated Learning

The Situated Learning Theory promotes learning in authentic context and promotes problem-based learning and context awareness (Lave & Wenger, 1991). Situated cognition involves cognitive processes that occur as an individual interacts with their environment. Therefore, the individual's learning experience and the context in which learning takes place are interdependent (Choi & Hannafin, 1995). According to the situated cognition perspective, context establishes the foundation for learning. Context is inseparable from an individual's cognitive functions within a learning activity and involves the physical and conceptual structure of the activity, the inherent purpose of the activity, and the social and historical constructs in which the activity is embedded (Choi & Hannafin, 1995). Because of this, educational mobile games are a powerful method of modeling systems by situating the learner within a real-life setting.

Many games often rely on a narrative to situate activity rather than perceptual mimicry. Narration context determines the game structure by providing a storyline that frames the virtual environment in which players can construct knowledge through play. Inclusion of narrative within mobile games helps to provoke curiosity and motivation to solve problems presented in the real-life contexts and furthermore, acting in these virtual worlds will elicit situated cognition that will shape students' conceptual understanding of subject matter in fundamental ways so they are able to transfer information outside of the learning environment.

Flow

There are certain components that must be incorporated into mobile games for effective delivery. The most effective gameplay that a user can experience is when they reach a psychological state of "flow", in which the user is completely absorbed or engaged in a goal-driven activity that all unrelated activity is neglected (Webster, Trevino & Ryan, 1993). Csikszentmihalyi's flow theory considers a person to have reached a flow state when he or she has fallen into a deep focus and has successfully channeled all of their emotions and attention to an activity they are engaged with (Csikszentmihalyi, 1990). When individuals enter flow state, they become more intensely captivated in the activity and become increasingly aware of their own mental processes that they are able to focus a majority of their attention to the task at hand and filter out irrelevant thoughts. For example, a study by Sites and Potter (2018) aimed at studying the impact of flow and music in video games. They found that participants who experienced "generative music" talked about flow more than participants who had more a "traditional" soundtrack in the videogame. In addition, a study by Özhan & Kocadere (2019) finds that increased flow allows for academic success because flow allows for more motivation. A study conducted by Webster, Trevino, & Ryan (1993) on flow in human-computer interactions found that the motivational impact of flow is influential in nurturing the learning process. Learning, therefore, becomes a product of motivation toward achievement to such an extent that one is entirely engrossed in the guided discovery process that the game experience affords. This active participation in a problem-based activity successively facilitates students' acquisition of target content either implicitly or explicitly within the context of the game. Thus, it is essential for mobile games to incorporate specific elements that induce flow behavior.

Webster and colleagues maintain that flow, within computer interactions specifically, is comprised of three dimensions: control, attention focus, and cognitive enjoyment. The convergence of these three elements within mobile game enables individuals to reach flow state and generates a positive effect in players' knowledge acquisition. Their findings illustrate that individuals who experience feelings of control over their environment display increased exploratory behavior. Mobile games epitomize the dimension of control by affording individuals autonomy within the virtual world. Imparting players with the freedom and control to choose which paths to navigate and explore equips them with a sense of agency, increases their motivation and precipitates flow behavior conducive to learning.

Drawing upon Csikszentmihalyi's flow theory, Webster describes how, as individuals enter flow state, they become more intensely captivated in the activity and become increasingly aware of their own mental processes that they can focus a majority of their attention to the task at hand and filter out irrelevant thoughts. To catalyze the onset of focused attention, the design characteristics of adventure-simulation games should be tactically organized so that it is perceptually easy to isolate the central activity, ensuring that the individual's play is not susceptible to any external or internal stimuli that may cause interference. Minimizing extraneous distractions helps make evident the target activity and accentuates the pertinent aspects that warrant attention.

Contrary to Csikszentmihalyi's theory, Webster proposes that curiosity and intrinsic interest are interdependent elements of flow, rather than discrete dimensions, that contribute to what he labels as cognitive enjoyment. Cognitive enjoyment is propelled by curiosity and interest and is characterized as involvement in an activity for its own pleasure or enjoyment. Webster details that clear goals and unambiguous feedback are antecedents of cognitive enjoyment, and consequently, overall flow.

It is critical that goals within games are clearly defined; ambiguous goals lack direction and elicit confusion. Joseph, Deepika, Pal, Tunks, & Mehta's (2019) study, found that inquiry-based learning environments promote motivation. Furthermore, in relation to motivation, Garris, Ahlers, and Driskell (2002) advocate that well-specified and challenging goals are likely to lead to enhanced motivation and performance. They suggest, "Games should employ progressive difficulty levels, multiple goals, and a certain amount of informational ambiguity to endure an uncertain outcome," to obtain an optimal level of challenge. In accordance with Garris et al., Webster argues that games must provide challenges that match the individual's skill level to increase the likelihood of flow and adjust to maintain this balance as a player's mastery evolves to keep the player in a flow state once it is reached. Webster stresses that difficulty of tasks should be within the individual's achievement capability stating, "Flow occurs when an activity challenges an individual enough to encourage playful, exploratory behaviors without the activity being beyond the individual's reach." Well-defined, level appropriate goals within mobile games will make it easier for players to monitor their progress toward achieving those goals, enhance overall fluidity of play, and contribute to cognitive enjoyment.

Feedback is influential in motivating learning and is a vital component of cognitive enjoyment in flow theory. Feedback allows players to evaluate their performance in the process of striving to achieve their goals and gives them the opportunity to strategize ways to minimize the discrepancy between their current level of performance and goal criteria. This goal-feedback discrepancy triggers greater attention and motivation for the activity (Garris et al., 2002). Mobile games should consistently provide concrete feedback to users, enabling them to test their predictions and learn from their decisions within the virtual world. In addition to contributing to motivation and flow experience, immediate feedback permits players to self-assess their actions and hypotheses and may lead to concept formation and discovery of better problem solutions. It is important that mobile games are structured in such a way that they integrate the dimensions of control, focused attention, and cognitive enjoyment that are essential in producing flow behavior, which in turn stimulates the learning process by enhancing the effectiveness of content delivery.

AFFORDANCES OF MOBILE GAMES FOR EDUCATIONAL GAMES

One way to understand how mobile devices contribute to learning is analyzing its technological affordances and capabilities. Uther (2019) says that questions such as "Why mobile?" and "'How can mobile learning be evaluated?" are questions that continue to puzzle us even now. Mobile technology is unique in that it is portable and allows access to information for learning in different contexts. It is important to consider is that while mobile devices are essentially computers, they should not be viewed as simply "portable versions" of static machines (Naismith, Lonsdale, Vavoula, & Sharples, 2004). They are key differences in learning from static machines like computers versus more portable technologies like cellphones (Sharples, Taylor, & Vavoula, 2005). Therefore, games designed for mobile platforms should take into consideration the mobility of users by making learning a constructive and social process. While many people envision technology as being mobile, Sharples argues that first and foremost, it is the learner that is mobile, not the technology. He says that learners are continually on the move and "opportunistically appropriating whatever technology is ready to hand as they move between settings (p. 5)." The small size and weight of cellphones enables them to be transferred from one site to another. The freedom to move about rather than being fixed to a location is what distinguishes mobile gaming, and mobile learning in general, apart from computer or console gaming. Because of this distinction, mobile gaming is primarily associated with informal learning. Mobile Learning theory relies heavily on frameworks for informal learning or learning that takes place outside of formal institutions or classrooms as learning in interwoven in everyday activities. In addition, mobile gaming enables data exchange and collaboration to take place face to face with different individuals and groups. The portability of mobile games allows students to collaborate much more effectively face-to-face when used in the classroom, compared to having to collaborate using computers, and in addition, are far less expensive than computers overall (Park, 2011).

In addition to portability, mobile technology adds new dimensions to learning such as social interactivity, context sensitivity and connectivity (Klopfer & Woodruf, 2002). Naismith et. al (2004) suggest taking an activity-based approach, because mobile devices enable individuals and groups to perform activities that are different than those desktop computers. Naismith et. al (2004) adds that mobile use "affords a highly personal experience in authentic context" (p. 12). Mobile devices can also gather real time data from the location or environment during a unique time. Furthermore, users can be connected to a shared network. In addition, scaffolding of learning material is also possible for individuals with different needs.

To an even greater extent, VR technology has the power to add enhanced dimension and richer context in the form of mobile learning. VR has the capacity to immerse a player into not just an authentic context that mirrors real life phenomena, but can accomplish this from any perspective, scaling from the micro-level (ex. interacting with atoms and molecules) to the macro-level (ex. examining and manipulating galaxies in the universe) and with concepts ranging from concrete to abstract. This function is especially beneficial for practicing or learning in scenarios that are typically high cost, high risk, or nearly impossible to replicate by providing a safe space to learn where the actions one takes or mistakes one makes are buffered within a low risk environment. This immersion can provide a player deeper exploration, greater level of interactivity, and sense of presence within the virtual world. Hedberg & Alexander (1994) define immersion, fidelity, and active learner participation as the three features that distinguish a 3D Virtual world, with high levels of each being the recipe for a reliably effective learning experience. VR allows players to be immersed within any place with no distractions and explore the environment at their own pace led by their own curiosity. This 3D immersion stimulates a player's sight, touch, and sound, which elicits a sense of self inside a given scenario. Interaction between the player and the virtual world creates a sense of presence through body ownership and agency within the space. In addition, virtual worlds also provide rich social contexts by creating an authentic social representation, 'social fidelity', 'immediate discourse', and 'social presence' (Whitelock et al,.1996).

Although there are numerous benefits of mobile games, the use of mobile devices is not without its limitations. Many mobile applications require internet access to run; others are downloadable software application. But this brings us to the issue of storage. Mobile devices typically have finite memory storage and limited battery life, exponentially less than that of computers. Other limitations of mobile phones are their small screens and possible violation of personal security Kukulska-Hulme (2007). In the case of VR technology, the primary reason it is less prevalent than mobile phones or tablets is cost. VR technology such as the Oculus Rift, has been highly expensive for the average gamer, thus limiting its access to a select few who can afford it leisurely. Only since 2017 recently has Oculus VR released more affordable versions Oculus Go and Oculus Quest. Oculus and Samsung's release of Gear VR has allowed players to transform their Samsung mobile phones for VR use. Still, many would argue that another limitation of VR is that it is primarily a 1:1 device, meaning only one person can play and learn at once. While this may be true per VR device, there are ways to make VR's use 1 to many. One of the ways to accomplish this is to insert a camera inside a 3D space and program the camera to film the avatars and then project the environment onto a 2D screen. The remainder of the class or participants can learn from following along.

AFFORDANCES OF MOBILE GAMING FOR SOCIAL GOOD

One area in which there has been considerable interest to employ mobile phones for social work is the gaming industry and the nonprofit sector. From a theoretical viewpoint, games for social change can also fall into the category of serious or persuasive games. The primary purpose of these games is other than pure entertainment. While many more games for social change exist for web and desktop platforms than their mobile counterparts, nonprofits are slowly beginning to realize the potential that mobile platforms hold for games. Going "mobile," however, does not necessarily mean simply replicating the desktop version of the game. Organizations do not want to develop games that cannot be used by the intended population due to restrictions or technological access. Therefore, it is not without careful consideration of the various factors, such as audience or learning goals, that organizations should decide on matters of platform use.

In harnessing the potential of mobile phones for learning, games for social change exist as a subcategory of games in which the primary purpose is to educate and empower. Therefore, understanding social change games requires important consideration of learning theories that are applicable to mobile learning in particular. Unlike static desktops, the mobile is portable and therefore theories of contextual and situated learning are highly applicable (Naismith et al., 2004). Understanding how exactly these learning theories can be applied in games, allows for better designs and dispersions of mobile games. This paper presents a brief overview of the learning theories tied to mLearning and considers mobile phone and game design principles to understand the ways in which developers to can fulfill the educational purpose of social change games to the fullest potential. Using use case scenarios of current mobile games and games for social change, especially those related to humanitarian aid, the paper attempts to understand the learning theories, challenges and design principles at play.

Background: Games for Social Good

There is much debate about serious games and their purpose. Many claim that these games are meant to be "fun," while others say that they should be engaging and successful in their educational goal. Suzanne Seggerman, the president and cofounder of Games for Change, says that there is a general sense that games are only good for "fantasy and escapism, and that there's an inherent conflict between real-world content and game-play" (Schreiner, 2008, p. 12). She argues this idea is mistaken as games hold much more potential. According to Michael and Chen (2006) serious games can teach, train and educate through the reinforcement, motivation and blended paradigms. Many may think that embedding learning into game design can take away the fun from the game, but there exist many other ways where both learning and fun can coexist in social change games. For example, Abt (1975) in his book Serious Games says that the educational purpose of games does not have to be in the "game design," but it can be an extraneous factor. Learning in fact does not have to completely take place during playing the game, but in effect can occur afterwards. However, there is opportunity for improvements in serious game design as noted by Westera (2019). The author argues that serious games can be more effective if there are opportunities for reflection and feedback on learning rather than performance.

Categorizing Mobile Games for Social Good

One of the many challenges facing the gaming community is categorizing the types of social games that exist. This is not only because social games cover a broad range of subjects, they also elicit different responses and experiences from various players. Two important categories that Schriener (2008) suggests include awareness-raising games and transformation games. These two categories of games are different in their approach to design, user experience and goals. Awareness-raising games touch on salient social issues and are often "low budget, grassroots creations distributed virtually. For example, Ayiti: The Cost of Life, developed by GameLab and GlobalKids is about a Haitian family making decisions about work, healthcare and education. The players of the game come to understand the effects of poverty on education in Haiti (Ferri, and Fusaroli 2009). Unlike awareness-raising games, transformative games are deeply intricate and "the focus is on creating concrete change in the real world" (Schreiner, 2008, p. 16). The game Peacemaker is an example of a transformative game that revolves around the Israeli-Palestinian crisis. The game includes real footage and facts about the event and incorporates them into the game. These are just two of the many categories that social games can fall into, but these also seem the most plausible and tangible for game developers to understand in the context of the larger gaming industry.

BENEFITS OF EDUCATIONAL GAMES

Games for learning have become increasingly popular within the last decade, and more and more industries including military training, business professional development, healthcare, and education are attempting to harness the motivating and influential characteristics of games to foster learning. The U.S. National Institute of Health and the Institute for Education Sciences has allocated millions of dollars to funding the development of educational games.

Mobile games have the potential for being an extremely valuable and productive tool for education. Incorporating educational computer games that children can use regularly in the classroom that are aligned with school curriculum creates favorable conditions for learning and improves the learning process. Because players' actions within the game are based on their natural tendencies toward learning, learning occurs during the process of play within the authentic setting that the video game affords (Rosas, Nussbaum, Cumsille, Marianov, Correa, Flores, & Salinas, 2003). The learning process is catalyzed through the immersion effect that the game provides. According to Hubbard (1991), the immersion effect occurs when a player submerges him or herself within the game environment and progressively allocates their cognitive resources toward obtaining a target goal (as cited in Rosas et. al., 2003). This immersion theory is similar to the effects that a player encounters when he or she reaches a state of "flow".

Research has shown that educational gaming has produced significant learning benefits. Digital game-based learning strategy was effective in promoting students' problem-solving skills and learning motivation (Yang, 2012; Rashid, Salleh, & Noor, 2019; Anastasiadis, Lampropoulos, & Siakas, 2018). Simulation games have shown to enhance children's spatial abilities (De Lisi & Wolford, 2002; White, 2019) and three-dimensional computer games have been used to enhance achievement and motivation in geography learning (Tüzün, Yılmaz-Soylu, Karakuş, İnal, Kızılkaya, 2009).

Empirical evidence supports the positive outcomes on school achievement, cognitive abilities, motivation, attention and concentration through the use of computer games for instruction. Several studies indicate that computer games support the acquisition of problem-solving skills, promotes strategic planning and self-regulated learning, and caters to and assists in developing different learning styles. Even besides games, simply technology facilitated instruction can boost motivation and collaborative learning as found by Yang, Kuo, Ji, McTigue's (2018) study on technology and reading instruction. Results from a study conducted by Lou, Abrami, and d'Apollonia (2001) indicate significant positive increases in achievement for students learning with computer games compared to traditional instruction;

and similarly, McFarlane, Sparrowhawk, and Heald (2002), reported increased performance in algebra and reading comprehension from exposure to computer game instruction. Favorable measures of student motivation for computer games over traditional teaching methods have been cited in the literature, and motivational increases were directly related to children's attention and concentration.

Though there is limited empirical research of mobile gaming in the classroom, congruent findings have emerged evaluating the efficacy of mobile games for learning. A study comparing the performance scored on an end of unit exam of a group of students exposed to a mobile game to students who continued with project-based instruction found that the students in the mobile game condition on average scored significantly higher than students in the project-based instruction condition (Huizenga, Admiraal, Akkerman, & Dam, 2009)

Many educational games lack the fun factor that sparks curiosity, engagement and motivation because they primarily focus on the instructional elements, but regardless of how effective an instructional game has the potential to be, learning will not occur unless the player is motivated to play it. A strategy of game designers is to emphasize the fun aspects of games rather than the learning components. Instead, they disguise target material within the more stimulating activities of the game in hopes to foster learning, unintentionally. The concept of incidental learning is introduced as a possible mechanism to mediate learning, achievement and enjoyment. Learning is considered incidental when knowledge is acquired without being explicitly presented. Rosas et. al. (2003) defines it as, "the unintentional or unplanned learning that results from activities not overtly educational and occurs through observation, repetition, social interaction and problem-solving during activities that involve implicit meanings." Rosas and colleagues evaluated the effects of incidental learning through the use of educational video games in the classroom and found a positive influence on learning, motivation and overall classroom dynamics due to game implementation.

Mobile Gaming in and out of the Classroom

Mobile games can be harnessed to support educational efforts for subject matter practice, inquiry, strategic thinking, problem solving, and many other curricula. Mobile games offer a useful framework for teaching critical thinking because their mobile nature can provide an authentic model of a problem in its natural setting and situate the learner within a meaningful context that will spark inquiry and encourage exploration within a problem-based environment. McTigue & Uppstad (2019) also suggest that teachers tell the students that the games in the classroom are different from the games they play at home, at the "goal is to learn." The authors suggest that saying this to the students is important to foster competence. Kiili (2005) suggests, "Games provide a meaningful framework for offering problems to students." Games in and of themselves embody problems that necessitate problem solving, manifested by goal directed behavior. Games typically comprise of a series of sub-tests, or "quests", that the user must complete in the process of achieving the overarching goal reflect problem solving processes and can be conceptualized as a larger problem being composed of smaller causally linked problems that must be solved in order to progress. Mobile games have the potential to immerse students in problem situations where they must play using the tools and resources provided in the virtual world to solve complex problems. Exploratory mobile games provide a flexible learning environment that allows players to endeavor to experiment with different solutions to a problem, aiding in broadening their conceptual knowledge of the subject matter.

Utilizing mobile games in the presents a captivating method to keep students engaged with the content, and is a way to ensure that students are receiving consistent feedback on their performance, enabling them to reflect on learning. Because mobile games are typically designed to provide comprehensive data on specific players, game log data can be used as a way to formatively assess student learning. Log data can be used to monitor student progress, diagnose problems, provide individualized support and targeted scaffolding, and differentiate classroom instruction according to student needs. Mobile games are a very powerful tool in supporting student learning.

da Silva, Silva, Juniora, Goncalves, & Viana (2014) exemplifies how mobile applications can harnessed for classroom use. They utilized Algodoo, a 2-D physics sandbox application for the iPad. This mobile application targets grades 4-12 and was developed by Algoryx. This application enables students to construct systems of environments with interacting physical characteristics. They discuss Algodoo's how ease of use and utility in executing qualitative and quantitative analysis of movement and manipulation of velocity, trajectory, equations of motion aid in conceptual understanding of physics content when used in the classroom.

Researchers compared the effectiveness and satisfaction of two interventions teaching 8-10 year old students about the water cycle, one method of delivery was an iPhone game and the other was traditional classroom instruction. Although both interventions showed similar results in terms of student performance, the iPhone game was found to be more satisfying and resulted in higher motivation outcomes (Furio, Juan, Seguí, & Vivó, 2015).

Mobile games are being used more and more for excursions outside of the classroom. Game developers and researchers in Italy created a whole class, multiplayer, treasure hunt game for mobile learning at archaeological sites. This game was meant to address students' lack of interest and excitement when visiting ancient ruins, especially the ruins that have decayed to the point to where their original aesthetic has ceased to exist (Ardito, Buono, Costabile, Lanzilotti, & Pederson, 2007).

Similar engagement outcomes were found in a controlled study of the iPad fractions game Motion Math. Researchers discovered that playing Motion Math every day for 5 days produced a significant increase in fractions test scores compared to scores of students in the control condition. Results also indicated that the game enhanced children's self-efficacy and attitudes toward fraction problems (Riconscente, 2013).

Tablet math games have shown to be beneficial in supporting a remedial math class comprised of students whose teachers identified as needing extra help with addition facts. Carpenter, Pagar and Labrecque (2013) conducted a study to examine the effects of the Teachley: Addimal Adventure tablet game for mathematical learning on student accuracy, speed, and strategy use. They found that student speed was reduced from pretest to posttest, answering more facts per minute at posttest, and found an improvement in students' strategy execution. Teachely: Addimal Adventure is now offered as part of a suite of math tablet games created by the same developers which also includes Teachley: Mt. Multiplis, a game focused on teaching the distributive property of multiplication.

Tablet games are also being used to address cognitive factors such as executive functioning. New York University's CREATE lab in partnership with the University of California, Santa Barbara have created a suite of tablet games that aim to increase player's executive functions. Their games Gwakkamole, All you can ET, and Crush Station aim to address subskills of executive functions which include the following listed respectively: inhibitory control, task switching, and updating.

We are now seeing VR experiences now being offered inside and outside of the classroom. Subsequently, the U.S. Department of Education has been investing a ton of money into funding VR learning games. One of these government funded games is Mission US: Time Snap, a game that leverages VR's immersive qualities to transport students to four critical periods in U.S. history, Boston 1770, the Mason Dixon Line in 1851, the Great Plans in 1872, and the Lower East Side of New York in 1907 (American Social History Project, 2018). VR technology has shown to be popular in the science domain. "Tablecraft" harnesses VR for chemistry learning by situating players in a laboratory that allows them to interact with and experiment with the periodic table of elements, discovering and creating different compounds as they play (Melnick, 2018). Similarly, Schell Game's HoloLAB Champions utilizes Oculus Vive to immerse learners in a gamified lab environment where they complete challenges using tangible laboratory equipment to earn trophies and, in the process, develop proficiency in using laboratory tools (Schell Games. 2018).

In early 2019, Immerse announced the release of the first ever VR language learning platform that connects learners and tutors in the virtual space (Immerse, 2019). Immerse supports that the optimal language learning recipe is to have "real conversation with real people in real places", and VR let's their product do just that - transport learners into authentic environments related to the target language, such as learning French from a French tutor in virtual Paris, or English from an American in New York City (Immerse, 2019). Opponents of VR often highlight its isolating qualities, but Immerse demonstrates the uniting power of VR's ability to transcend borders and build bridges between learners and educators across the globe. As VR begins to gain traction, we are curious to see how it can be leveraged to connect learners beyond the confines of a traditional classroom.

BENEFITS OF SOCIAL GAMES

Social games are valuable, because they represent current social and political issues in ways that engage the player with intrinsic forms of engagement. Educational games can also be persuasive in nature, because there is an agenda to educate the player and take a side on an issue such as homelessness, poverty or war. Some frameworks for social change games have been proposed. For example, Swain (2007) notes that there are core design, theory and activism principles that need to be embedded in social games in order to for them to be successful. Some of these include integrating subject matter specialists, partnering with like-minded organizations, embracing wicked problems (cultural and social problems that are difficult to solve) and measuring the transference of knowledge. He says that humans can learn more effectively through games because it offers active learning and games are able to communicate a "shared understanding of a complex phenomenon." David L. Michael and Sandra L. Chen in their book *Serious Games: Games That Educate, Train, and Inform* (2005) mention that serious games for education are able to draw students by methods such as deductive and inductive reasoning as well as problem solving (as mentioned by Barbosa et al., 2014).

Learning From Current Games for Social Good

Numerous games for social change exist on web and desktop platforms. A closer look at these games can help give preliminary direction as that social change games should take for going mobile. Many existing social games are tied to doing deeds and charitable work for those in need. The game design techniques used by these types of games are mostly tied to awareness-raising games and transformative games as described above. However, important to consider is that these games also draw on a variety of learning theories related to mobile technologies such situated, collaborative, behaviorist and constructivist. The following will draw on existing web and mobile games for social change to discuss the types of contributions they present to the field of education and social change.

Many web games have the potential and are better suited for becoming mobile games without much revision. The United Nations Food Programme's game, Free Rice, was founded by John Breen in 2007. While there have been many requests by users online to make this game mobile, it has still not officially been released in the iOS or Android markets. The game draws on behaviorist learning, by allowing users to learn about art, vocabulary, geography, math and literature through drill-feedback systems. In promotes "fun" in the game as not only past time (submission), but also presents game as a challenge. For every question a player gets right, Free Rice donates ten grains of rice to combat world hunger. The company has already donated over 91 million grains to date as it attempts to feed and educate. Some argue that Free Rice offers is a band-aid solution that is not going to fix all the problems that exist with world hunger (Cox, 2011). Deep learning is not encouraged, but Cox notes that the game works at a deeper social and psychological construction as it is not only about teaching social good, but developing learning and self-satisfaction through game mechanics. In terms of design principles for mobile games, this game would be well suited since it does not require many screens to achieve the goals of the game and can be executed in a simple, clean user interface, much like the website.

While games like Free Rice allow more people to help those in need, mobile games for social change have also been extremely fitting and appropriate for the opposite audiences in developing nations. To illustrate, Half the Sky is an organization that has distributed three mobile games in developing nations such as India, Kenya and Tanzania. 9-minutes is a mobile game that allows you to experience nine months of pregnancy in nine minutes (Kristof & WuDunn, 2012). The goal is to allow users to understand the risks of pregnancy and to promote safe conception. The United States Agency of International Development (USAID) funded the game and targeted India as a target audience due to her high neonatal mortality rate. In 2012, there were 64 female deaths per 1000 live births compared to 59 male deaths in India (Dasgupta, 2012). Many of these neonatal deaths are due to sex-selective abortion, poverty, and poor maternal health. USAID conducted an evaluation on the effectiveness of the mobile game platform and the results noted measurable shifts in knowledge and attitudes towards promoting safe pregnancy. The study also specifically acknowledges that users of mobile games are becoming more salient and widespread, further emphasizing the importance of exploring these game avenues that target social good (Dasgupta, 2012).

The audience is an important element of consideration when choosing a platform for games, both web and mobile. For example, Half the Sky is currently trying to reach communities in East Africa and India. The reason they chose to develop mobile games rather than games on other platforms, is due to the increasing use and penetration of mobile phones in global networks. Many people have access to a phone, but not a computer or a fixed telephone line. In order to play the games, the phone cannot be a feature phone or a smartphone and must be a certain resolution. When Games for Change went to India to test their apps, they discovered that the majority of people in India were savvy with mobile technology. Important to note is that the game takes on a challenge through drama and applies situated learning theory by asking females in India to use this during their pregnancy, thus giving it authentic context. It is thus important to consider audience characteristics. While such mobile game initiatives may make sense to be implemented in developing nation and users in America and Europe on a web platform, if design calls for a web base or a larger game (Dasgupta, 2012).

Many mobile games are in attempts of getting implemented into classrooms and curriculum, but it is difficult to get teachers and educators aware of such games and their potential for learning. Get Water! is a game that aims to educate people about the scarcity of water and lack of access to clean water in developing nations. It is based around a girl named Maya, who must leave school to go get clean water (Epstein, 2013). Many game dynamics such as skills and supplies supplement the game. Another game that has been making its way into schools is My Life as A Refugee Against All Odds, an app for both iOS and Android phones created by The Office of the United Nations High Commissioner for Refugees or UNHCR (2012) . UNHCR deals with displaced communities. The player is placed in the shoes of a refugee and the game allows them to experience what refugees must go through as they are displaced, whether it is finding shelter, food or water. In terms of design, the game uses static images and quotes to represent different situation and is a clean design. Behaviorist learning theory is also implemented in the story as users navigate the challenges by answering questions, which present rewards of punishments depending on the choices you decide to make. Despite the learning opportunities this app offers, one of the challenges that the app developers and UNHCR face is distribution of the game to schools and teachers and then getting people to share the game (British Broadcasting Corporation, 2012).

More recently, UNICEF launched the TAP Project, web-based game that lack of access to clean water in many countries. For every ten minutes that the person does not use their phone, UNICEF donates a days' worth of water. Over 250 million minutes have been played and at this time, donations are inactive, but people can support the project by playing. One of the many conveniences of the Tap Project is that users need not download an app or game to participate. Instead, users simply go to a website to play the game. As time passes, users receive notifications and little facts about water use and sustainability. The project is simply getting users to understand that phone checking has become almost as vital as having access to clean water. Some argue they have "merged our good deeds without consumerist tendencies" and larger nonprofits like UNICEF have taken on a "slacktivist" approach by the "trivialization this humanitarian platform." Again, like Free Rice, the TAP Project does not aim to solve the problem of water access, but does raise awareness about water access issues. According to UNICEF, 768 million people do not have safe, clean water to drink and everyday 1400 children die from consuming unsafe water or from lack of adequate sanitation facilities. Therefore, the challenge to effectively use social change games to create real world behavioral changes still exists.

DESIGN FOR MOBILE GAMES

While learning theories provide a conceptual basis to highlight the affordances mobile technology can bring to education and game design, design principles of mLearning apps should also be considered for effective teaching. Design principles of mobile learning are largely different from web-based learning tools. The importance of mobile design was supported by a IBM – Columbia University research study that discovered that there was a dramatic increase in both user satisfaction after altering mobile interface design (Ahmad and Orton, 2009). The study further suggests some design principles for effective mobile learnings. First of all, it is important to remember that unlike the web counterparts, you do not need everything on the mobile, just enough relevant information. By taking such a holistic point of view one must realize that you want to give your audience three screens of fewer as more clicks leads to more distraction and cognitive load. Similarly, content chunks must be made, and shorter sentences utilized. In addition, images should be used sparingly as they take long to download. Another important aspect to consider is whether the learning activity is going to use the mobile web browser or be a locally installed app. Mobile web-based application are more accessible to a variety of smartphones, but have limited functionality. On the other hand, locally installed apps can take advantage of built in features and shortcuts. In order to make educational games effective, these mobile design principles should not be neglected.

Mechanics-Dynamics-Aesthetics (MDA)

One way to look at this design mobile games effectively is to consider the Mechanics-Dynamics-Aesthetics (MDA) framework for game design. The premise of the framework is that the way in which a player discovers a game is different from the way a designer conceptualizes a game. For example, a player first discovers the mechanics, rules and regulation of the game as they learn how to jump or shoot. Next, they discover dynamics, which refers to the system or how the player learns to manipulate the game to kill, collect or bluff. The aesthetics or presentation of the game refers to the look and narrative such as an interesting story or intriguing music. Hunicke, LeBlanc, and Zubek (2004) further propose that "fun" is not an aesthetic, because it is not descriptive enough.

Instead they offer eight terms to describe aesthetics.

1) Sensation: game as a sense of pleasure, 2) Fantasy: game as make-believe, 3) Narrative: Game as drama, 4) Challenge: game as obstacle course, 5) Fellowship: game as social framework, 6) Discovery: game as uncharted territory, 7) Expression: game as self-discovery and 8) Submission: game as pastime (p. 2).

Integrated Model of Multimedia Interactivity (INTERACT)

The Integrated Model of Multimedia Interactivity (INTERACT) presented by Domagk Domagk, Schwartz, and Plass (2010) emphasizes the reciprocal interaction between the learner and the mediated information through a multimedia learning system. Interactivity is represented in the feedback loops within the model that connect the principal components of the multimedia learning system. This view suggests that the learning is facilitated through interactivity between the affordances of the multimedia environment and the characteristics and cognitive processes of the learner.

Multimedia Theory

Mobile game designers should monitor the amount of information they present at once because too much information can result in less understanding. It is important to limit the amount of simultaneously presented multimedia stimuli on mobile screens as a cognitive load reduction strategy to prevent detrimental effects on learning.

Presenting excessive stimuli for multiple modalities in a game, can impose serial bottlenecks in learners' information processing systems. Anderson (2009) defines these bottlenecks as "points at which it is no longer possible to continue processing everything in parallel". At the onset of a bottleneck, our cognitive mechanisms undergo a serial selective process of deciding which pieces of information to attend to (Anderson, 2009). When designing a mobile game, use caution if including supplemental text or auditory input.

Presenting students with games that make use of several modalities, even to accommodate for individual learning preferences, can be detrimental to the learning objective of a given game. Research by Neisser and Becklen (1975) demonstrate that people can focus their attention within their visual field to process what they are interested in. For students who prefer to process information textually, they would endogenously control their goal-directed attention to reading the supplemental narrative, however, simultaneously presenting a more salient graphics/video can draw attention away from reading the words to the action on the screen. This is an example of stimulus-driven attention. The split-attention effect occurs when visual attention is divided between on-screen text and animation; this results in poorer performance on retention and transfer of presented information (Mayer, Heiser, & Lonn, 2001).

In a multimedia game situation, adding on-screen text to accompany the visual stimuli that is already supplemented with auditory stimuli can impose a high cognitive load. High cognitive load can severely reduce instructional effectiveness. A study analyzing the effects of multimedia video presentations incorporating on-screen narration in both summary and full text form provides evidence that student learning was impaired by the addition of redundant on-screen text (Mayer et al., 2001). These results are compatible with research conducted by Kalyuga, Chandler, and Sweller (1999) compared several combinations of multimedia stimuli. The following combinations were tested: diagrams with printed text, diagrams with speech, and diagrams with both text and speech. They found that learners performed worse when presented with a combination of all three inputs compared to receiving only diagram and speech. Similarly, Moreno & Mayer (2000) discovered that adding auditory material to multimedia instructional message detracted from student learning. Educational game designers should restrict stimulating competing multiple modalities concurrently for this can overload information processing channels and reduce effective cognitive resources toward integration, memory, and transfer; and thereby interfering with the learning of target material.

Design-Based Theory for Mobile and VR

The brief overview of the learning theories and frameworks above demonstrates that the mobile technology can provide the similar learning potentials as those presented by computers or other educational tools. In addition, these learning theories are also applicable to VR experiences. Mobile devices are generally small in size and portable and therefore offer many unique possibilities for learning that are different from desktop. Similarly, many VR experiences are portable and allow for entire new learning contexts for learners to be immersed in. Guidelines for effective policies for implementing mobile technology indicate that organizations should look at factors such as the cost model and infrastructure. Mobile learning can fit training niches and leverage technologies that people already use such as SMS. One should also study the requirements of those involved in using technologies and consider whether the learning content provided is applicable for the intended audience. When designing a VR experience, it is important to have a framework in mind for how one might want to build and create a genuine VR experience that would meet the intended learning goals.

Existing research on designing VR experiences references traditional design principles such as ADDIE or Merrill's First Principles of Instruction (Merrill, 2002) as Hanson and Shelton (2008) mention. According to Hason and Shelton, as a part of the analysis phase, it is important to research VR technology and potential uses, as well as, learn about your audience and how they may use the VR environment. In the design phase, one would consider the various "components" of the VR lesson. Hanson and Shelton suggest "specifying how the virtual world will be reified, the amount of immersive qualities to be included, the level of sensory referents and feedback, and the degree of user interactivity planned for the VR application" (p. 129). It is equally important to recognize and consider the limitations of VR before the development phase. Hanson and Shelton suggest running a pilot test to understand the types of limitations and how you may want to rectify them. Hardware and software issues also need to be taken into consideration, as well as cost and resourcing as VR technology is not as ubiquitous. In terms of design considerations, they also suggest thinking through how to build the different components of the VR lesson and create an outline for the course or game. They state the main goal should be allowing for the "environment to be problem-based" (p. 122).

Strategically enriching the VR experience is important to ensure the game or environment is meeting the intended learning goals. Chen and Teh (2013) suggest ways in which instructional models for VR can be enhanced. Their ideas are borrowed from important contributors of instructional design principles like Gagné, Merrill and Mayer. In a study to evaluate a VR based learning system where beginner car drivers learned about traffics regulations, Chen and Teh applied many of Gagné & Merrill's (1990) design principles. For example, they took into consideration the objectives and integrative goals, as well as, the detailed scenarios that were to be played out for the learner to resolve. In addition, Mayer's (2002) principles and ideas about spatial continuity and modality were also tested. For example, Mayer says that learners learn better when words and letters are presented closer together than when farther apart. Similarly, it is better to present learners with animation and narration together than with animation and text where learners must read.

At a more micro-level, Chen and Teh (2013) also evaluate areas of real-life simulation including environmental richness and collision detection. They provide insight into aspects of VR technology that can be considered in design. For example, they added virtual objects in the training experience to "enhance its attractiveness and realism" (p. 703). Two separate instances of the design were tested, one with virtual objects such as trees, buildings and bushes, while the other with only roads and signs pertinent to the learning of traffic rules and regulations. Results indicated that there was no clear outcome as to whether adding these virtual objects affected the learning outcomes. Collision detection was also added to one design, while not to the other. The detection, when active allowed did not allow the user to collide with virtual objects and was prevented when a use came in close contact with a virtual object. When the feature was not active, users "navigated along the road side" and thus Chen and Teh conclude that adding such as feature helps prevent disorientation for users (p. 707). The study indicated that their participants all concurred that learning in a VR experience was beneficial to them as opposed to learning in a real life setting on the road where they may not have the same opportunity or time to negotiate and solve the problems they encountered. Thus, Chen and Teh argue that such environments are beneficial for constructivism and allowing learners to build on previous experiences.

CHALLENGES IN DESIGNING SOCIAL CHANGE GAMES

One of the major challenges that the gaming industry faces in making games for social good is accuracy and authenticity. An analysis of social change games these challenges still need to be addressed also when adopting games into the mobile community. For example, dealing with serious socio-political content is tricky because of its complex nature. Dahya (2008) suggests that the poor functionality of game mechanics, limited playable elements and fragmented story construction that are not embedded in gameplay are some of the challenges that persuasive games face. Ayiti: The Cost of Life does not represent the full complexity of the poverty in Haiti because the game does not explain the origins of poverty in the country (as presented by the United Nations (UN)).

Furthermore, games like Darfur is Dying, developed by University of Southern California after consulting with humanitarian aid workers who had spent considerable time in Sudanese region, still presents similar challenges. The game involves Darfuri family who has been displaced and players go through various experiences, such as the process of getting water and having to avoid being captured by the Janjaweed militia. Even though the developers did consult with humanitarian aid workers, the game still uses language and stereotypes that are not entirely accurate. For example, the game villainizes Arabs by categorizing all people of Arabic descent into one militant group. Despite these challenges and structural problems that hinder learning and playability of such games, however, serious games have reached out to a large audience. Therefore, it would be helpful to continue analyzing the challenges and successes of such games.

CHANGING PERCEPTIONS AND MARKETABILITY

A main challenge that social games face, regardless of platform, is finding a place in the context of the larger gaming industry. This is largely due to the fact that social games have more of an educational purpose than other more "fun" games. However, Ari Mubarak, the designer of Peacemaker, talks about how games were never only about fun. He uses chess as an example to explain how games used to be tiring and demanding and that the origin of games was supposed to be a "teaching device." He says that entertainment and fun were embedded just to attract kids (Schreiner, 2008).

Proponents of educational video games of all types find support for their view in theories promoting the underlying concept of children's play. Play is an integral part of children's cognitive and social development and the natural structure of games promotes play behavior. Supporters argue that using these games as instructional tools creates a learning environment that reflects students' interests and yields more meaningful learning. The versatile nature of educational technology is also beneficial for teachers by allowing them to experiment with alternative methods of instruction corresponding to individual student needs and the opportunity to deliver subject matter in various contexts to support comprehension that traditional instruction does not afford (Rosas et. al., 2003).

Although studies support that video games have a beneficial impact on motivation, cognitive abilities and performance outcomes, there remains much speculation regarding its impact on learning and the anticipated consequences of implementation. Conventionally, school and play have been conceptualized as discrete entities, however, as Rosas describes, this was not always the case. Throughout infancy and early childhood, learning is through playing is encouraged. As a child advances from kindergarten to later elementary and on through high school, it is custom to disassociate the two because more and more time is allocated to instruction than play behavior. The child begins to understand that the nature of play as they have experienced it does not fit within the confounds of instruction, traditionally. Subsequently, teachers typically view them as independent and tend to separate school and play. This school-play dissociation and the unwillingness to deviate from conventional practices is the source of resistance to adopting computer games for classroom use.

Resistance is present on both classroom and school level implementation. Rosas reports that resistance on the level of the classroom originate from "(1) teacher's perceptions of games as merely entertaining and not as useful instructional tools, (2) teachers' lack of knowledge and skills with respect to computer assisted instruction, and (3) insufficient developments of effective educational hardware and software".

He also notes that issues on computer game implementation on the level of the school system stem from the penetration of computers in the educational environment, teacher resistance towards a technology-centered learning environment, and poorly structured educational games. In addition, many video game opponents caution against their use based on several research findings that suggest that video games elicit aggressive tendencies in children and reflect gender stereotypes that favor masculinity (Rosas et. al., 2003).

The topic of integrating VR gaming into classrooms is even more of an uphill battle. The cost and lack of efficacy research creates a large hurdle for VR game penetration in schools. Often a prerequisite for adoption in traditional education settings is proof of impact, but reliable and valid impact is difficult to research without wide scale adoption that is generalizable. One obstacle is that sufficient funding is necessary to conduct research of this sort, another is that the results needs to be disseminated and PR around the beneficial impact of VR on learning would need to be spread in order for perceptions to change.

Most published research that has been conducted utilizes computer programs, due to limited access of mobile devices in schools. We have seen a rise in schools that are providing computer carts for students or even adopting 1:1 student technology in classrooms, so we suspect more scientific publications on the use of tablets in the classroom. Tablets are more prevalent in schools than ever before, yet it is not uncommon for school firewalls and platform plug-ins to limit access to learning games. It is also the case that mobile phones, which statistically may actually prove to be more prevalent in schools than tablets or computers, are more often than not, prohibited for use on school grounds. Mobile phone regulations are in place in schools for the purpose of inhibiting classroom disruption. However, the authors of this chapter advocate that the tech savvy generation of today, who have been exposed to mobile phones since birth, have trouble "disconnecting" during the school day just as adults of today have trouble "disconnecting" from their online presence when they are at home. Harnessing this generation's expertise in navigating technology by having them apply it to a learning activity that will simulate their technology needs in a meaningful way, such as playing mobile game could produce favorable results. This learner-centered delivery just might be the way educators can finally speak the students' language, instead of the other way around.

This chapter aims to shed light on the affordances, limitations, prevalence, market demand, and challenges that mobile games face. We explore the learning theories that lay the foundation for their educational and social impact. Current examples demonstrating the wide use of mobile technology for educational purposes to teach a variety of subjects ranging from history, to science, to literacy, and social good is discussed. As designers, developers and educators of mobile games, it is important to observe how learning theories have been applied to the design and development of mobile games to ensure that learners, through gameplay, receive effective learning experiences.

REFERENCES

Ahmad, N., & Orion, P. (2010). *Smartphones make IBM smarter, but not as expected.* American Society for Training and Development.

Ajegbomogun, F. O., & Oduwole, O. K. (2017, October). Social media trends and collaborative learning for scholarly research among postgraduate students. In *European Conference on e-Learning* (pp. 9-17). Academic Conferences International Limited.

American Social History Project. (2018, September 14). *Mission US: Time Snap – Can VR Enhance the Teaching of History?* Retrieved from https://ashp.cuny.edu/news/mission-us-time-snap-can-vr-enhance-teaching-history-0

Anastasiadis, T., Lampropoulos, G., & Siakas, K. (2018). *Digital Game-based Learning and Serious Games in Education.* Academic Press.

Anderson, J. R. (2009). *Cognitive Psychology and Its Implications* (7th ed.). New York, NY: Worth Publishers.

Ardito, C., Buono, P., Costabile, M. F., Lanzilotti, R., & Pederson, T. (2007, November). Re-experiencing history in archaeological parks by playing a mobile augmented reality game. In *OTM Confederated International Conferences on the Move to Meaningful Internet Systems* (pp. 357-366). Springer. 10.1007/978-3-540-76888-3_58

Barbosa, A. F. S., Pereira, N. M. P., Dias, A. F. F. J., & Silva, G. M. F. (2014). *A New Methodology of Design and Development of Serious Games.* Retrieved from Hindawi Publishing Corporation: http://www.hindawi.com/journals/ijcgt/aip/817167/

Bargerhuff, M. E., Cowan, H., Oliveira, F., Quek, F., & Fang, B. (2010). Haptic glove technology: Skill development through video game play. *Journal of Visual Impairment & Blindness, 104*(11), 688–699. doi:10.1177/0145482X1010401103

Bodily, R., Leary, H., & West, R. E. (2019). Research trends in instructional design and technology journals. *British Journal of Educational Technology, 50*(1), 64–79. doi:10.1111/bjet.12712

British Broadcasting Corporation. (2012). *UN turns players into refugees in a new app*. Retrieved from BBC Technology: http://www.bbc.com/news/technology-18553601

Bruner, J. (1966). *Toward a Theory of Instruction*. Cambridge, MA: Harvard University Press.

Carpenter, K., Pagar, D., & Labrecque, R. (2013). *Teachley: Addimal Adventure: Bridging Research and Technology to Help Children Foster Strategy Development*. Conceptual Understanding, and Number Fact Fluency.

Chen, C., & Teh, C. (2013). Enhancing an instructional design model for virtual reality-based learning. *Australasian Journal of Educational Technology*, *29*(5). doi:10.14742/ajet.247

Choi, J. I., & Hannafin, M. (1995). Situated cognition and learning environments: Roles, structures, and implications for design. *Educational Technology Research and Development*, *43*(2), 53–69. doi:10.1007/BF02300472

Clark, R. E. (1983). Reconsidering research on learning from media. *Review of Educational Research*, *53*(4), 445–459. doi:10.3102/00346543053004445

Collins, A., Neville, P., & Bielaczyc, K. (2000). The role of different media in designing learning environments. *International Journal of Artificial Intelligence in Education*, *11*(1), 144–162.

Cox, K. A. (2011). *Helping to Feed and Educate*. Undergraduate Works, Award-Winners & Notable.

Csikszentmihalyi, M. (1990). *Flow: The psychology of optimal experience*. New York: Harper & Row.

da Silva, S. L., da Silva, R. L., Juniora, G., Goncalves, E., & Viana, E. R. (2014). *Animation with Algodoo: a simple tool for teaching and learning physics*. arXiv preprint arXiv:1409.1621

Dahya, N. (2009). Serious Learning in Playful Roles: Socio-political games for education and social change. Academic Press.

Dalgarno, B., & Lee, M. J. (2010). What are the learning affordances of 3-D virtual environments? *British Journal of Educational Technology*, *41*(1), 10–32. doi:10.1111/j.1467-8535.2009.01038.x

Dasgupta, P., Tureski, K., Lenzi, R., Bindu, K., & Nanda, G. (2012). Half the Sky Movement Multimedia Communication Initiative: An Evaluation of the 9-Minutes Mobile Game and Video. Washington, DC: C-Change/FHI 360.

De Lisi, R., & Wolford, J. L. (2002). Improving children's mental rotation accuracy with computer game playing. *The Journal of Genetic Psychology*, *163*(3), 272–282. doi:10.1080/00221320209598683 PMID:12230149

Domagk, S., Schwartz, R. N., & Plass, J. L. (2010). Interactivity in multimedia learning: An integrated model. *Computers in Human Behavior*, *26*(5), 1024–1033. doi:10.1016/j.chb.2010.03.003

Epstein, E. (2013). *Get Water! A Mobile Game that Advocates for Change*. Retrieved from Girls Globe: http://girlsglobe.org/2013/01/24/get-water-a-mobile-game-that-advocates-for-change/

Ferri, G., & Fusaroli, R. (2009). *Which Narrations for Persuasive Technologies? Habits and Procedures in Ayiti: The Cost of Life*. Academic Press.

Furió, D., Juan, M. C., Seguí, I., & Vivó, R. (2015). Mobile learning vs. traditional classroom lessons: A comparative study. *Journal of Computer Assisted Learning*, *31*(3), 189–201. doi:10.1111/jcal.12071

Gagné, R. M., & Merrill, M. D. (1990). Integrative goals for instructional design. *Educational Technology Research and Development*, *38*(1), 23–30. doi:10.1007/BF02298245

Games, S. (2018, October 22). *'HoloLAB Champions' Now Available on Oculus Rift*. Retrieved from https://www.schellgames.com/blog/schell-games-adds-oculus-rift-support-to-its-award-winning-virtual-reality-chemistry-lab-game-hololab-champions

Garris, R., Ahlers, R., & Driskell, J. E. (2002). Games, motivation, and learning: A research and practice model. *Simulation & Gaming*, *33*(4), 441–467. doi:10.1177/1046878102238607

Halupa, C. (2015). *Pedagogy, Andragogy, and Heutagogy*. Retrieved from https://www.researchgate.net/publication/297767648_Pedagogy_Andragogy_and_Heutagogy

Hanson, K., & Shelton, B. (2008). *Design and Development of Virtual Reality: Analysis of Challenges Faced by Educators*. ITLS Faculty Publications. Retrieved from https://www.researchgate.net/publication/43952612_Design_and_Development_of_Virtual_Reality_Analysis_of_Challenges_Faced_by_Educators

Hedberg, J., & Alexander, S. (1994). Virtual reality in education: Defining researchable issues. *Educational Media International, 31*(4), 214–220. doi:10.1080/0952398940310402

Huizenga, J., Admiraal, W., Akkerman, S., & Dam, G. T. (2009). Mobile game-based learning in secondary education- engagement, motivation and learning in a mobile city game. *Journal of Computer Assisted Learning, 25*(4), 332–344. doi:10.1111/j.1365-2729.2009.00316.x

Hunicke, R., LeBlanc, M., & Zubek, R. (2004, July). MDA: A formal approach to game design and game research. *Proceedings of the AAAI Workshop on Challenges in Game AI*, 4.

Immerse. (2019a, February 1). *Immerse Emerges from Stealth to Launch First Ever Virtual Reality Language Tutoring Platform* [Press release]. Retrieved from https://www.immerse.online/press-kit

Immerse. (2019b, February 1). *The Immersive Method: A New Era of Language Learning* (White paper). Retrieved from https://app.hubspot.com/documents/4379388/view/34399933?accessId=919a7a

Joseph, B., Deepika, R., Pal, A., Tunks, J., & Mehta, G. (2019). Intrinsic vs. Extrinsic Motivation in an Interactive Engineering Game. *Journal of Advances in Computer Engineering and Technology.*

Kalyuga, S., Chandler, P., & Sweller, J. (1999). Managing split-attention and redundancy in multimedia instruction. *Applied Cognitive Psychology, 13*(4), 351–371. doi:10.1002/(SICI)1099-0720(199908)13:4<351::AID-ACP589>3.0.CO;2-6

Kiili, K. (2005). Digital game-based learning: Towards an experiential gaming model. *The Internet and Higher Education, 8*(1), 13–24. doi:10.1016/j.iheduc.2004.12.001

Klopfer, E., & Woodruff, E. (2002, January). The impact of distributed and ubiquitous computational devices on the collaborative learning environment. In *Proceedings of the Conference on Computer Support for Collaborative Learning: Foundations for a CSCL Community* (pp. 702-703). International Society of the Learning Sciences. 10.3115/1658616.1658795

Knowles, M. S. (1978). *The adult learner: A neglected species* (2nd ed.). Houston, TX: Gulf Publishing Company.

Kozma, R. B. (1991). Learning with media. *Review of Educational Research, 61*(2), 179–211. doi:10.3102/00346543061002179

Kristof, N. D., & WuDunn, S. (2012). *Mobile Games: Reaching the Hardest to Reach*. Retrieved from Half the Sky Movement: http://www.halftheskymovement.org/pages/mobile-games.html

Kukulska-Hulme, A. (2007). Mobile usability in educational contexts: What have we learnt? *The International Review of Research in Open and Distributed Learning*, *8*(2). doi:10.19173/irrodl.v8i2.356

Lave, J., & Wenger, E. (1991). *Situated Learning: Legitimate Peripheral Participation*. Cambridge, UK: Cambridge University Press. doi:10.1017/CBO9780511815355

Li, M., Sareh, S., Xu, G., Ridzuan, M. B., Luo, S., Xie, J., ... Althoefer, K. (2016). Evaluation of pseudo-haptic interactions with soft objects in virtual environments. *PLoS One*, *11*(6), e0157681. doi:10.1371/journal.pone.0157681 PMID:27352234

Lou, Y., Abrami, P. C., & d'Apollonia, S. (2001). Small group and individual learning with technology: A meta-analysis. *Review of Educational Research*, *71*(3), 449–521. doi:10.3102/00346543071003449

Mayer, R. E. (2002). *Multimedia learning*. Cambridge, UK: Cambridge University Press.

Mayer, R. E. (2005). Cognitive theory of multimedia learning. In R. E. Mayer (Ed.), *The Cambridge handbook of multimedia learning* (pp. 31–48). New York: Cambridge University Press. doi:10.1017/CBO9780511816819.004

Mayer, R. E., Heiser, J., & Lonn, S. (2001). Cognitive constraints on multimedia learning: When presenting more material results in less understanding. *Journal of Educational Psychology*, *93*(1), 187–198. doi:10.1037/0022-0663.93.1.187

McCann, S. (2017). Mobile Learning in Workforce Development: Cultivating Creativity on Action Learning Teams through Higher-Order mLearning. In D. Mentor (Ed.), *Handbook of Research on Mobile Learning in Contemporary Classrooms* (pp. 139–159). Hershey, PA: IGI Global. doi:10.4018/978-1-5225-0251-7.ch007

McFarlane, A., Sparrowhawk, A., & Heald, Y. (2002). Report on the educational use of games: An exploration by TEEM of the contribution which games can make to the education process. Cambridgeshire, UK: Academic Press.

McTigue, E. M., & Uppstad, P. H. (2019). Getting Serious About Serious Games: Best Practices for Computer Games in Reading Classrooms. *The Reading Teacher*, *72*(4), 453–461. doi:10.1002/trtr.1737

Melnick, K. (2018, July 5). *VR Chemistry Game 'Tablecraft' Makes Learning The Periodic Table Fun*. Retrieved from https://vrscout.com/news/vr-chemistry-game-tablecraft/

Mentor, D. (2015). From App Attack to Goal-Oriented Tablet Use. In Tablets in K-12 Education: Integrated Experiences and Implications (pp. 1-21). IGI Global. doi:10.4018/978-1-4666-6300-8.ch001

Moreno, R., & Mayer, R. (2000). A coherence effect in multimedia learning: The case for minimizing irrelevant sounds in the design of multimedia instructional messages. *Journal of Educational Psychology*, *92*(1), 117–125. doi:10.1037/0022-0663.92.1.117

Naismith, L., Lonsdale, P., Vavoula, G. N., & Sharples, M. (2004). *Mobile technologies and learning*. Academic Press.

Neisser, U., & Becklen, R. (1975). Selective looking: Attending to visually specified events. *Cognitive Psychology*, *7*(4), 480–494. doi:10.1016/0010-0285(75)90019-5

Özhan, Ş. Ç., & Kocadere, S. A. (2019). The Effects of Flow, Emotional Engagement, and Motivation on Success in a Gamified Online Learning Environment. *Journal of Educational Computing Research*, 0735633118823159.

Pask, A. G. S. (1976). *Conversation Theory: Applications in Education and Epistemology*. Amsterdam: Elsevier.

Rashid, N. A. M., Salleh, S. M., & Noor, N. M. (2018). The Role of Game Elements in Improving Jawi Skills through a Mobile Game 'G-Jawi'. *International Journal of Interactive Mobile Technologies*, *12*(7), 20–30. doi:10.3991/ijim.v12i7.9636

Riconscente, M. M. (2013). Results from a controlled study of the iPad fractions game Motion Math. *Games and Culture*, *8*(4), 186–214. doi:10.1177/1555412013496894

Rosas, R., Nussbaum, M., Cumsille, P., Marianov, V., Correa, M., Flores, P., ... Salinas, M. (2003). Beyond Nintendo: Design and assessment of educational video games for first and second grade students. *Computers & Education*, *40*(1), 71–94. doi:10.1016/S0360-1315(02)00099-4

Schreiner, K. (2008). Digital Games Target Social Change. *IEEE Computer Graphics and Applications*, *28*(1), 12–17. doi:10.1109/MCG.2008.4 PMID:18240782

Sharples, M., Taylor, J., & Vavoula, G. (2005). Towards a theory of mobile learning. *Proceedings of mLearn 2005, 1*(1), 1-9.

Sites, J. D., & Potter, R. F. (2018). Everything Merges with the Game: A Generative Music System Embedded in a Videogame Increases Flow. *Game Studies, 18*(2).

Skinner, B. F. (1968). The Technology of Teaching. New York: Appleton-CenturyCrofts.

Statista. (2019). *Mobile Gaming Industry - Statistics & Facts*. Retrieved from https://www.statista.com/topics/1906/mobile-gaming/

Statista. (2019). *Mobile phone gaming penetration in the United States from 2011 to 2020*. Retrieved from https://www.statista.com/statistics/234649/percentage-of-us-population-that-play-mobile-games/

Statista. (2019). *Number of mobile phone gamers in the United States from 2011 to 2020 (in millions)*. Retrieved from https://www.statista.com/statistics/234635/number-of-mobile-gamers-forecast/

Swain, C. (2007). *Designing Games to Effect Social Change*. DiGRA.

The Office of the United Nations High Commissioner for Refugees. (2012). *Could you survive?* Retrieved from My Life as a Refugee: http://mylifeasarefugee.org/index.html

Tüzün, H., Yılmaz-Soylu, M., Karakuş, T., İnal, Y., & Kızılkaya, G. (2009). The effects of computer games on primary school students' achievement and motivation in geography learning. *Computers & Education*, *52*(1), 68–77. doi:10.1016/j.compedu.2008.06.008

Uther, M. (2019). Mobile Learning—Trends and Practices. *Education in Science*, *9*(1), 33. doi:10.3390/educsci9010033

Webster, J., Trevino, L. K., & Ryan, L. (1994). The dimensionality and correlates of flow in human-computer interactions. *Computers in Human Behavior*, *9*(4), 411–426. doi:10.1016/0747-5632(93)90032-N

Westera, W. (2019). Why and How Serious Games can Become Far More Effective: Accommodating Productive Learning Experiences, Learner Motivation and the Monitoring of Learning Gains. *Journal of Educational Technology & Society*, *22*(1), 59–69.

White, L. T. (2019). *Video games could teach spatial skills lost to a society dependent on devices*. Academic Press.

Whitelock, D., Brna, P., & Holland, S. (1996). What is the value of virtual reality for conceptual learning? Towards a theoretical framework. In P. Brna, A. Paiva & J. A. Self (Eds), *Proceedings of the European Conference on artificial intelligence in education (EuroAIE)* (pp. 136–141). Lisbon: Edicões Colibri.

Yang, X., Kuo, L. J., Ji, X., & McTigue, E. (2018). A critical examination of the relationship among research, theory, and practice: Technology and reading instruction. *Computers & Education*, *125*, 62–73. doi:10.1016/j.compedu.2018.03.009

Yang, Y. T. C. (2012). Building virtual cities, inspiring intelligent citizens: Digital games for developing students' problem solving and learning motivation. *Computers & Education*, *59*(2), 365–377. doi:10.1016/j.compedu.2012.01.012

Chapter 9
Machitia:
An Educator-Focused Liberation Platform for Education

Joél-Léhi Organista
Columbia University, USA

ABSTRACT

Machitia is an educator-focused mobile app prototype where educators create, collaborate, and share lesson plans. These lesson plans embed the following liberating and transformative theoretical frameworks and pedagogies called in this chapter "circles of liberation": (1) Dis/ability critical race theory (DisCrit), (2) biliteracy, (3) culturally sustaining pedagogy, (4) radical healing, (5) critical pedagogy, (6) proficiency-based learning, (7) queer theory, and (8) decolonizing theory. After introducing those frameworks, a mapping of currently existing educator-focused platforms prelude the review of mobile technology theoretical frameworks Machitia's design incorporates. Then, the discussion turns to how all the circles of liberation and mobile technology theoretical frameworks manifest as features within Machitia. By the end of the chapter, learners and educators will have a sense of the various possibilities of, and the need for, an education-focused liberation platform.

INTRODUCTION

The word "machitia" in the Mexican indigenous Nahuatl language means, "to teach someone something". By starting with centering an indigenous language and word, this chapter seeks to decolonize the intersectionality and possibilities of lesson plans, pedagogies, mobile technology, transformation, as well as liberation. At the

DOI: 10.4018/978-1-5225-9351-5.ch009

core of Machitia is focus on an education for liberation platform, where educators create, collaborate, and share lesson plans. Through this core task, educators are empowered and enabled to further their own lifelong learning, facilitate peer to peer constructive criticism, labor alongside each other in the education justice movements, and, most importantly, build community among the various intersectionalities that are found within the community of educators dedicated to the emancipatory potential in teaching and learning.

The aim of this chapter is to consider that transformative and liberating theoretical frameworks and pedagogies are neither static or monolithic, but fluid and directly connected to each other. Such frameworks and pedagogies must manifest themselves beyond the classroom and actualize within the lifelong learning, teaching of an educator through their lesson plans on mobile technologies, and community building. Just as knowledge is constructed and deconstructed, an educator focused mobile application (app) prototype has been constructed as a platform, in order to invite, engage, and lead how education for liberation is a constant process of transformation of learning and unlearning. The chapter will offer reports from the user testing in addition to the usability testing conducted by using the interactive mobile app prototype.

OVERVIEW OF CIRCLES OF LIBERATION

Prior to any discussion on the specific mobile technology theoretical frameworks informing Machitia's design, it is vital to understand what liberating, transformative theoretical frameworks and pedagogies the author will draw from, as well as how technology can facilitate the process of education for liberation. Since the purpose of Machitia focuses on liberating and transformative lesson plans, it is important to articulate what that means. There are eight theoretical frameworks and pedagogies that have been identified that collectively make up what a liberating and transformative lesson plan has the potential to be. We call these within Machitia, "circles of liberation": (1) Dis/ability Critical Race Theory (DisCrit), (2) Biliteracy (3) Culturally Sustaining Pedagogy, (4) Radical Healing, (5) Critical Pedagogy, (6) Proficiency Based Learning, (7) Queer Theory, and (8) Decolonizing Theory.

Dis/ability Critical Race Theory

DisCrit combines aspects of Critical Race Theory (CRT) and Disability Studies to propose a new theoretical framework that incorporates a dual analysis of race and ability. CRT has at least five tenets that "inform theory, research, pedagogy, curriculum, and policy: (1) the intercentricity of race and racism; (2) the challenge to dominant ideology; (3) the commitment to social justice; (4) the centrality of experiential knowledge; and (5) the utilization of interdisciplinary approaches" (Yosso, 2005, p. 73). While there are different ethnicities, the word racism is used here as it has been purveyed as a trope in society, even though scientific fact indicates that there is only one race. The word racism in this chapter is used to refer to forms of bias and discrimination experienced by people because of their ethnicity.

While there are seven tenets of DisCrit:

1. DisCrit focuses on ways that the forces of racism and ableism circulate interdependently, often in neutralized and invisible ways, to uphold notions of normalcy (Annamma, Connor, & Ferri, 2016, p. 11);
2. DisCrit values multidimensional identities and troubles singular notions of identity such as race or dis/ability or class or gender or sexuality, and so on (ibid, p. 11);
3. DisCrit emphasizes the social constructions of race and ability yet recognizes the material as well as the psychological impacts of being labeled as raced or dis/abled, which sets one outside of the western cultural norms (ibid, p. 11);
4. DisCrit privileges voices of marginalized populations, traditionally not acknowledged within research (ibid, p. 11);
5. DisCrit considers legal and historical aspects of dis/ability and race, additionally how both have been used separately and together to deny the rights of some [individuals] (ibid, p. 11);
6. DisCrit recognizes whiteness and Ability as Property and that gains for people labeled with dis/abilities have largely been made as the result of interest convergence of white, middle-class citizens (ibid, p. 11);
7. DisCrit requires activism and supports all forms of resistance (ibid, 2016, p. 11).

Biliteracy

Biliterate, bilingual, and bicultural lessons focus on using two-dimensional practices when teaching language. The language experience approach can be one of many approaches to reading, writing, listening, and speaking naturally. Biliterate education should be incorporated within any and all lesson plans to aid knowledge construction and deconstruction of tropes. Machitia will provide the starting place of resources to help educators use "evidence-based strategies for fostering biliteracy in any classroom" (Delbridge & Helman, 2015) which will aid sustaining culturally sensitive pedagogy.

Culturally Sustaining Pedagogy

Culturally Sustaining Pedagogy (CSP) builds on the literature of culturally relevant teaching and culturally responsive pedagogy. Culturally relevant teaching has three components: (a) the teachers' conceptions of themselves and others, (b) the way classroom social interactions are structured, and (c) teachers' conception of knowledge (Thompson, 2014). Then, culturally responsive pedagogy describes educators who: (1) are socio-culturally conscious, (2) are favorably disposed to diversity, (3) see themselves as cultural brokers in educational institutions, (4) understand and embrace constructivist as well as deconstructivist views of knowledge, teaching, and learning, (5) know about the lives of their students, (6) design instruction to draw on students' strengths and to address their needs (Thompson, 2014).

All these foundations come together and are expanded to become CSP which "seeks to perpetuate and foster—to sustain—linguistic, literate, and cultural pluralism as part of the democratic project of schooling" (Paris, 2012, p. 93). Shifting from culturally responsive to culturally sustaining, allows the focus to not be on the "translation of schooling into culturally responsive materials for the purpose of achievement, but positions education as the vehicle for cultivating, celebrating and sustaining cultural knowledges that have otherwise been targeted for extinction" (Smith, Tuck, & Yang, 2019. p. 16) which could also lead to radical healing.

Radical Healing

Radical Healing is a lens that is critical and liberating in addressing both ourselves as educators/learners, and those we are experiencing a formal educational lesson. While Social Emotional Learning (SEL) has become a foundational theory that has been established with educators, SEL is still in the process of providing all children and adolescents with the opportunities to learn, acquire and practice the social-emotional competencies needed to succeed in life. Its components are: 1) Self-awareness, 2) Self-management, 3) Social awareness, 4) Relationship skills, 5) Responsible decision-making skills (Oberle, Domitrovich, Meyers, Weissberg, 2016). Radical Healing Framework (RHF) challenges both SEL and Trauma-sensitive practices in a critical lens through a five framework: Culture, Agency, Relationships, Meaning, and Achievement, referred to as "CARMA" (Ginwright, 2016). CARMA also includes identity exploration, action, social connectedness, understanding positionality and power. Students experience trauma within and outside of school, for which trauma sensitive awareness and practices could help.

Trauma-sensitive practices and learning environment are about a classroom, school, or any other teaching location in which each student is healthy, safe, engaged, supported, and challenged. The trauma-sensitive strategies toward practice are organized around four primary themes: self-awareness, relationship, belief, and "live, laugh, love" (Morgan, Pendergast, Brown, Heck, 2015). All of the aforementioned, in many school contexts, still require a social transformation as the schools act as a microcosm of following social structures that can be traumatic.

Critical Pedagogy

Critical Pedagogy is both the theory and practice of social transformation that connects education to social justice. Critical pedagogy became popular because of Paulo Freire's book, Pedagogy of the Oppressed (1970). His book calls for problem-posing education which is education for freedom, and it emphasizes that both teachers/educators and students/learners must see each other in a partnership "in the fight for their own liberation" (p. 53). He identifies five stages in the process that leads to action and reflection on that action: (1) Identify a problem, (2) Analyze the problem, (3) Create a plan of action to address the problem, (4) Implement the plan of action, (5) Analyze and evaluate the action. Liberatory pedagogy roots in understanding that knowledge is not a product, but a process that can be utilized to build proficiencies. Proficiency based learning counters the deficit and dehumanizing manner lesson plans as well as pedagogies can be operationalized if they focus solely on memorization, performance, and evaluations. Sometimes termed as competency or mastery-based learning, proficiency on a content and the application of knowledge should be aligned with the rest of the liberation and transformation pedagogies. Once this happens the learner is being empowered to take what they are learning outside the classroom and into their lives.

Queer Theory

Queer Theory centers around "anyone who self-identifies as a member of the LGBTQAI+ community. That includes but is not limited to lesbian, gays, bisexual, transgender, agender, asexual, pansexual, intersex, genderqueer, aromantic, androgynous, bigender, bicurious, androsexual, gynosexual, demisexual, polyamorous, genderfluid, and questioning individuals" (Berila, 2016, p. 6). It challenges heteropatriarchy and gendernormativity.

It is important to keep in mind how one learns about oppression and its role towards deconstructing and dismantling systems of oppression. Anti-oppressive pedagogies like all the circle of liberation recognize knowledge production as a political practice where power dynamics are always in play. There is an intersectionality between other anti-oppressive pedagogies, including contemplative pedagogy and a feminist queer pedagogy, where they recognize "that knowledge is contextual, a focus on the generative nature of teaching practices, an embrace of multiple perspectives, and a recognition that the 'life of the mind' is inevitably connected to our bodies" (Berila, 2016, p. 8). Queer theory should intersect with other circles of liberation. An example of this is how queer anti-oppressive pedagogy and mindfulness practices within radical healing allow for liberatory and transformative experiences (Berila, 2016).

Decolonizing Theory

Decolonizing Theory is "informed by Indigenous theory, history, epistemology, and futurity…[it] emphasize[s] the ways that colonization and decolonization are time-specific and land-specific" (Smith, Tuck, & Yang, 2019. p. 12). Decolonizing the mind reminds us that "coloniality shapes and severs human and non-human relationships across land, nation-state, water and time" and intersects with social justice education that contributes to "broad implications for pedagogy, curriculum, schooling, education policy, and social movements" (p. 13). Three demonstrations of decolonizing theory include (1) Red Pedagogy, (2) Decolonial Participatory Action Research, and (3) Culturally Sustaining Pedagogy for Indigenous Students.

Red pedagogy draws upon critical pedagogy becoming a pedagogy "that attends to decolonization in its material politics in order to recognize and nurture Indigenous practices of present and future changes" (Smith, Tuck, & Yang, 2019. p. 15). Decolonial participatory action research attempts to accomplish ground- and grassroots-knowledge production and most importantly, "draws limits to this territory by refusing to hand over anything and everything to the academic enterprise, by drawing attention to power's code of ethics" (ibid., p. 16). Culturally sustaining pedagogy for Indigenous students can manifest in endless ways. Some examples of this and how intersectionality with the other circles of liberation include: "literacies of Land" (p. 39), "paddling as pedagogy" (ibid., p. 40), place-based knowledges engaging "Xicanx epistemology that engages Indigenous as well as attempting to decolonizing pasts, presents, futures" (p. 40). Additionally the "fluidity of oceanic boundaries and movements to unsettle the naturalization of land-as-territory with stable landscapes readily available for cartographic borders" (ibid.,p. 41), "Indigenous educational institution-building beyond the parameters of nation-state" (ibid., p. 41), "Queering Indigenous Education" that clarifies the "inseparability of land sovereignty and body sovereignty" and "restoring traditional understanding of bodies to land, fluidity and tradition become complementary" (ibid., p. 42), "community ideologies of disability" (ibid., p. 44) and reshaping them, "incorporating elder input, science input, and policy input into climate change problem-solving" (ibid., p. 44), and "remembering of language as a restorative practice to counter the dismemberment or disembodiment enacted by residential schools and the reserve system...pedagogies of land-based language education that triangulate stories, ontologies, and place with Indigenous language" (ibid., p. 44).

Figure 1. A Diagrammatic representation of Intersectionalities showing decolonizing theory and the other seven circles of liberation

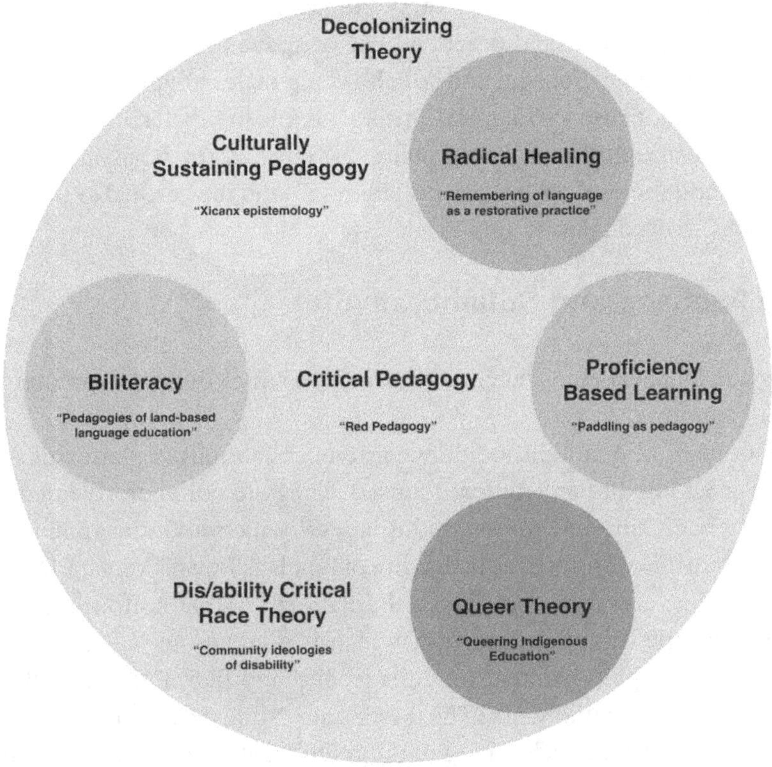

Practicing decolonizing theory in our teaching and learning should naturally be intersectional with the other circles of liberation as figure 1 shows. Decolonizing is the:

acknowledgment that lands and waters are polluted everywhere, and so to treat water as 'pure' and sacred and others as impure is against Indigenous callings for defending the water. Likewise, to treat normative bodies as capable and pure enough to be defenders of the waters, and others as too ill, to contaminated, too disconnected, is a form of ableism and queer-phobia. To meet lands and peoples where they're at is to engage with waters/bodies as they are...these practices may not be free of colonial contamination, but they are practices of harm reduction, of healing (Smith, Tuck, & Yang, 2019. p. 45).

MAPPING EDUCATORS FOCUSED PLATFORMS

Imagine an educator named Terry. Terry creates amazing lesson plans, but does not have any easy, simple, and practical way or space to share and collaborate on lesson plans they create. Sharing and collaborating on lesson plans should not be so frustrating. Educators use various platforms available to fulfill their need of finding, sharing, and collaborating on lesson plans. In figure 2, the horizontal axis ranges from "hard to collaborate" to "easy to collaborate", and the vertical axis ranges from "student focus" to "educator focus".

Student Focused and Collaborative

Surveying the landscape of educator focused apps, highlights the greater need for the Machitia app.

The bottom right quadrant includes various collaborative platforms educators use in their teaching, but are student focused. They are collaborative platforms we could call "whiteboard and screencasting apps," which allow anyone in that live space to insert different multimedia documents such as PowerPoints, PDF, pictures, audio, video files, and drawings. The only platform in that bottom right quadrant that is not that type of app is Nearpod, which is a "ready-to-teach K-12 lessons" interactive learning platform that requires technology like iPads and other tablets for every student, in order to work. The lessons are "slideshows" that are interactive, and capture data on student's learning. It is easy to create and share "ready-to-teach" lessons, but they are not lesson plans and it is not focused on educators' collaboration.

Figure 2. Graph of educator focused platforms

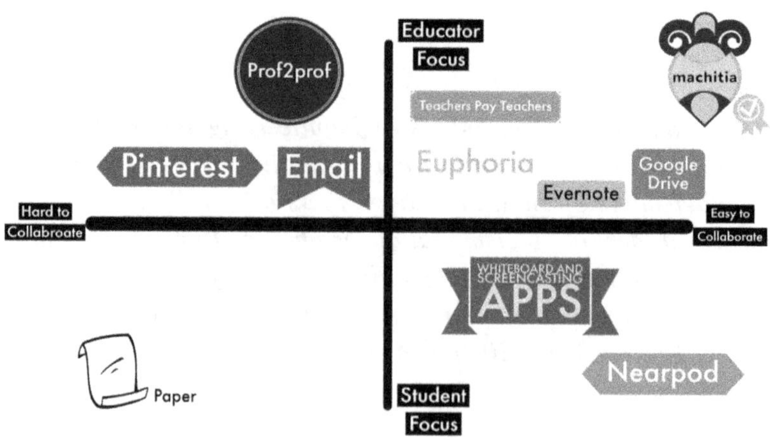

Student Focused and Not Collaborative

The bottom left quadrant is paper, which is not a digital platform, but important to put into perspective what educators use for both themselves and youth learning. Paper is used for creating lesson plans and is sometimes easy to share, however it is never ideal for collaboration.

Educator Focused and Not Collaborative

The top left quadrant includes platforms that millions of educators use to share lesson plans and resources, but sharing is not collaboration. Emailing someone your lesson plan is not a collaborative act, but is nonetheless a real interaction that happens often between educators seeking to help other educators and educators seeking to learn from other educators. Prof2Prof is a "professor-centric" platform where mainly higher education educators share syllabus and reading recommendations. Again, the focus here is sharing and not collaborating.

Pinterest is the last major platform in top left quadrant focused on educators sharing, over educators collaborating. In an article published in 2015 titled "There's a big hole in how teachers build skills, and Pinterest is helping fill it", the author quotes educators stating "often, in-person [professional development] does not offer strategies we can immediately use in our classroom… Pinterest does" (Cummings). The article positions Pinterest as the main platform educators have adopted for sharing images of lesson plans. Pinterest usage has risen steadily over the years, yet educators have become a vital group for the platform to the point that Pinterest gave educators their own "hub" on the site (Cummings, 2015). That hub includes over 30 boards and more than 163,000 followers. Educators do not need to be following that hub or boards to engage in their own created boards. Cummings shares that "education pins per day climbed from 500,000 to about 1.3 million in [2015]...popular lesson plans on Pinterest can be posted and reposted thousands of times" (2015). Posting and reposting pins are all behaviors of sharing, but not collaborating.

Educator Focused and Collaborative

The last quadrant in the figure 2 is the top right one, which includes platforms that vary in how educators focus on educator collaboration. Eduphoria is the most comprehensive tool in this quadrant. It has three integrated apps that allow for lesson planning, monitoring student progress in assessments, and provides a collaborative platform for educators and administrators. "Forethought" is the main app within Eduphoria that allows for adaptable, responsive, intertwined, team focused lesson plan creation, sharing, and collaboration. The only very real drawback of this comprehensive platform is that it is expensive, only school/district wide licensing, and the collaboration can only happen within that paying school/district, which excludes everyone without the licensing of a school/district.

Teachers Pay Teachers (TpT) is popular and known educator focused platform. Just as the name suggests, educators create lesson plans, materials, and resources available for free or for sell. TpT claims that more than 2 out of 3 educators in the US have used a resource, that over 5 million educators used TpT in the last year, and that over 1 billion resources have been downloaded from TpT (About us, 2019). Content for sell range from zero dollars to hundreds of dollars depending on the amount of resources provided within a bundle. It is common practice for new educators to spend their own money in buying resources not only from TpT, but wherever they can find what they need. The main reason TpT is on the quadrant for collaboration is because educators can rank and post comments on the quality of the resources. The majority of the comments are constructive and positive, but they are still limited because educators cannot truly collaborate with the created lesson plan on sale. They would either need to buy it, adapt it, and make it clear what they improved on by themselves, and post it on TpT again. The thread of the original lesson plan will not be officially or easily correlated to the new lesson plan posted. Therefore TpT is closely on the border leading to "hard to collaborate".

The two other platforms in this educator focused collaborative quadrant are well known platforms for non-educators as well. Evernote and Google Drive use a cloud space for documents that can be easily created, shared, and collaborated on. It is simple to control who has access to view, comment, and edit a document created within those two platforms. The documents allow for live collaboration across devices and users. Both are not initially educator focused, but educators have adopted these and other "cloud spaces" or "virtual flash drives" to share and collaborate on lesson plans.

Educators in the ethnic studies movement use Google Drive, but have come across the stress of how this tool has no frameworks specific to enhancing, reminding, and reinforcing liberating, and transformative pedagogies. It is a great tool, but it was neither built by educators for educators. Organizations have also come across the lack of organizing their resources they produce and communicating within their own members how their creation of resources, in this circumstance that being lesson plans, should have certain baselines of frameworks and pedagogies. A great example of this are summer camps that create lesson plans for their activities, but sometimes forget to keep in mind how they can use a platform for sustainability. They reinvent the wheel every year because they forgot to create a digital copy in a platform that allows for easy sharing and collaborating. Additionally, educational consultants currently only have cloud spaces and email as a place for providing specific and constructive feedback to their clients like individual teachers and sometimes whole school districts. There is not an exclusive platform that is specific to educators supporting their continual growth, which is necessarily accessible, affordable, and more importantly, rooted in applying transformative and liberating frameworks.

Machitia is on the furthest top right corner of figure 2 because it strives to be educator focused and streamline collaboration to be an experience that is not only easy, but a practice of transformation and liberation. This is possible by understanding and designing the platform from theoretical frameworks that solve the gaps the platforms mapped lack and enhance the nature of the circles of liberation.

MOBILE FRAMEWORK

There are six broad mobile technology theoretical frameworks informing Machitia's design. Those frameworks are: (1) Constructivist Theory, (2) Collaborative Theory, (3) Social Connectedness Theory, (4) Behaviorist Theory, (5) Learning and Teaching Support Theory, and (6) Informal and Lifelong Theory. When describing the following theories, the "learner" in this context is the users, usually k-12 educators because they are the ones that would use the app and are "learners" engaging in activities within the app like creating lesson plans.

Constructivist Theory

The first theory is the Constructivist theory that centers on "activities in which learners actively construct new ideas or concepts based on both their previous and current knowledge" (Naismith, 2006, p. 2). This happens through the core task of educators applying their own knowledge on constructing a lesson plan in their private space in the app from "scratch" or building off previous lessons. Constructing lessons is the core task in the app and is the center of where the rest of the interactions, activities, engagement, and learning stem branch out from.

In figure 4 you can see constructivist theory with units or clusters of lesson plans the educator creates having a space to articulate the narrative of how the lesson plans come together and build upon each other. The circles of liberation will also be visible on both the label area of the lesson plans and the units. Similarly, to video playlists or music playlists on other platforms; Machitia allows the user to create shelves within their library of any combination of lesson plans they create, and others create that are available to the public. No matter what is created in Machitia, the credit goes to the creator.

Figure 3. Intersectionalities of mobile technology theoretical frameworks and features in Machitia

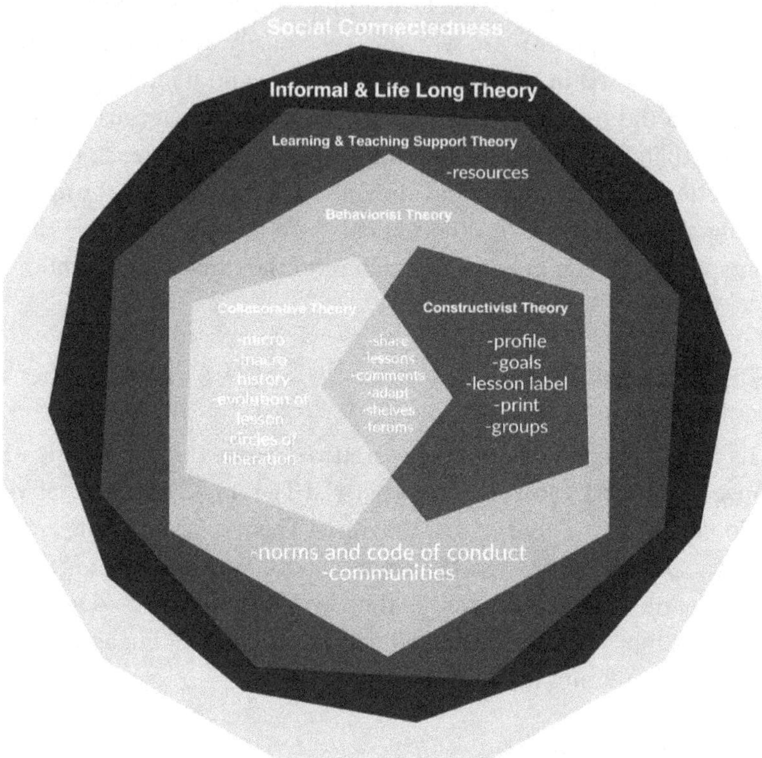

In figure 5 it shows the creator can easily choose which options they want to allow for viewing, sharing, commenting, and collaborating for their lesson plans, units, and shelves. Comparable to the Snapchat feature which notifies the user if someone screenshot a picture they sent or uploaded on their story; Machitia will notify users anytime someone duplicates your lesson plan in order to adapt it. This is of course if the user allows it to be duplicated. An icon will show up on any duplicated lesson plan linking it to the original. Lesson plans will have a couple of prepared templates users can choose from or create their own. Constructing theory is central to the core task of creating anything in Machitia.

Collaborative Theory

The second board theory is Collaborative theory, which is "activities that promote learning through social interaction" (Naismith, 2006, p. 3). This comes to play once the educator decides to change the private setting of the lesson to a shared and collaborative setting. This settings of who can view, comment, and edit the lesson, in order to enable the peer-to-peer educator collaboration can be adjusted at anytime.

Figure 6 demonstrates how the platform is unique and like nothing that is out there, because it allows that the collaboration to happen in the micro and macro level. From the micro level of commenting on any element of the lesson specifically, to the macro level of creating groups, and communities working on whole units, lessons, and curricula. Users can upload images and videos on any aspect of the lesson plan, including at the end where you can showcase final products of what happened in the lesson after it was taught. This allows for a collage of examples of possible outcomes in result to teaching that lesson. Through any level of collaboration, it is easy to see the history and evolution of a lesson plan updating its content because of the positively constructive feedback. The unique element of Machitia compared to the other platforms that have been mapped out is the way collaborative theory is designed in the mobile technology.

Social Connectedness Theory

The collaborative practice also creates a community of practice with elements of social connectedness highly present (Mentor, 2016). The sense of affiliation as educators share lessons went beyond just saving time and effort, but created virtual communities of practice who share the same work requirements. Additionally, educators can now also experience social connectedness through the app's various virtual engagement media, such as from the online social networking channel and instant messaging options in the app. Similarly, as with online social media, the app

Figure 4. Menu screen, My Lessons screen, and Home screen in Machitia

provides the user with a connected presence in addition to on-the-go accessibility (Mentor, 2018).

The mobile phone affords quick virtual check-ins with family and friends. In short, the consequences of using and interacting with, and on the app, offers a new, professional, interpersonal means to stay in contact (Mentor, 2016) while feeling socio-emotionally supported in a job that can feel devoid of community support. The previous examples provided hinge on the tacit existence of the cultivation of intended and unintended social connectedness. the latter signify the production and value of the successful maintenance of reciprocal professional relationships and support. In short, the app offers a means to relate to networks of like-minded contacts, thus enhancing feelings of socio-emotional support, especially important in an educational environment.

Behaviorist Theory

Behaviorist theory is the fourth framework, which is "activities that promote learning as a change in learners' observable actions" (Naismith, 2006, p. 2). This is manifested when the creator of the lesson plan is engaging with the comments and feedback on their lesson plan which "provides the reinforcement" needed to update the lesson plan. Likewise, the educator can comment on any public lesson plans to provide the same "stimulus" for the creator of that lesson plan, in turn promoting that behavior of changing content, pedagogy, and approaches possible in future implementation of a lesson plan.

Figure 5. My Library screen, lesson label features, create lesson screen in Machitia

Machitia has ways the educator can articulate tips of applying the circles of liberation to the specific lesson plan being constructed built into the lesson plan structure as seen in figure 6. This is another activity that promotes learning for observable actions, which in this audience is the pedagogy of educators in the classroom and outside the app. Within the terms and conditions of Machitia, it includes norms and a code of conduct that brings together all the circles of liberation. Users must follow that in order to be allowed on the platform. Users can be reported by other users on specific norms that have been violated. If found that the report is true, the reported user will have restorative practices as options to continue their learning and be permitted back to the space they have been removed from.

Learning and Teaching Support Theory

As seen in figure 3 the fifth theory builds off the past three, because it highlights the way Machitia is the first platform to provide a space for educators to apply within the app what they learn from resources shared within the community. Learning and Teaching Support theory is understood as "activities that assist in the coordination of learners as well as resources for learning activities" (Naismith, 2006, p. 4). This comes in the form of both the collaborative direct feedback as comments within the public lessons plans, as well as the peer led creation of spaces, communities and forums. Additionally outside the lesson plans, the provision of resources on anything educators want to coordinate around. Examples of topics of resources educators can

Figure 6. Circles of Liberation, comment section, and create lesson sections in Machitia

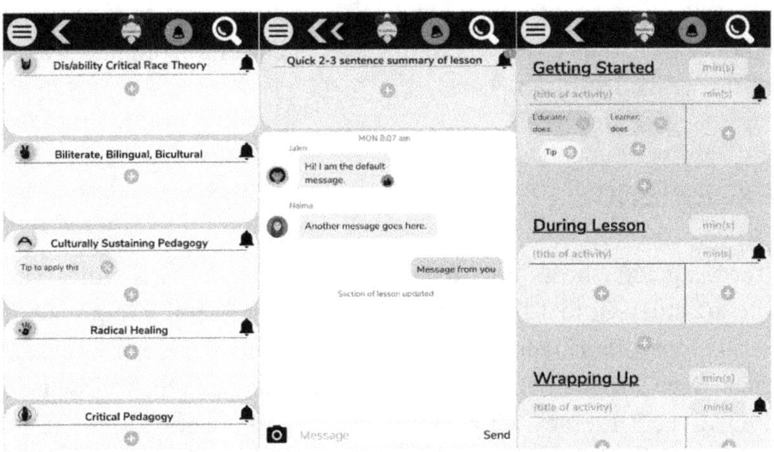

collaborate around are the eight circles of liberation, classroom decor, wellness, parent engagement, trainings, events, and activism.

One feature that educators as user testers have requested is having a one-click printing feature, which allows them to have all their lesson plans to be easily accessible on paper, digitally, and offline. Like in other mobile technologies, there will be bi-monthly challenges to grow and learn, which is mainly to expose users to content, frameworks, pedagogies, resources, and practices they might otherwise not come across.

Informal and Lifelong Theory

The last broad mobile technology theory brings together all of the theories previously mentioned as demonstrated in figure 3. The Informal and Lifelong theory are "activities that support learning outside a dedicated learning environment as well as formal curriculum" (Naismith, 2006, p. 4). As mentioned in the brief review, through the core task of creating, sharing, and collaborating on lesson plans, educators are empowered then enabled to further their own lifelong learning. While facilitating peer to peer constructive criticism, and most importantly build community among the various intersectionalities that are found within the community of educators dedicated to the emancipatory potential in learning.

Since the platform is targeted and catered to educators who are on there to improve the own teaching, which requires being a lifelong learner themselves; Machitia, has endless opportunities and potential for an educator to become both empowered and a learner as well. Within the user's profile they create, they have the option to articulate the level they know, understand, and can apply all the circles of liberation within their pedagogy and interactions. This allows for others to engage in empathic ways to any comments, feedback, questions, and engagement of a user because others can actually see where the user is coming from and wanting to go in terms of their own learning.

USER EXPERIENCE

Imagine Terry again, who is now using Machitia for the first time. They create their profile articulating that they are gender non-conforming, high school health teacher, Latinx, polyglot fluent in Spanish, English, Japanese, Portuguese, Nahuatl, and American Sign Language (ASL). Terry is well versed in the following circles of liberation: DisCrit, Biliteracy, Culturally Sustaining Pedagogy, Critical Pedagogy, Proficiency Based Learning, Queer Theory, and Decolonizing Theory. However,

Terry is not proficient in Radical Healing Framework. Terry can make that known on their profile page and say that they are still new to radical healing framework, but is available as a resource and support for the other circles of liberation.

The first thing Terry does is create a lesson plan on Machitia after creating a profile. The lesson plan is titled “Sleep”, which is a lesson in the unit of physical health. On the third image in figure 5, it shows the layout where Terry can start to fill out the label of the lesson plan with information on the grade, subject, and length. Additionally, Terry fills out the quick summary of the lesson and tags, so others can find the lesson once Terry makes the lesson public or shares it to groups and individuals. Afterwards, Terry adds what the objectives are and any preparation needed for the lesson.

The third image in figure 6 shows how Terry can break down the lesson plan with a beginning part titled “Getting Started”, middle part called “During Lesson”, and ending titled “Wrapping Up”. Terry adds in the Getting Started section an activity that lasts 5 minutes long titled “Warm Up”. At the beginning of every lesson Terry always has a warm up ready on the whiteboard for students to complete, in order to get them started in thinking about the topic or reviewing what was taught before. Terry can continue to add as many activities in the Getting Started section and add attachments, go into detail about what the educator does, the learner does, and any tips to applying any of the circles of liberation. The During Lesson section is the core of the lesson plan. Similar to the beginning section, Terry can be as detailed as they want on how the plan is written up. The Wrapping Up section is where Terry can articulate how the lesson plan can end.

Once Terry has finished everything, they want to create by themselves, they can invite specific individuals and groups to collaborate in viewing, commenting, and editing the lesson plan. Terry has an educator friend named Joshua that also teaches health, but in another state. Terry adds Joshua as a collaborator who can edit and comment on the lesson plan. Joshua starts by commenting on the lesson plan on sections he really likes. He notices that the worksheet Terry uploaded to an activity is outdated and he recently created a new worksheet for that exact activity. Joshua comments that information and uploads the more worksheet he created. Once Terry allows the lesson plan to be public to either the whole Machitia platform or just to certain groups, educators can see all the comments, updates, and history of the lesson plan. This includes seeing the options of the original worksheet Terry uploaded and the newer version by Joshua.

Terry decides that both have created the lesson plan to a level they feel comfortable to share with a group of health teachers they are a part of. Terry makes the lesson plan public with that group. Farima is part of that group and is able to comment on the lesson plan. She starts with providing feedback on potential new tags that can be added, in order to make it easier for other educators to find this lesson plan. Farima is also well versed in Radical Healing Framework and noticed in Terry's profile that he is seeking to learn how to apply this in their pedagogy. Farima's comments focus on suggestions as to how Terry can apply Radical Healing in the lesson plan. Terry then changes Farima from a view and comment only collaborator to now also be able to edit the lesson plan. Farima is now free to directly add the tips to the lesson plan, rather than just comment about them in the comment sections.

Throughout the school year, Terry receives notifications on comments from educators all over the country on how great the lesson plan is, how they have implemented it, and images of results from the activities in the lesson plan educators upload. Terry allowed the lesson to be duplicated. This means when an educator like Farima who now teaches pre-school, can duplicate Terry's original lesson plan and adapt it. Farima does this because she feels lots of the content already there can just be adapted for younger kids and does not want to type everything again. There is an icon on this duplicated lesson plan that links it to the original lesson plan. The same collaborating process happens with this new lesson plan. The creators and owners of a lesson plan can turn off the features for others to view, comment, edit, share, and duplicate a lesson plan. Every user can also control the amount, pace, and type of notifications they receive on a lesson plan, comment section, direct message, group message, and page. Terry, Joshua, and Farima are just examples of creators, community, users, and educators that can be on Machitia. There are endless opportunities and ways to create, share, and collaborate on Machitia.

CONCLUSION

Reflecting on the mapping of educator focused platforms and their degree of allowing accessible collaboration, it becomes clear that there is a need for Machitia. There is an urgency for a space where knowledge, resources, best practices, and ideas rooted within transformative and liberating frameworks and pedagogies can be created, collaborated on, and shared. Machitia embodies the circles of liberation in the space it constructs. It goes beyond lesson plans, but into the actualization of how community building happens in a liberating and transformative manner. Machitia is truly the only educator focused platform that provides the space for organic and constructive collaboration. By centering educators, the circles of liberation, and the mobile technology theoretical frameworks, Machitia is a unique platform that enables education for liberation and transformative learning for educators and learners.

FUTURE RESEARCH

The future and possibilities of Machitia are endless. The more diversity there is of educators on Machitia that have different intersectionalities of identities, lived experiences, subject matter expertises, and age group focus, the more empowering the platform becomes. All of which requires various research approaches as well as short and long term research foci. Machitia enables both informal and lifelong theory and learning and teaching support theory to the educator, focused through a new central space and platform for educator empowerment and learning.

Through user feedback and engagement, Machitia will improve as constructivist, collaborative, social connectedness, behaviorist, learning and teaching support, as well as informal and lifelong theories manifest in the platform. This includes, but is not limited to templates of lesson plans, showcasing new circles of liberation, moderating comment sections, communities, and forums. Furthermore, what is highlighted in your profile, new ways to collaborate, and using restorative practices in addressing any harm caused between users will also be addressed. In the future it will be important to explore how to take advantage of in-person gatherings of platform users, in order to bridge the gap between the digital space and the physical plane to support and enhance teaching, learning and transformative experiences happening through Machitia. Every educational space should facilitate as many circles of liberation no matter if it is a mobile platform, an actual in person space, as well as the digital world we traverse.

Future research on how the platforms mapped in figure 2 can be enhanced by bringing both the circles of liberation and mobile theoretical frameworks even closer, is another possible important step, since the educators using Machitia might use those platforms as well. In this regard, Machitia will explore the Learning Tools Interoperability (LTI) framework to provide users the ability to seamlessly connect between those platforms and Machitia. The research, theoretical frameworks, and practices in Machitia are applicable to the application of teaching and learning spaces, opening up additional spaces for future research. For example, using the data analytics from the app, a quantitative approach could help the developers of the app, as well as the platform users if shared responsibly while respecting FERPA, PIP, and individuals' privacy. While a qualitative approach could also help gather feedback for enhancements of the app, or input or sharing from platform users in specific districts or distally located, regarding specific lesson plans, projects or subjects. A mixed method approach of various research tools, could obviously also add a means of triangulating or crystalizing the data. There is both an invitation and challenge to implement what this chapter shares in such both physical and digital spaces and platforms, with much more research to track, monitor and evaluate the success of the app as well as the goals and objectives of what it sets out to achieve.

REFERENCES

About Us. (2019). Retrieved March 5, 2019, from https://www.teacherspayteachers.com/About-Us

Annamma, S. A., Connor, D. J., & Ferri, B. A. (2016). Disability critical race studies (DisCrit): Theorizing at the intersections of race and dis/ability. In J. D. Connor, B. A. Ferri, & S. A. Annamma (Eds.), *DisCrit: Disability studies and critical race theory in education*. New York: Teachers College Press.

Berila, B. (2016). Mindfulness as a healing, liberatory practice in queer anti-oppression pedagogy. *Social Alternatives*, *35*, 5–10.

Cummings, M. (2015, April 2). *There's a big hole in how teachers build skills, and Pinterest is helping fill it*. Retrieved from https://slate.com/human-interest/2015/04/pinterest-and-teachers-how-the-site-is-filling-a-gap-in-teacher-training.html

Delbridge, A., & Helman, L. (2015). *Evidence-Based Strategies for Fostering Biliteracy in Any Classroom*. New York: Springer Science & Business Media.

Freire, P. (1970). *Pedagogy of the Oppressed*. Academic Press.

Ginwright, S. (2016). *Hope and Healing in Urban Education: How Urban Activist and Teachers are Reclaiming Matters of the Heart*. New York, NY: Routledge.

Mentor, D. (2016). EMxC3= e&mLearning Cultivating Connected Communities: Sustainable Workforce Talent Development. In *Handbook of Research on Mobile Learning in Contemporary Classrooms* (pp. 240–259). IGI Global. doi:10.4018/978-1-5225-0251-7.ch012

Mentor, D. (2018). Micro to macro social connectedness through mobile phone engagement. In *Encyclopedia of Information Science and Technology* (4th ed.; pp. 6184–6194). IGI Global. doi:10.4018/978-1-5225-2255-3.ch537

Morgan, A., Pendergast, D., Brown, R., & Heck, D. (2015). Relational ways of being an educator: Trauma-informed practice supporting disenfranchised young people. *International Journal of Inclusive Education*, *19*(10), 1037–1051. doi:10.1080/13603116.2015.1035344

Naismith, L. Lonsadale, P., Vavoula, G., & Sharples, M. (2006). *Literature Review in Mobile Technologies and Learning*. Academic Press.

Oberle, E., Domitrovich, C., Meyers, D., & Weissberg, R. (2016). Establishing systemic social and emotional learning approaches in schools: A framework for schoolwide implementation. *Cambridge Journal of Education*, *46*(3), 277–297. doi:10.1080/0305764X.2015.1125450

Paris, D. (2012). Culturally sustaining pedagogy: A needed change in stance, terminology, and practice. *Educational Researcher*, *41*(3), 93–97. doi:10.3102/0013189X12441244

Smith, L. T., Tuck, E., & Yang, K. W. (2019). Indigenous and decolonizing studies in education: Mapping the long view. New York, NY: Routledge, an imprint of the Taylor & Francis Group.

Thompson, S. (2014). Encyclopedia of Diversity and Social Justice. Academic Press.

Yosso, T. (2005). Whose culture has capital? A critical race theory discussion of community cultural wealth. *Race, Ethnicity and Education*, *8*(1), 69–91. doi:10.1080/1361332052000341006

Zembylas, M., & Keet, A. (2018). *Critical human rights, citizenship, and democracy education: Entanglements and regenerations*. London: Bloomsbury Academic.

KEY TERMS AND DEFINITIONS

Behaviorist Theory: Activities that promote learning as a change in learners' observable actions.

Biliterate: Strengthening bridges between languages in their reading, writing, listening, and oracy.

Collaborative Theory: Activities that promote learning through social interaction.

Constructivist Theory: Activities in which learners actively construct new ideas or concepts based on both their previous and current knowledge.

Critical Pedagogy: A collective process that utilizes dialogical learning approaches which are critical of the underlying systems and structures of oppression, systemic in their inquiry into both theory and practice, participatory in involving communities in transformation, and creative in bringing into play cultural productions to re-read society.

Culturally Sustaining Pedagogy: Teaching that perpetuates and fosters linguistic, literate, and cultural pluralism as part of schooling for positive social transformation.

Decolonizing Theory: Indigenous theory, history, cosmologies, axiologies, epistemology, and futurity that is people-specific, time-specific, land-specific, in critiques of settler colonialism, borders, conceptualizations of antiblackness, heteropatriarchy, gendernormativity, and ableism.

Disability Critical Race Theory: Critical race theory and disability studies intersecting in a new theoretical framework that incorporates a dual analysis of race and ability.

Informal and Lifelong Theory: Activities that support learning outside a dedicated learning environment and formal curriculum.

Learning and Teaching Support Theory: Activities that assist in the coordination of learners and resources for learning activities.

Machitia: In the Nahuatl language it means to "teach someone something" and in this chapter refers to the mobile app showcased.

Pedagogy: The art of teaching.

Proficiency-Based Learning: The focus on mastery and being proficient over the deficit and dehumanizing focus on solely memorization, performance, and evaluations.

Queer Theory: A fluid framework of challenging heteropatriarchy, and gendernormativity, and its relation to intersectionalities.

Radical Healing: Interdisciplinary approach to healing bringing in psychological trauma, civic engagement, cultural sustainability, and social justice frameworks.

Social Connectedness: Concept often been used to characterize degrees of interpersonal trust, attachment security, social competency, and a sense of belonging, online and in the real world.

Chapter 10
M-Learning Apps for Film Studies:
An Approach Towards Self-Determined Learning

Jeanny Vaidya
Columbia University, USA

ABSTRACT

While there are many educational apps for traditionally taught subjects such as mathematics and science, more specialized curriculum has largely been left unexplored in terms of m-learning. Film studies, an academic discipline that deals with the theoretical, historical, and critical underpinnings of film, is one such subject that has very few mobile applications. This chapter explores creating a mobile application to teach basic approaches to film interpretation and in addition, considers a heutagogical approach in design. Benefits of m-learning include increased delivery options for multimedia, context-based learning support, and the prospects of a more fulfilling learning experience. This chapter provides direction for implementation and evaluation techniques for an introductory film studies module on film noir, which can be integrated into a mobile format to make film theory more relevant and accessible.

DOI: 10.4018/978-1-5225-9351-5.ch010

M-LEARNING APPLIED TO SPECIALIZED CURRICULUM

As a result of the rapid growth of educational technology, many apps now exist for core subjects like mathematics (math) and science, yet less traditional subjects (media studies and film studies, to name a few) have largely been left unexplored in the field of mobile learning, otherwise known as M-Learning. Interestingly, literacy, despite being a core skill to modern society, is also considerably under-represented. A study on educational apps at Apple's App Store revealed that early learning and STEM (Science, Technology, Engineering and Math) apps are popular, while literacy and art/creativity apps are sparse (Shuler, 2012). Shuler says that this presents an opportunity for developers to fill the gap in the market. The study likely considered literacy based on core reading and writing skills, hence it alludes to an even greater lack of the aforementioned less traditional disciplines related to the literacy umbrella, which may not even be categorically represented in the study. Furthermore, the Shuler study also reported that 80% of apps target children and found that while the percentage of apps made for children were increasing, apps for adults have been in decline. Taken together, there exists a notable absence in apps geared towards less traditional subjects and adult learners. What follows is a preliminary investigation of these gaps in the M-Learning community. In order to examine the potential of M-Learning outside of traditional domains, we will investigate how it might be applied to teaching film studies. Emphasis will be given to demonstrating how M-Learning contributes to the pedagogy of the contents and skills connected to film analysis.

One of the key considerations of this paper is how a heutagogical approach can be integrated into mobile education for film learners. Heutagogy is the "study of self-determined learning" (Hase & Kenyon, 2000). To further clarify, Mulrennan (2018) says "Heutagogy is a net-centric teaching method in which learners are highly autonomous and self-determined." In regard to the app proposed in this paper for film learning, this may mean that students are not only learning to identify film tropes, meanings and genre classifications through a mobile application, but also reflecting on why such classifications exist. Students may come up with their own assessments and content within the application. To make this inquiry more manageable, we will first explore the ways in which an introductory module for film studies, specifically Film Noir, could be designed and integrated into a mobile format. However, it should be known that film studies are but one of countless disciplines and Film Noir merely serves as a sample genre. The app, more broadly, would teach basic approaches to film interpretation, and our central aim is to incorporate M-Learning affordances in order to render film studies more relevant and accessible. This feat could conceivably be achieved for any discipline. A secondary objective of this investigation presents film studies as a necessary part of reflective media studies skills.

Background: M-Learning

Mobile Learning or M-Learning is a relatively new field that emerged as a result of many technological and educational advancements. Crompton (2014) outlines these advancements by noting that the creation of the first web browser, digital camera and graphics calculator in the 1990s subsequently allowed for multimedia and portable digital devices (PDAs) to be utilized in educational settings. Crompton further argues that the introduction of web 2.0 in the new millennium, represented an important shift in the Internet from "read-only" to innovative "read-write" web platforms (p.12). This birthed new forms of learning environments, sparking a revolution in communication and forever altering the way in which we engage with technology. More importantly, it gave users the opportunity to interact with and actively contribute to online content. Soon after came the development and rapid advancement of portable smartphones and tablets with unique capabilities. The infinite knowledge on the Internet and the plethora of mobile applications immensely changed the educational technology landscape (Crompton, 2014). Furthermore, Mentor (2018) says that "mobile learning offers micro-moments of learning in a multitude of multimodal stimulating means cultivating opportunities for transformation."

Modern learning environments are highly digital and have been transformed as a direct result of developments in mobile technology. Yet an underlying issue with mobile devices is that they are generally built for the mass consumer market. This makes it difficult to leverage such technology for teaching and learning purposes. Small screen sizes, assessment of learning outside of classrooms, support for learning across contexts, as well as, conceptual differences between e-learning and m-Learning are just a few of the challenges (Mehdipour & Zerehkafi, 2013). Some research even points to showing that mobile devices in academic contexts are not very favorable as there are difficulties in multitasking while learning as well as the pedagogical challenges in creating instruction (Pedro, de Oliveira Barbosa, das Neves, 2018). However, research and implementation of mobile devices in classrooms and the workplace have shown success at various levels (Mentor, 2016; 2018). Benefits of M-Learning include increased delivery options for multimedia, context-based learning support and the prospects of more fulfilling learning experiences (Mehdipour & Zerehkafi., 2013).

It is important for M-Learning advocates to identify M-Learning's unique advantages over other types of technology enhanced learning. Many compare M-Learning to eLearning and Mass Online Only Courses (MOOCs), because both can leverage collaborative and multimedia capabilities (Waard et al., 2011). M-Learning is also seen as a "new stage" in the development of distance learning and e-learning (Georgiev, Smrikarov & Georgieva, 2004; Mentor, 2018). However, what lends to the distinct nature of mobile technology is, not surprisingly, portability, which allows for contextualized and location-based learning support. It ensures that users can flow between learning spaces without location being a constraint. Historically, M-Learning has been focused on new developments, such as miniaturization of technology (in terms of size and portability). Only more recently has this focus shifted to the mobility of the user for "seamless" learning support (Specht, 2014).

When taking into consideration the mobility of the learner, context-based learning plays an important role. Context can be seen as "the formal or informal setting in which a situation occurs; it can include many aspects or dimensions, such as environment, social activity, goals or tasks of groups and individuals [and] time (year/month/day), personal and social activity, resources, and goals and task structures of groups and individuals." (Brown et al., 2010, p. 4). In M-Learning, context becomes highly valuable as it determines the ways in which users leverage M-Learning affordances. For example, user context can be derived on mobile devices using sensors and other information seeking technology. Cross-context support can also be augmented using cloud-based services, which allow for multiple devices and seamless learning to occur (Wong & Loon, 2011). An example of successful seamless learning can be found in a study promoting mobile journalism among a group of Ecuadorian high school students from New York. Participants collaborated in a filmmaking sessions to talk about issues of discrimination, racism, hunger and bullying, to name a few (Casale and Mentor, 2018). The authors say that "teaching of digital film via hypermedia offers collated means and incorporates the use of online environments for the real-time teaching and learning of films. With the 'inter-web' paradigms shifting from end-users to active participators, the milieu of the web have increased human ICT use and interaction" p. 146. The authors' research is highly valuable in supporting the idea that film learning and teaching can go hand in hand with mobile learning in a technologically rich era.

In sum, changes in technology have led to the rapid development of mobile devices that offer novel learning environments, which can be formal or informal. Learning affordances include higher access to information, personalized, and social learning across time, location and context. Learning is not only context aware, but also augmented, as information can be combined across multiple devices for seamless learning (Specht, 2014).

THE TRANSFORMATIONAL POTENTIAL OF MEDIA

Virtually every topic of inquiry has the possibility of having mobile technology to support learning the topic. With respect to M-Learning efficacy, it is important to understand the motivations underlying a user's decision to use a given app. With that in mind, motivation to learn film theory can come from various sources, whether it be *intrinsic* or *extrinsic*. A simple, but profound motivation for learning how to analyze film, for instance, can come from the notion that cinema has a transformational potential. Film studies often falls under the umbrella of media studies, a concentration that explores the nature and influence of media. Tisdell (2008) describes media literacy, essentially the ability to consume media through a critical and informed lens, to be an important tool for transformative learning in subjects surrounding diversity and equity. Film studies, more specifically, has also been tied to the concept of transformative learning, a term coined by Mezirow (1998), which in essence emphasizes critical reflection. Tisdell (2008) argues, "Media has the power to challenge one's assumptions about structural power relations, both through informal learning, as when one sees a movie or television show for entertainment, as well as through formal learning in a critical media literacy class" (p. 62).

APPROACHES TO FILM STUDIES

One may question why it is necessary to learn film theory; after all, story structure and dialogue are usually deliberately designed such that users can derive meaning without learning film history or jargon. Film analysis using film theory is very different from talking about movies from a simple referential point of view, wherein only plotlines and characters are described as seen in many movie reviews. Jacobs (2012) describes that one approach to film interpretation is to understand that film has both *explicit* and *implicit* content. For example, the "moral of the story" or the "socio-political attitude" can be explicitly communicated through a dialogue in the movie. He further says that *implicit* content is less obvious, but can be inferred from less overt instances, such as how characters grow or develop. They can also point to relationship dynamics that are not explicitly stated by characters.

Audiences can have different approaches to analyzing film. Each individual may see a film differently based on their own personal experiences. Learning film theory provides some fundamental, underlying tools to unpack and interpret films in a manner that helps individuals make coherent sense of content. Learning film theory is also a useful for understanding and predicting the incorporation of tropes in newer films. Holmberg, Moore & Peters (2007) say "scholarly theories imply a systematic ordering of ideas about the phenomena of our field of inquiry and are usually of two kinds. One is concerned with understanding, the other with explanation and prediction" (p. 429). Below are some of the major approaches to film analysis that should be considered when talking about film.

Film Theories: Structuralism, Realism and Contextualism

One of the major approaches to film analysis is *structuralism*. It argues that film can be analyzed using key structural elements (Jacobs, 2013). Similarly, *formalism*, also falls under *structuralism*, and looks at the ways in which meaning is created from structure and form. For example, this means that narrative analysis is presented through formal consideration of elements such as character, setting, story structure, editing and camera movement (Jacobs, 2013).

Another popular approach to film analysis is *realism*, which looks at how reality is portrayed in the movie. For example, sometimes filmmakers are more focused on creating certain moods and emotions (Jacobs, 2013). In essence, they use cinematic techniques to more accurately portray reality. The realist approach still considers formalist theory such as editing, mise en scène, setting and characters (Jacobs, 2013).

Film can also be examined through a contextual lens. According to Jacobs (2013), a *contextual* approach considers film in a broader context and examines a wide variety of other approaches including *cultural*, *psychological*, *dualist*, *Marxist* and *generic* theories. For example, a *cultural* approach takes into consideration the time, culture and place that the movie was made. A *psychological* approach explores sexual symbolism and representations of the subconscious, ego and superego. A *dualist* approach looks at opposing dynamics (for instance, male vs. female, good vs. evil, rural vs. urban, to name a few). *Marxist* theory considers the class struggles of characters and events as well as contexts of an oppressive government. A *generic* approach compares films of the same genre while looking for symbolic motifs that are present throughout the films (Neale, 2000). All of these approaches, broadly speaking, can be considered contextual elements. As should be evident, there is no shortage of methods for an audience to derive meaning from a film.

Time and Film

An interesting approach to film analysis is taking a closer look at "time" in films and film theory. Unlike some other art forms that are more static, film is one that encompasses and appeals to a variety of sensations, for the most part, visual and auditory. Donaldson (2016) suggests that if time differentiates film from other arts, then it requires a closer look at the experience films offer. Matilda Mroz's book

Temporality and Film Analysis, as reviewed by Donaldson, is an exploration of the many facets of time when analyzing film. Mroz argues that we have the tendency to break down and analyze film, but that the "temporality of film" pushes against the ability to do so. We are always trying to find meaning in film, but that meaning is constantly changing in the face of time. Time that is spent watching a film and even time spent analyzing a movie is constantly in flux and pushes against our constant desire to put films in fixed boxes while categorically creating meaning out of it. For example, repeated viewings of a film can change the interpretations we have of characters, events and even the creator's intended meaning. Nevertheless, this does not imply that film analysis is not an important part of learning film theory. Mroz's argument proves valuable to consider as we analyze film and think about how one should be teaching it. Perhaps the greatest takeaway from Mroz is that our interpretations are never definitive and this is because time is an important factor causing the constant shift (Donaldson, 2016).

Film Studies in Formal and Informal Settings

As mentioned, film studies, an academic discipline that deals with the theoretical, historical and critical underpinnings of film, can be considered a sub-discipline of media studies. The discipline is popular in higher education; however, less importance has been placed on learning it before higher education or even outside of formal educational settings. A typical higher education course on film studies utilizes text, lectures, discussion and the film content itself. For example, an Introduction to the Study of Theory and Film course at Columbia University School of Arts typically covers learning film language, cinema history and approaches to film criticism and theory, as well as, exploring the work of major directors (Kim, 2015). An important note is that unlike many other educational topics that may appeal to many age groups (based on level of difficulty), film studies has traditionally been a field popular among adult learners. (Admittedly, given the ubiquitous popularity of film, if film studies were available to early learners, it would likely be popular for younger ages as well).

While people can read a variety of books and websites that talk about film criticism, a coherent take on the field may primarily only present within communities of formal education or film societies. Self-directed or informal learning environments for media studies is therefore still lacking. Informal environments refer to instances wherein the user does not have to rely on an instructor or organization and happens as a result of "daily work-related, family or leisure activities," such as watching television or listening to a radio (Halliday-Wynes & Beddie, 2009, p. 2). For users to gain the most out of their informal learning environments, it requires more than just the passive consumption of media.

Therefore, vastly important to consider is the ways in which media studies can become a part of everyone's informal learning experience. While e-learning platforms and books exist for people to gain this knowledge, there is rarely a central space that replicates an introductory film course to learn film theory and critique, to understand how story structure and meaning are created visually. To reiterate, mobile technologies act as an interactive platform, which allows for embedded multimedia content that is portable and accessible. There are few mobile apps that teach about film, and out of those that do, most do not incorporate film theory. These apps are structured in a way to provide either glossary items or trivia questions about movies, but do not necessarily provide space for collaboration or interaction to unpack and understand the media content itself. The rarity of film theory knowledge among popular masses and the lack of apps that deal with approaches to interpreting film calls for a deeper look at the ways in which mobile technology can be leveraged for learning film theory to enhance the informal learning experience of movie watching.

A MOBILE APPLICATION FOR FILM STUDIES

Now that we have a basic understanding of why we might consider building an app to teach film studies, we are ready to conceptualize a prototype mobile application. For the purpose of testing, the scale of the application will only focus, as mentioned, on one genre of film. A full application would consist of various genres and movements such as westerns, detective, action/adventure, science fiction, gangster or movements such as neorealism or the new wave films. Staying true to the primary purpose of the app is paramount. While teaching various genres may be the long-term goal, the proposed prototype has one function: to teach film noir. The utility of the app depends on how well it accomplishes that very simple goal. Sounds easy enough, however, to teach film noir, we need to know a lot about how adults learn. We also need to determine features of mobile technology that best support the conditions for real learning to take place.

Let's work backwards (sometimes it really is better to start from the end). A user just finished exploring the film noir app. What, specifically, did the user get out of it? First and foremost, the user should come to know how to define film noir. *Film Noir is a style of Hollywood crime dramas that emerged in the 1940s and 1950s.* Great, what else? Hopefully, s/he can now easily identify the genre in terms of tone, style, and cinematography. For instance, *low-key lighting, visual motifs and unique characters like the anti-hero and the femme fatale represent distinguishing features of film noir.* In essence, the user should also have a better understanding of film theory and jargon and perhaps most importantly, a growing appreciation not just for film noir but also for film analysis more broadly. If the user's engagement with the app accomplishes all that, we can say that the app is supporting their learning and engagement.

With that in mind, the prototype of the app will define film noir and introduce learners to common elements found in film noir movies from the 1940s and 1950s so that they can identify them in more modern noir films (neo-noir or even tech-noir). Along the way, it will teach the basic theories and aspects of film noir (motifs, lighting and characters) and interweave some historical information about influences on film noir to help learners see the lineage of film noir. In the end, learners should be able to identify elements of film noir in newer films and gain a deeper understanding and appreciation of the genre and story construction across time periods.

Audiences of a Film Theory Application

Taking into consideration the complexity of film analysis as well as the aforementioned fact that it is traditionally a subject taught in higher education, the target audience are likely be adults. But to a certain degree, in this instance, the specification of that demographic is unimportant. The app targets movie enthusiasts that wish to gain a deeper appreciation for genre, form and content (a desire that is not necessarily reduced to age demographics), but can also be used by teachers of a language, social studies or any other relevant subject or discipline. Specifically, the app is targeted for non-academic use or self-directed learners, and there are many aspects to consider in the app development process in order to make it more appealing to a general audience, outside of an academic setting.

As mentioned earlier, film studies is an important facet of media studies in higher education where formal classes incorporate resources from textbooks and lectures. However, for those who do not have access to such courses, a resource such as an app would greatly help them learn the basics of film interpretation. Expected users of this app would be individuals who want to gain a deeper understanding of genres and film theories. Even though the user may not have any prior knowledge of film studies, at a basic level, they most likely understand the importance of media studies.

The app primarily targets beginners to film studies because it offers basic compact, multimodal content on film noir and assumes no prior knowledge of film theory. While a baseline for motivation for learning media studies might already exist, this motivation will hopefully increase after using the app. With respect to mass communication theory, audiences are generally split between two groups, *active* and *passive*. Biocca (2012) says that active audiences are critically engaged with content, being selective, intentional and involved. Intentionality refers to the "schematic processing and structuring of incoming information (p. 53). Selective attention and perception also reflects in their retention of information. On the other hand, the passive audience is described as "gullible, anomic, vulnerable, victims" who do not critically engage with the content (Biocca, 2012, p. 53). Users for this app do not necessarily fall into either extreme of the passive-active spectrum; however, the goal is that they move towards becoming more active viewers as a result of this app.

As indicated earlier, challenges may arise due to differences in perception of film. For example some audiences may perceive a movie as noir while others as horror. Users are expected to have different viewing and learning experiences based on their level of knowledge of genre specific films. A user may also have affinity with a particular genre already. Therefore, another goal of the app is to teach "genre-knowledge conventions" that allow audiences to understand each genre more deeply (Beach, 2015). Increased experience of viewing the genre will hopefully help them build critical thinking skills such as a deeper understanding of storylines, identifying symbolism or even inferring the function of setting to explain characters' behaviors. For example, one might recognize that an eerie music in a horror movie signals potential danger for characters (Beach, 2015).

Learning Contexts for Film Studies

Space and time considerations are important for designing a mobile application for film learning, because learning may not confined to an academic community or classroom. Rather learning may be very individualistic and relies on the self-motivation and interest of the learner. It also means that the app should have resources available anywhere, at any time. The app could be more effective if users use the app after having watched a particular movies covered in the app in order to contextualize the learning, but is not necessary. This is because users should be able to pick up on film theory, elements and questions in the evaluation portion of the application from just watching the video clip, going through the glossary or reading the information provided about the specific genre.

For example, many movie goers may not realize that newer movies often incorporate elements of noir. This app would be beneficial to help them learn about the narrative and visual aspects of a recent movies they have watched. If a viewer who just watched *Shutter Island* (2010) found the film to be very confusing, especially Leonardo DiCaprio's character, he or she can use this app to learn about film noir. Subsequently, they can unpack the movie to better analyze DiCaprio's complex character and even perhaps come to some conclusions as to how the film's director was influenced by classical noir films. Therefore, identification of genres and visual styles is important, not only for the sake of categorization, but also because it is a tool that can help learners gain a deeper appreciation of newer film.

Learning will take place between film viewings and the learning space provided by the app. This lends to a more immersive and seamless environment. The app allows users to watch a movie before or after diving into the learning environment. Some of the external factors that this app relies on is mobile data and network speeds.

EXISTING MOBILE PROJECTS ON FILM THEORY

Important to consider are the types of app in the market that teach film. Currently, those that teach film are minimal in functionality for a comprehensive approach to studying films. *Film, Tv and Video Production* by SageMilk (2011) is a study guide app geared towards helping film and video production professionals learn jargon that will assist them with getting a position in the film production world. The app includes multiple choice quizzes and a glossary. Another similar app is *Film Making – Learn Basics* by KoolAppz (2011), which encourages users to make their own films. The app acts as a guidebook that provides information for story and production elements. These elements include learning how to get great lighting in your film or how to promote your film. While being available in mobile format increases accessibility, a more interactive approach in missing, such as assessment pieces for the user to check for understanding.

One particular app called *Close- Up Film Language Glossary* developed by Neue Wege des Lernens e. V. *(2014)* targets learning film theory. The app includes a glossary and visual markers that teach users about film language. The visual markers serve as impressive theoretical models. The app includes many film techniques that are neglected such as composition and color in films. The text is supplemented by icons and visual markers that help users "visualize" the film techniques. Users can also store their favorite terms. The app is well designed and includes components that promote critical thinking and engage the user actively.

While the available film learning applications are extremely useful for novices, they do not provide a comprehensive approach to film theory. It would however, be naive to think that an app could replace an entire course without the built in functionality of connecting to other peers and providing sufficient materials and direction. The apps mentioned above serve mainly for the purpose of references with little flexibility for accounting for student goals. Furthermore, let us also acknowledge that it is difficult to have an entirely heutagogical approach in many of these apps, because the opportunity for self-reflection is not present in a more scaffolded instructional approach. However, such learning apps in and of themselves can also be considered andragogical in approach. While pedagogy is faculty-centered education, andragogy is student-centered education (Knowles, 1970). A student may be interested to learn about film theory with a self-directed approach and use an app like *Close- Up Film Language Glossary* to understand film language. Similarly, searching the app store and negotiating which aspects of film they want to learn about and subsequently, choosing an app to download, supports an andragogical approach.

CONTENT ANALYSIS FOR NOIR

The World of Film Noir is an example app designed for use by movie enthusiasts who wish to learn more about the noir and neo-noir films. Noir was chosen as a genre to explore because many of the structural elements of noir are replicated in newer movies.

Noir is a style of cinema that emerged in the 1940s. In 1946, French authors Raymond Bord and Etienne Chaumenton coined the word "noir" to describe films that were made in the United States before World War II (Naremore, 1996). The films generally consist of a male protagonist getting lured and destroyed by a femme fatale. Subjects that were explored included the dark side of humankind and the use of light and shadow to represent good and evil as well as psychological and sexual relationships and dimensions of the male and female relationship.

The general design of the app is geared towards learning film theory through the different approaches to interpretation discussed earlier, such as *formalism*, *realism* and *contextualism*. The user is presented with three of four main concepts of a genre or movement through which all of these different schools of theory will be interpreted to explain most of the genre. In the case of noir, characters such as the femme fatale and the antihero, as well as lighting are three of the most central concepts to noir films. However, there are other elements that users should learn about such as the influence of German Expressionism on noir films. Important historical and contextual information will be intricately embedded in the app structure. The majority of the explanation is done through text, videos, short quizzes, images and discussion questions. Discussion questions range anywhere from asking the user to analyze a specific scene to identify symbols in a film sequence.

- Glossary section with appropriate icons that teach users common tropes and symbolism in noir.
- Quiz assessment to help users build confidence. Immediate feedback is provided.
- Discussion questions to allow for users to think about structural elements in terms of settings, characters, photographic composition, camera movement, editing choices, etc.

- Videos that illustrate the particular characteristic (lighting or characters) of noir.

The app incorporates specific video clips in the app. The app assumes that users will be able to grasp enough information about the subject being explored, such as lighting or the femme fatale, without having watched the whole movie. The app is more effective if the users have watched the film before using the app, but there is no way to ensure that this will happen, especially outside of a structured film course.

Learners will be evaluated using a quiz on how well they can identify film noir elements, distinguish between film noir and non-noir movies, as well as how well they can explain those elements and distinctions. Important to note are that quiz questions are deliberately short (unlike traditional evaluations) to keep the user engaged. The point of the quizzes is just to make sure that users are on target with what they are learning.

A central part of this app will be allowing students to reflect on what they have just watched and what they learned. In addition, reflections would be made through comments that are public for other app users to read. Furthermore, a central part of this app is understanding how jargon, motifs and characters related to film noir have been transformed throughout history and how they exist in present time. The opportunity for this discovery will be a part of the discussion that learners would have in the comment section. In addition, there would be weekly discussion questions given by commenters, which would allow for a more heutagogical approach. A local viewing option, which allows learners to look up contemporary film noir or related film viewings, would also allow for a heutagogical approach as students are in control of what films they want to view.

Overall, the content in the app will help users develop a deeper understanding of story construction. To do this, the app teaches common elements of film noir such as jargon, motifs and characters. Furthermore, it provides adequate content and practice so that users can observe the influence of film noir in newer movies or neo-noir films. Through the use of video clips, the app makes learners proficient in using these terms when analyzing movies to see how they apply to different movies regardless of whether it is a classical noir film or a newer film that applies noir techniques.

APPLICABLE LEARNING THEORIES

The main overarching theory that the film learning app supports, is the theory of informal and lifelong learning (Futurelab et al., 2004). This theory posits that people can acquire skills, knowledge and values outside of formal learning environments or more structured environments such as institutions and schools. Informal learning occurs in daily experiences from surrounding environments and mass media. To demonstrate that informal learning is valuable, this film theory application incorporates various movie clips from YouTube. Many people learn from channels on YouTube instead of attending school for hobbies or professions they want to pursue. Similarly, the availability of already uploaded film clips is extremely useful for this type of informal learning. The app is able to take available content and curate it to teach individuals about film noir. The app also directs users to local screenings, so that they can learn more about special viewings of noir and neo-noir movies that they can attend. In this way, the app recognizes and authenticates the diversity of informal learning situations that film lovers can benefit from.

The app also supports the Multimedia Theory (Mayer, 2009) and the theory of Retention and Transfer (Edge, Searle, Chiu, Zhao & Landay, 2011). The Multimedia Theory is based on the notion that combining images/visuals with text and audio, maximizes the learning experience rather than simply presenting text alone. It has also been noted that presenting both verbal and visual information together helps retention. With an intentional pedagogical goal, texts combined with pictures and videos are part of the design to improve the learning process. In addition, the glossary was enriched with icons to provide not only texts but also visual stimuli to visual and kinesthetic learners. For example, the glossary feature also utilizes icons as opposed to simply listing film terms found in film theory textbooks. The activities proposed to the app users to learn about film noir also draw upon the theory of Retention and Transfer. The theory of Retention and Transfer (Edge et al., 2011) is based on the idea that testing helps learners retain more information than simply studying it. This is the reason that the app has small quiz questions interlaced throughout to accompany the video clips. While some may think that only formal learning has assessment, it is important to remember to include assessment in informal learning as well. Assessment in informal learning settings does not have to have to be as stringent as in formal environments.

The film learning app also draws on the Constructivist theory (Futurelab et al., 2004). This theory argues that learners construct new ideas and concepts based on their previous or current knowledge. The design of the app has been intentionally constructivist in nature. After using the film noir app, users are not only able to identify common elements of film noir such as jargon, motifs and characters, but they are able to observe the influence of film noir in newer movies or neo-noir films. Again, identification is not the main purpose of learning about film, but it does lead to a deeper understanding of the story constructs and character analysis. Fedora hats, private detectives, smoking, amnesia and a femme fatale that lures the hero into committing crimes - are all characteristic of noir films. Suppose an individual has learned these terms and motifs through using the app and has also watched the movie *Laura* (1944). At another time, the user may watch *Shutter Island* (2010) and begin to see the similarities between Dana Andrews' character in *Laura* (1944) and DiCaprio's detective role in *Shutter Island* (2010). In this way, the app empowers users to use their knowledge of film noir to analyze and understand these new movies and come to some creative analyses and conclusions.

Being wireless is one of the greatest advantages of mobile technology as we are not tied to a specific location for learning. In this sense, content adaptability is highly important in M-Learning. The point of the film theory app was to make a system which is highly adaptable to the needs of movie goers, but at the same time less predictable. When personalized information is received, students tend to value the given content. In this sense, the idea of "local viewing" was introduced as a means to make the learning experience more personal.

Considering a Heutagogical Approach

An important consideration for this app is understanding how adults learn best. Knowles (1970) says that adults are self-directed learners who have the knowledge and experience to learn in ways that are internally motivated (as cited by Halupa, 2015). Because of this, learners can develop learning environments that are suited to their learning goals. In andragogy, teachers can involve students in assessing their goals and interests to help come up with objectives and provide resources. Heutagogy, on the other hand, takes student centered learning a step further by allowing students to take a larger role in the learning process. As noted by Halupa (2015), heutagogy is more suited for adult learners in areas such as higher education courses, because adults are motivated to take a more active role in their learning. Halupa says "In true heutagogical approaches the teacher provides the material, but the students decide how to negotiate the learning process." Similarly, Hase & Kenyon (2000) explain that currently, because we have access to a plethora of information, the traditional methods of teacher-based learning is not enough to account for the ways in which students needs for "immediacy of learning" can be met. They suggest that there are ways in which the we can address such learner needs through by examining and conveying the "deficiencies of the pedagogical and andragogical methods." For example, "negotiated assessment" is also crucial to heutagogy (Blaschke, 2012). Therefore, a more flexible curriculum to accommodate different learning goals and needs as well as learner directed questions is important.

With the advent and use of more mobile technologies in education, in some areas of film education, there has also been a simultaneous desire to consider a more heutatogical approach (Cochrane, Antonczak, & Wagner, 2013). A 2009 to 2012 study on post-web 2.0 pedagogy explores the ways in which learning film can be enhanced to include student-generated content, mobile learning, collaboration and even student-generated contexts (Cochrane et al., 2013). The study focused on a higher education Film and Television course which explored the challenges and advantages of integrating Wireless Mobile Devices (WMDs) in education courses. In addition, central to the study was exploring how WMBs could support collaboration, reflection and interest to generate a more creative and motivating environment for students. The students in the study leveraged social constructivist e-learning tools such as Google Plus, YouTube and Twitter. Social constructivism is the idea that "knowledge is defined as the meaning that is negotiated through social interaction within a community of learners. This meaning is obtained by the participation of individuals as they become initiated into their community of learners." (Barbour and Rich, 2007, p. 3). The study explores the ways in which a teacher focused course evolved into a more student focused one. The students in the course even generated their own content and contexts for assignments and assessments. A key concept of heutagogy includes "double loop learning" as conceptualized by Argyris & Schön (1996). Ervin and Slife (2019) say double-loop should "help students critically examine values and assumptions of relevant topics." This implies that learners are not only looking at problems, their outcomes, but also reflecting on the process and how it affects their own values (Argyris & Schön, 1978, as cited by Cochrane et al., 2013). A core part of the study was exploring ways in which the course could incorporate student reflection and this eventually led to the opportunity for students to make changes and input to the course itself.

As discussed before, the idea that an app should be heutagogical in and of itself is challenging.It would be difficult to create an entirely heutagogical approach, but allowing for heutagogical principles such as "double loop learning" and student chosen contexts and assignments to flow in would prove beneficial. It is important to note, that in this day and age, we should be considering heutagogical approaches to learning and even learning that happens through technology. Heutagogy relies on the idea of the constructivist theory, the idea that students "obtain knowledge by filtering new knowledge through their own personal experiences" (Halupa, 2015). Learners adapt to their experience of learning by doing and they adapt to their learning experiences and make adjustments accordingly. Previous experiences help form and shape their new experiences. A heutagogical approach for the app may be considered in a variety of formats as discussed below and will allow students to contribute their own knowledge and shape their own learning.

FUNCTIONALITIES OF THE WORLD OF FILM NOIR

The World of Film Noir is a prototype for the various types of film theory apps that can exist. The example app was built on Proto.io as part of a mobile learning app development graduate course (Mentor, 2018). Proto.io is a web-based collaborative software that was chosen because it allows development of fully interactive apps without requiring coding. Sound, video, images and text can be easily added, and the app is quite intuitive to use. The prototype works on touch screen devices. Video can also be embedded in the prototype using YouTube for ease of access/availability of content. The app requires that the mobile device be connected to the Internet for functionality and streaming purposes. M-Learning allows contextual learning to take place between viewing movies on the screen and using the app to make sense of film theory. It also supports learning on the go and informal learning theories.

- **Overview (Homepage):** Includes a small quote about film noir to encapsulate the genre. Film music also plays. A stop/start button is included for music. An iconic picture of an anti-hero. The double navigation at the bottom points to the three characteristics of noir that will be explored including lighting, femme fatale, anti-hero as well as links to the glossary, neo noir and local viewings. Navigation is also available at all times from the sidebar.
    - **Element 1:** Lighting Overview. Describes the importance of lighting in noirs. Different types of lighting include chiaroscuro, eye lighting and the effect of lighting on venetian blinds.
    - **Example 1:** Chiaroscuro. Describes the influence of German Expressionism and Renaissance period on film noirs for the mood that lighting creates. Links to specific German Expressionist movies that came about due to World War I and described dissonance and evil that Germans felt. Specific renaissance paintings that display light and shadow are serve as examples. Example of how structural, contextual and realist approaches to film theory can be applied into analysis.
    - **Movie Example:** German Expressionism. Includes an explanation of the historical and cinematic influence of German expressionism on noirs. Includes a scene from *The Cabinet of Dr. Caligari* (1920) as an example.
    - **Quiz:** Pertains to the influence of art forms on noir films. Only one right answer/multiple choice.
    - **Comments:** Students can comment on the video, quiz questions or other aspects of the subject.

  - **Element 2:** The femme fatale overview. Includes an explanation of a femme fatale and three distinct movies: *Laura* (1944), *Out of the Past* (1947), *Sunset Boulevard* (1950) serve as examples.
  - **Movie Example:** *Laura* (1994). Movie overview of Laura's character as a femme fatale. A scene depicting Laura that describes her role in the film. At the top is an icon for a quiz question.
  - **Quiz:** Pertains to Laura as a femme fatale. Only one right answer/ multiple choice.
  - **Comments:** Students can comment on the video, quiz questions or other aspects of the subject.
  - **Element 3:** The anti-hero overview. Includes an explanation of an anti-hero and three distinct movies, *Maltese Falcon* (1941), *Double Indemnity* (1944), *Sunset Boulevard* (1950) serve as examples.
  - **Movie Example:** Sunset Boulevard (1950). Describes the anti-hero in noir films and gives an example of a noir film, Sunset Boulevard (1950) with an archetypal anti-hero. A scene depicting the anti-hero and his role in the film. At the top is an icon for a quiz and discussion question.
  - **Discussion:** Discussion question pertains to what is viewed in the scene for *Sunset Boulevard* (1950). Open ended question.
  - **Comments**: Students can comment on the video, quiz questions or other aspects of the subject.
  - **Glossary:** Includes a search function. However, the icons represent certain characters, settings, symbols, editing and lighting techniques present in film noirs.
  - **Glossary item Anti-Hero:** Description of anti-hero with bolded words to describe main characteristics that link back to other items in the glossary. Importance of icons are noted for retention and transfer. Images and text are placed together.
- **Local Viewings:** Alert the user of neo-noir films that are out in theaters right now. Movie location, synopsis and times included.
- **Neo-Noir films:** Describe what neo-noir films are and link to the some classic modern examples that viewers might have watched. Helps users who are interested in seeing these neo noir films and be able to apply what they learned.
- **User Contribution:** An area for users to comment and share what they have learned in the app and even outside of the app, in external contexts. They can also create assessments and ask questions to help other learners.

- **Comments and Feedback:** This is a section of the app which allows users to leave feedback on how they would like to change or add to the application in order to help their learning.

User Flow and Use

A user would download and open the app and land on the homepage. Here, they see a quote describing the film noir style. They can also see the three main characteristics of film noir, a glossary section, a section about neo-noir and local viewings. The user can also turn on some film music if they wish. If they first decide to learn more about lighting in film noir, they can read about how deep shadows, venetian blinds and low key lighting portray the idea of good and evil. They can also decide to read more about low key lighting, also known chiaroscuro. Users can learn about the historical influences of the Renaissance paintings and German expressionism on lighting in film noir. They can also watch a clip from a 1920 silent German film, *The Cabinet of Dr. Caligari*, to visually experience the light and effects. Users can then go back to see if they can answer a quick question the renaissance and German influence on film noir. They can also explore the Anti-Hero page and the Femme Fatale page where they will see similar movie clips and questions. The user also notices the dialogue icon on the Anti-Hero page which prompts them to a discussion question, one that they must answer to test their own knowledge of what motifs are associated with an anti-hero. They could also go to the glossary page to understand why voice-over it is heavily employed in noir films.

One day the user decides to check out the Local Viewings page and notices that *Looper* (2013) is playing at the local theater nearby. They read up on what neo-noir films are about and is interested to go see *Looper*. The user clicks on the movie to see viewing times and invites some of their friends to watch the movie. During the movie, the user has to explore the ways in which this movie relates back to the motifs and jargons of film noir and how the genre has evolved over the years, thus allowing for a heutagogical approach to learning. Figure 1 outlines the hierarchical flow of the application.

Implementation

An app such for film learning can be implemented via an iterative design process and user testing. For example, for the sample film noir prototype, adult movie goers would be asked about their user experience. When we talk about usability, we should not only consider whether the app can be used without any issues or technical

Figure 1. A hierarchical flow of the application.

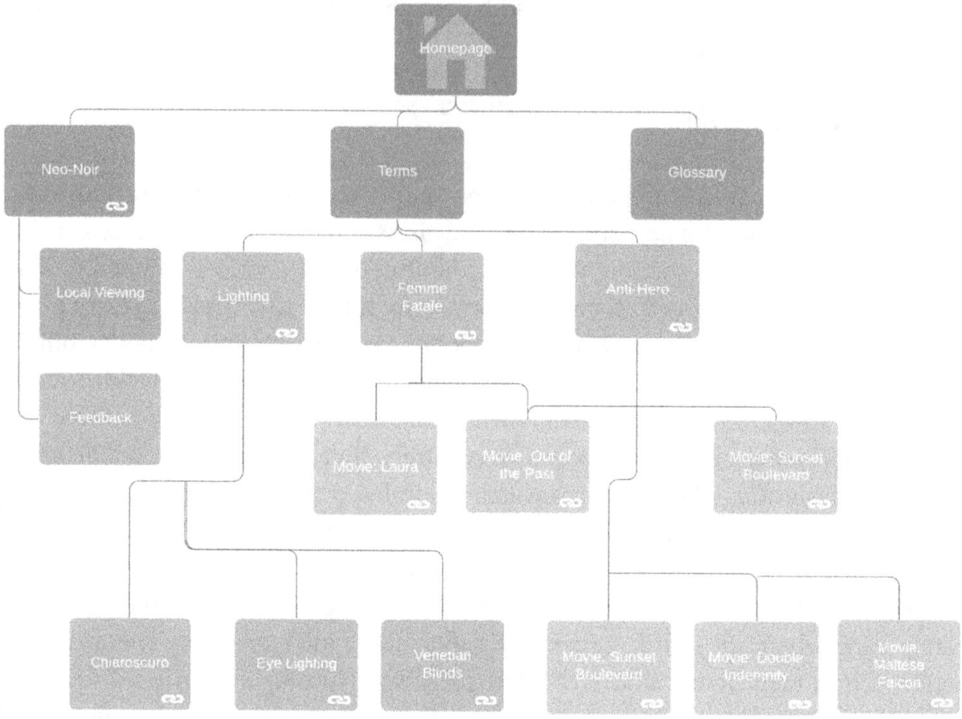

problems. It is vital to understand, through testing, whether movie audiences would be motivated to use and benefit from such an app and whether it would augment their film analysis skills. Some further questions to consider are:

- When is the best time for users to use the app? (For testing purposes as well).
- Are short video clips the best medium for presenting examples of film theory?
- Are the short discussion/quiz questions beneficial for the user? If so, to what extent?
- What kind of motivation is necessary for users to learn from this app?
- What additional features can be added during the iterative design process?
- How can the app best incorporate heutagogy practices that allow for student directed learning, self-determined learning and self-reflection?

Some other challenges to consider during implementation are more technical and deal with compression and feed. If users do not have good data connection, it could deter them from using the application. Furthermore, there are legal issues surrounding copyright to consider. Many of uploads of movie clips on YouTube may not be legal, so there would need to be a way to obtain properly copyrighted information and videos.

Another challenge, from a pedagogical and sustainability stance, is the ability for the app to engage users of all backgrounds and levels in film studies. As mentioned above, the film noir app would be for beginning film students, but how can the content be challenging enough for them to want to continue using it? These are questions to consider during the iterative process of designing such an app.

Evaluation

When evaluating an app, qualitative and quantitative approaches can be useful. Proper analysis should include in-person user testing, interviews and surveys, as well as performance data analysis. Surveys come in handy because they can help determine the overall appeal and motivation for using the app, as these may differ from whatever our assumed motivations were. Furthermore, surveys are useful in that they can include things like: rate of *effectiveness* (degree to which learner is able to reach learning), *usability* (degree to which app allows user to learn without hindering their performance) and *satisfaction* (degree to which user is satisfied by what they are learning from the app). To measure such things, scales are often employed (i.e. from best to worst, most to least true on a scale from 1 to 10). Again, for this app, learning goals for users should be based around the following: 1) identification of genre, 2) knowledge of film theory and jargon, and 3) appreciation for further application and analyses. Hence, the data collected would ideally be supportive of this. Considerations might also include functionality and navigation. Here are some goals for evaluation that should be considered:

Overall Appeal and Motivation

- How appealing is the idea of learning Film Theory through a mobile application?
- In what instances would you use such an app?
- User background and interests.

Effectiveness Content a Differentiation in Learning

- Does the content supplement a user's understanding of noir and neo-noir?
- Does a genre based approach work to enhance structural and contextual elements of film analysis?

- Are the videos, images, text suitable and appropriate for each element that is being taught?
- Which content was the most useful to meet learning goals? Which content was the least useful?

Usability User Friendliness and Performance

- Does the UX/UI render usable in terms of functionality and navigation?
- How easy is the hierarchical flow of information to follow intuitively?

Satisfaction and Transformative Learning Potential

- Does the user feel like that have learned about noir elements, history, art and filmmaking in a cohesive manner?
- Does the user feel that they can apply what they have learned to other films or settings?
- What other functionalities should be added or subtracted to enhance learning experience?
- Does the user feel that they played a sufficient role in shaping their learning?

In terms of design, evaluation should shed light on both aesthetic appeal and user experience. For instance, users may find thnde interface to be simple and clean or busy and complex (hopefully not the latter). Users may offer insights about theme consistency and the extent to which the app was or was not intuitive. In the case of this app, our hope would be that users report the learning experience was straightforward, and that the content was broken down into digestible bites. However, sometimes things are not as clear as we might think. For example, hierarchical flow of information and design (colors, fonts, images) might have to be revised once evaluation is conducted based on user experience.

Until an app is tested, it is hard to objectively gauge strengths and weaknesses. However, for the purposes of testing, it is important to realize one's own assumptions while designing an app. For example, to appeal for more commercial use, the film noir app would need to be designed a bit differently. The app might have to completely eliminate talking about classical film noir and instead focus on neo-noir or movies people are more familiar with. Another assumption during the design process was that users will want watch to short video clips in order to learn about films. Designing apps requires rethinking such assumptions and figuring out alternatives proposals if those assumptions prove ineffective during evaluation.

FUTURE RESEARCH

Future research in the area of film studies and mobile technologies calls for a look at a more comprehensive heutagogical approach. Şahin & Mentor (2016) have explored the value of m-assessments in contemporary classrooms. They argue that "the strengths of m-assessment suggest using m-assessment for representing real-life learning scenarios. Therefore, more studies using m-assessment in realistic learning scenarios are needed." In film learning, the incorporation of m-assessments can be truly valuable, especially if students are able to define and contribute to shaping the m-assessments. This paper has explored m-assessments in terms of small quiz assessments in more real-life learning situations, such as going to the theater. However, further research should be conducted in terms effectiveness. Further comparisons should be made with existing assessments in film studies courses, while keeping a heutagogical approach in mind.

CONCLUSION

The rapid changes in functionality and portability of mobile devices call for a closer look at the ways in which this technology can be leveraged for teaching and learning of film and media studies. Given the importance of learning critical media and the scarcity of apps that teach it, M-Learning is valuable medium through which critical thinking skills can be learned.

When considering the prototype for *The World of Film Noir*, some learning considerations that were examined in the analysis of M-Learning affordances did not fully fit into the app design. For example, this app concentrated on using short video clips as a primary means of communicating and unpacking film noir conventions. However, effective collaboration was still missing in the sense that the affordance of cross-contextual support that mobile technologies possess was not fully leveraged. This chapter proposed considering a more heutagogical practices that allowed for student feedback to shape learning and content. Future of film studies and other areas of media studies should think about leveraging technologies such as live twitter feeds or chats viewable on television or live streaming. Incorporation of such technology would be ideal for seamless and cross-contextual support in M-Learning and film studies.

REFERENCES

Barbour, M., & Rich, P. (2004). Social constructivist e-learning: A case study. In *E-Learn: World Conference on E-Learning in Corporate, Government, Healthcare, and Higher Education* (pp. 1754-1759). Association for the Advancement of Computing in Education (AACE).

Beach, R. (2015). *CI5472 Teaching Film, Television, and Media*. Retrieved April 19, 2015. Retrieved from http://www.tc.umn.edu/~rbeach/teachingmedia/module7/3.htm

Biocca, F. A. (2012). Opposing Conceptions of the Audience: The Active and Passive Hemisphere's of Mass Communication Theory. *Communication Yearbook*, *11*, 51–80.

Blaschke, L. M. (2012). Heutagogy and lifelong learning: A review of heutagogical practice and self-determined learning. *The International Review of Research in Open and Distributed Learning*, *13*(1), 56–71. doi:10.19173/irrodl.v13i1.1076

Brown, E., Börner, D., Sharples, M., Glahn, C., De Jong, T., & Specht, M. (2010). *Location-based and contextual mobile learning*. A STELLAR Small-Scale Study.

Casale, R., & Mentor, D. (2018). Digital Narratives of Immigrant Youth: Cultivating Mobile Activism and Mobile Journalism Skills. In Computer-Mediated Learning for Workforce Development (pp. 132-149). IGI Global.

Cochrane, T., Antonczak, L., & Wagner, D. (2013). Post-web 2.0 pedagogy: From student-generated content to international co-production enabled by mobile socia media. *International Journal of Mobile and Blended Learning*, *5*(4), 1–18. doi:10.4018/ijmbl.2013100101

Crompton, H. (2014). A diachronic overview of mobile learning: A shift toward student-centered pedagogies. In M. Alley (Ed.), *Perspectives on Open and Distance Learning: Increasing Access through Mobile Learning*. Athabasca University.

de Waard, I., Abajian, S., Gallagher, M. S., Hogue, R., Keskin, N., Koutropoulos, A., & Rodriguez, O. C. (2011). Using mLearning and MOOCs to understand chaos, emergence, and complexity in education. *The International Review of Research in Open and Distributed Learning*, *12*(7), 94–115. doi:10.19173/irrodl.v12i7.1046

Donaldson, L. F. (2016). *Temporality and Film Analysis* by Matilda Mroz (review). *Cinema Journal, 55*(2), 159-163.

Edge, D., Searle, E., Chiu, K., Zhao, J., & Landay, J. A. (2011). *CHI '11 Proceedings of the SIGCHI Conference on Human Factors in Computing Systems*. ACM.

Ervin, J., & Slife, N. (2019). *Double-Loop Learning: An Approach to Critical Thinking*. Academic Press.

Futurelab, N., Naismith, L., Lonsdale, P., Vavoula, G., Sharples, M., & Series, N. F. (2004). *Literature review in mobile technologies and learning*. Academic Press.

Georgiev, T., Smrikarov, A., & Georgieva, E. (2004). *m-learning: a new stage of e-learning*. doi:10.1145/1050330.1050437

Glossary: Learning Activity, New Learning - Transformational Designs for Pedagogy and Assessment. (2015, January 1). Retrieved from http://newlearningonline.com/learning-by-design/glossary/learning-activity

Halliday-Wynes, S., & Beddie, F. (2009). *Informal learning: At a glance*. Adelaide: National Centre for Vocational Education Research.

Halupa, C. M. (2015). Pedagogy, Andragogy, and Heutagogy. In *Transformative Curriculum Design in Health Sciences Education* (pp. 143–158). IGI Global. doi:10.4018/978-1-4666-8571-0.ch005

Hase, S., & Kenyon, C. (2000). *From andragogy to heutagogy*. Ulti-BASE In-Site.

Holmberg, B., Moore, G. M., & Peters, O. (2007). *The Theories and the Theorists: Why Theory is Important for Research*. Presented at the 4 th EDEN Research Workshop on "Research into Online Distance Education and E-Learning: Making the Difference," Castelldefels, Spain.

Jacobs, C. (2013). *Film Theory and Approaches to Criticism, or, What did that movie mean?* Retrieved from http://www.und.edu/instruct/cjacobs/FilmTheory&Analysis.htm

Kim, N. (2015, January 1). *Introduction to the Study and Theory of Film*. Retrieved January 1, 2015, from http://arts.columbia.edu/summer/film/course/study-and-theory

Kim (2015). *Introduction to the Study of Theory and Film*. Retrieved from http://arts.columbia.edu/summer/film/course/study-and-theory

Knowles, M. S. (1970). *The modern practice of adult education* (Vol. 41). New York: New York Association Press.

KoolAppz. (2011). Film Making – Learn Basics (1.0) [Mobile application software]. Retrieved from https://play.google.com

Martin, A., & Chamberlain, L. J. (2015). Technology-Driven Service Learning: The Instructor's Role and the Student's Experience. *EDUCAUSE Review*. Retrieved from http://www.educause.edu/ero/article/technology-driven-service-learning-instructors-roleand-students-experience

Mayer, R. (2009). *The promise of multimedia learning, Multimedia Learning* (2nd ed.). New York: Cambridge University Press. doi:10.1017/CBO9780511811678

Mehdipour, Y., & Zerehkafi, H. (2013). Mobile learning for education: Benefits and challenges. *International Journal of Computational Engineering Research*, *3*(6), 93–100. Retrieved from http://www.ijceronline.com/papers/Vol3_issue6/part%203/P03630930100.pdf

Mentor, D. (2016). EMxC3= e&mLearning Cultivating Connected Communities: Sustainable Workforce Talent Development. In *Handbook of Research on Mobile Learning in Contemporary Classrooms* (pp. 240–259). IGI Global. doi:10.4018/978-1-5225-0251-7.ch012

Mentor, D. (2018). TLC for MOOCS: Teaching and Learning Communities for Computer Programming. In Computer-Mediated Learning for Workforce Development (pp. 93-110). IGI Global.

Mentor, D. (2018). mClass Planet of the Apps: The Rise of Mobile Learning. In Computer-Mediated Learning for Workforce Development (pp. 196-215). IGI Global.

Mezirow, J. (1998). On critical reflection. *Adult Education Quarterly*, *48*(3), 185–198. doi:10.1177/074171369804800305

Mroz, M. (2013). *Temporality and Film Analysis*. Edinburgh, UK: Edinburgh University Press.

Mulrennan, D. (2018). Mobile social media and the news: Where heutagogy enables journalism education. *Journalism & Mass Communication Educator*, *73*(3), 322–333. doi:10.1177/1077695817720762

Naremore, J. (1996). American Film Noir: The History of an Idea. *Film Quarterly*, *49*(2), 12–28. doi:10.2307/1213310

Neale, S. (2000). Questions of Genre. In *Film and Theory: An Anthology* (pp. 157–178). Malden, MA: Blackwell Publishing.

Pedro, L. F. M. G., & de Oliveira Barbosa, C. M. M. (2018). A critical review ofmobile learning integration in formal educational contexts. *International Journal of EducationalTechnology in Higher Education*, *15*(1), 10. doi:10.118641239-018-0091-4

SageMilk. (2011). Film, TV and Video Production (1.0) [Mobile application software]. Retrieved from https://play.google.com

Şahin, F., & Mentor, D. (2016). Using Mobile Phones for Assessment in Contemporary Classrooms. In *Handbook of Research on Mobile Learning in Contemporary Classrooms* (pp. 116–138). IGI Global. doi:10.4018/978-1-5225-0251-7.ch006

Shuler, C. (2012). iLearn; An Analysis of the Education Category of the iTunes AppStore. New York: The Joan Ganz Cooney Center at Sesame Workshop.

Specht, M. (2014). Design of contextualised mobile learning applications. *Increasing Access, 61.*

Tisdell, E. (2008). Critical Media Literacy And Transformative Learning: Drawing On Pop Culture And Entertainment Media In Teaching For Diversity In Adult Higher Education. *Journal of Transformative Education*, *6*(1), 48–67. doi:10.1177/1541344608318970

Wege des Lernens, E. V. N. (2014). Close-Up Film Language (2.6.1) [Mobile application software]. Retrieved from https://itunes.apple.com

Wong, L.-H., & Looi, C.-K. (2011). What seams do we remove in mobile assisted seamless learning? A critical review of the literature. *Computers & Education*, *57*(4), 2364–2381. doi:10.1016/j.compedu.2011.06.007

Chapter 11
Utilizing Mobile Learning for Orphans Aging Out:
Orphan Youth Development in Peru With Universal Lessons

Tara Callen
Columbia University, USA

ABSTRACT

How can mobile e-learning be used to support and enhance the goals of working with youth aging out of an orphanage. This chapter describes the use of an ethnographic narrative approach to tell the stories of eight young women who were "aging out" from their orphanage where many of them had spent most of their lives. The chapter examined the way in which a collaborative art community could support the participants as they narrated their lives over a 16-month period through photo-journaling and social media outlets. The focal points of this study were community building via art making and building of personal aesthetic, community engagement, reflection on self-identity, cross-cultural art education, and shared experience via photo-art narratives. The authors also examined the role of collaborative art experiences in helping these young women form social connectedness with their peers. The project was also designed to help sustain the girls digitally and virtually in the present and future. The research project studied the identity development and how these functioned within a collaborative medium that supported the young participants as they moved out of their orphanage.

DOI: 10.4018/978-1-5225-9351-5.ch011

INTRODUCTION

The focal point of this chapter is on the elements of which e-learning and m-learning (e-&-m-Learning) for a project involving the creation of digital narratives of orphan girls aging out of their orphanage in Peru. The orphanage was administered and run by nuns. The study was designed using a qualitative research mode of inquiry that utilized a mixed methodology approach. The chapter is informed by an ethnographic narrative study tracking eight young women who were "aging out" (turning 18) or forced to leave their orphanage in Peru, where most of them had spent most of their lives. The research project examined the way in which a collaborative art community could create a sense of social connectedness, a sense of affiliation (Mentor, 2018) and to support the participants as they related their lives over 16-months through digital photo-journaling and through social media outlets.

The research also examined the opportunities of collaborative art experiences in helping these young women advance their agency, construct new identities and form relationships with their peers. One of the main questions that I sought to answer was how adolescent girls who have lived a large portion (if not their entire) lives within an orphanage might utilize an artistic outlet (such as digital photo journaling) as an aid to document their journeys of leaving home, school, and the place of their childhood to enter the unknown, outside world. These electronic and mobile journals could tell us about a topic so seldom addressed not just within their society, but from within the individuals themselves for the benefit of all societies.

E-&-m-Learning) modules are increasingly being utilized across educational platforms. The "virtual classrooms allow students and teachers to communicate synchronously using features such as audio, video, text chat, interactive whiteboard, and application sharing" (Yuktirat, Sindhuphak, and Krissana Kiddee (2018) p.152). As a researcher using the tools that e-&-m-learning offered, I was a bit of a novice when implementing the structure of what would best serve my participants. It was later in the study that I began to realize that many of the elements of my work did involve improvising on how to teach from afar and I began to make connections necessary to better support my participants during our ongoing work together.

The arena of art education would seemingly be one of the trickier subjects to teach using technology. Many formalist art educators shun the use of technology in their classrooms because it can in fact get in the way of the artistic process and sometimes be distracting (citation or citations needed to support this statement). There are many other reasons why technology might hinder rather than support creation, but if used sparingly or integratively in support of educational objectives (Mentor, 2015) and in the manner of facilitating certain lesson plans and ideas, technology can be incredibly helpful. In fact, more and more art museums are using iPads and other devices to educate museum-goers and students on their exhibitions and many of them are reporting great success in reaching a wider audience (Goeser, Caroline 2013; Şahin, & Mentor, 2016). Furthermore, when teaching seminars and other lesson plans from afar, the avenues that computer and mobile devices have opened up for teachers and students, possess much greater possibilities and more people across the globe are now gaining access to material that was not available to them in past years.

POPULATION STUDIED

The children and adolescents who make up the populace of most orphanages have few resources, and often they cannot rely upon family to house and provide for them. When they reach adolescent maturity at the age of 18, these young people are subsequently aged out of their homes and places of education to make room for incoming children. The term "aging out" means that when young people are finished with schooling and residency at the orphanage, usually at the age of 18 to 20 (unless they have special needs), they must leave because they are too old to qualify for care under the orphanage system.

Could introducing a way for the girls to chronicle this particular year of their life, one that is so monumental to them, help society better understand their struggles and lives as they launch into the real world? And how could I as a researcher living thousands of miles away construct a project that could support them, and be supported when teaching from afar to sustain its momentum over a long period of time. This is where e-&-m-Learning became tremendously helpful in my work overseas. Through first establishing in-person contact at the project site and then residing there for a little over a week so as to get to know my participants through interviews, observing them in their environment and teaching workshops about how to use the technology. Over the course of a fairly short period of time, I was able to open up communication with the participants using social media and their devices as tools utilized for correspondence and ongoing learning ventures.

When first preparing for this study, preliminary research on topics of social media and comparing different platforms that would be best suited for photography and video editing work led me to social media sharing platforms such as Instagram, Flicker, Starmatic, Tumblr, and Stipple, just to name a few, where people share their art work, writings, musings, and ideas. In the utilization of some of these sites, I commenced a project with the objective of supporting the girls in their art making while keeping them connected to one another and to myself as the teacher and researcher throughout the course of the study. Mentor says "Mobile phones and tablets have expanded learning engagement time exponentially. Supporting intentional and incidental learning episodes in people's daily lives, catering for more opportunities for learners to be active agents in their learning. Mobile technology enables everyday learning to be owned, naturally pursued, expanded on, monitored, and does so by blending seamlessly and unobtrusively into a user's life." (2018). While McCann, use MoLeNET's (2011, para. 4) definition of mobile learning, as "the exploitation of ubiquitous handheld technologies, together with wireless and mobile phone networks, to facilitate, support, enhance and extend the reach of teaching and learning." He goes on to quote (Shippee & Keengwe, 2014) who distinguishes mLearning through mobile devices from other forms of technology by their portability, ability to connect, and ability to reach the learner and the learner's needs. Furthermore, mobile learning is offering micro-moments of learning in a multitude of multimodal stimulating means cultivating opportunities for transformation" (Mentor, 2018) and can also be used for capturing collaborative dialogues and narratives digitally while on the go.

The foremost notion of creating a story with a small group of girls via one continuous thread that we all contributed to evolved as a shared experience. Which served as an opportunity for the girls to learn how to access technology and utilize digital narratives in creative and interesting ways. Picture taking, iPhone art, editing, scrapbooking, and video-making are all methods in which the girls were trained in order to begin the 16 months of recording and collecting images and story lines that encompassed who they were and where their lives would ultimately lead them. My own art practice also evolved, and I developed a mode of continuous communication between the girls and myself by contributing to several mutual social media platforms dedicated to art sharing in the form of photography, journaling, and individually profiling each subject. Through using unified, community-driven platforms, the dialogue certainly benefited from greater mobility, impetus, and flow.

BACKGROUND

The participants specific to this study were eight young girls, some who still resided at Hogares de Niñas, and some who have already left. This is an all-girls' orphanage, run solely by Catholic nuns, and is located in Puente Piedra, Peru, an outer lying territory of Lima. The very first conversations I had with the nuns and girls revealed quite a bit of uncertainty on everyone's part concerning the girls' future paths in life. The resources of the orphanage are quite sparse and not fully secure. In other words, they have few private funders and they cannot count on one sustaining source, patron, or even government aid. Additionally, the two universities that are really the only two realistic options for these particular girls have strict guidelines concerning whom they will accept. Their system is based only on grades and academic scholarship. If some of the girls were not entirely proficient in their studies, then most likely the opportunity to pursue a college education was not financially feasible or realistic. This left many lingering questions for all involved in terms of finding viable alternatives for some of these young women and the outcomes of their 16 months in transition. The opportunity to document the experiences of 8 adolescents transitioning out of an orphanage in Lima, Peru revealed much regarding their individual experiences over the established period of 16 months through their writing, photographing, and video-making which was shared over social media platforms and facilitated by implementing a technology-based lesson plan.

CHAPTER FOCUS

Computer and mobile technology have been utilized through many different outlets in advancing educational approaches. It was especially advantageous for my particular project work because of its ability to create a strong and consistent dialogue between myself and my participants. Dr. Muhammed Imran Yousif (2007) discusses in depth the "use of new technology for educational purposes" and how technology "has helped in covering the physical distances between teachers and students, to enable the flexible delivery of education at a distance, anyplace and anytime (p. 122). The key words here being flexible and at a distance, teaching from afar is solely enabled via technology and through some key trial and error lesson plans. For the sake of my project work, I quickly learned that not only would I rely on technology within my lesson plan, but I would need to rely upon it as the primary source of my data collection.

My initial aim was to offer opportunities for eight young women transitioning out of their orphanage to use art making (photography) as a way of reflecting on their new identities as they moved into uncertain futures. It is assumed that the arts provide some formative tools that not only help the individuals to develop further agency or add constructs to their identity/ies, but was also designed to provide an important collaborative activity establishing dialogue and connection among the participants. A central feature of the project was to teach new skills in creating photo-narratives to document personal experiences as a means of reflecting on a critical period of transition. Further support was garnered for this project by the ease of communication using technology and devices that were bought at a discount because they were third or fourth generation refurbished products. This point is an especially important one because one of the largest hindrances in the utilization of e-&-m-Learning in underserved communities is expensive technology devices and access to a secure and private internet connection. Most of us think of the internet as being something that is widely available for the most part, but in fact, it is not. There is great expense in first building the infrastructure to support broadband networks and then who exactly pays the fees for the service? Most schools and a majority of orphanages cannot afford such an added and seemingly superfluous expense so the case must be made to potential donors that the internet is an important expense to undertake. This is a difficult point to argue when basic human needs such as food and shelter are trying to be met.

PARTICIPATORY COMMUNITY RESEARCH MODEL

The relationship between the participants and researcher is referred to as Participatory Research, which involves a collaborative style of approach where ultimately action type research evolves out of a process that is fundamentally shared between all the group members involved (Jason et al., 2004). Participatory Research essentially marries the idea of an existing organizational framework to sustain an academic and community outreach approach geared toward prevention of certain social problems. Thus, the theory of "academic-community collaboration" (p. 88) is one that informed the entirety of this project.

It is important to remember that when conducting ethnographic research, the researcher is also placed in the role of a "participant-observer"; thus, the project was intended to be rooted within the community and the partnerships they formed. "The participant-observer attempts to generate the data from the perspective of the individuals being studied" (Wiersma & Jurs, 2009, p. 284); this point essentially means that as an observer that is working within a society and collecting data on their activities and personal profiles, ethnographic research must be "comprehensive, continuing and total. The observer attempts to record all relevant information in an unobtrusive way; consequently, observation is quite unstructured" (p. 284). The type of relationship that is fashioned here is one of mutual respect and learning that is in tandem, where observation occurs and is conducted by said researcher in a way that s/he is placed in the position of a "privileged observer" (p. 284). This concept is paramount to respecting multi-cultural populations and for conducting research that is ethical and respectful of the needs of the population being studied. Researchers and educators that are not directly from the community they wish to examine and who are considered "outsiders" must practice extreme care in the way they interview, ask questions, gather data, and observe their participants, because the relationship is one that is mutually beneficial in terms of co-learning.

For example, formal interviews may be too stressful on some participants or too formal in nature. In ethnographic and community-based participatory research, one must keep in mind that "casual and informal interviewing can be done when an occasion presents itself during observation. Questions might be asked of those being observed in attempt to clarify what is happening or in an attempt to capture those being observed" (Wiersma & Jurs, 2009, p. 286). These are poignant arguments in the case of the participants in this study because in observing the girls in their natural environment, a true attempt to create a narrative of their everyday lives evolved over time. It was necessary to spend time in their natural environment, amongst the influences and activities that affect their identities and contribute to how they come to make important decisions for their lives in the future.

A relationship of trust and respect is the pillar to forming long-term and viable support of such individuals, and occasionally structured interviews may work against the overall end goal. The researcher should not feel as if they are requiring that their participants engage in answering questions that they are uncomfortable with addressing directly. In other words, the population that I was specifically studying was quite delicate and vulnerable in a fundamentally multilayered way. I did not want them to feel as though they were being subjected to scrutiny they might have perceived as negative in any respect or that they would be made to feel even worse about their situations or not give truthful answers. The role of ethnographic and community-based participatory research is to "contribute to solving problems" while also "identifying the most important and key issues to be addressed" (Wiersma & Jus, 2009, p. 300) and to avoid placing undue stress on subjects.

The primary component of my methodology is what is referred to as Participatory Community Research, which is a "strategy that actively involves research participants in the development and implementation of the research process" (Jason et al., 2004, p. 17) and where "shared power between researchers and community participants is an essential element" (p. 10). The participants are essentially in harmony with the goals of the researcher, and the process becomes one that is coauthored and not solely written from the point of view of the so-called "expert in the field" (Jason et al., 2004). Participatory Community Research was strongly suited to my goals of engaging further social action and responsibility in identifying a certain topic that has yet to be thoroughly examined, thus giving a voice to those who have not had adequate representation in broader society. In addition, Participatory Community Research supports the idea that research from this perspective is non-elitist, and although the subjects are not "experts," so to speak, in the field of inquiry being examined, they are, in fact, representatives of their own voice and, therefore, have a right to establish the content and trajectory of the research.

Jan Irgen Karlsen (1991) furthers the above concepts by explaining the "process of change" as it relates to "participator-based" actions, which "means that those who are affected will also take part in defining which problems are to be treated." This compares to the "participator controlled" actions, which means, "that they will be in a position to participate in making decisions about what measures are to be implemented" (p. 146). The overreaching goals of Participatory Community Research are based upon two systems involving the role of the participants and their initiative to make change that is significant to their circumstances and that may benefit in the future those who are experiencing similar conditions.

As previously mentioned, an essential component of my methodology was the Participatory Community Research model, which allowed "strategies that actively involved research participants in the development and implementation of the research process" (Jason et al., 2004, p. 17) and where "shared power between researchers and community participants was an essential element" (p. 10). The participants were essentially in harmony with the goals of the researcher, and the process became one that was coauthored and not solely written from the point of view of the so-called "expert in the field" (Jason et al., 2004). Participatory Research was strongly suited to my goals of engaging further social action and responsibility in identifying a significant topic that had yet to be thoroughly examined, thus giving voice to those who had not had adequate representation in broader society. Here lay the overreaching goals of Participatory Community Research, which engages two systems based upon the role of the participants and their initiative to make change that is significant to their circumstances and that may benefit, in the future, those who are experiencing similar conditions.

This line of thought aligned seamlessly with my research paradigm, buttressing my argument that "the passive observer cannot learn anything rational about the inner dynamics and conditions" of certain social systems that are being examined, but "by becoming a member and making observations over a long period, the observer can succeed." Remember here that the "role of the research involves the dual aim of theorizing and taking action, with action based on theorizing" (Karlsen & White, 1991, p. 145). This perspective corroborates the intent to create a longstanding art network while partaking fully in the artistic process with my participants. It is through our shared actions and the resilient bonds we build with one another over time that we can initiate change as a group and perhaps contribute to the wellbeing of the next wave of girls that age out of this particular orphanage while adding significant data for orphanages globally.

SOCIAL CONNECTEDNESS

A significant portion of the study was to forge a stronger understanding of how a community operating against such odds would receive a completely foreign and new experience. As I initially concentrated on one single age group, I was surprised at how eager the nuns also were to join in on the experience. They too needed an outlet of expression; their journeys are rarely seen by the public because Catholic nuns are usually regarded by society as somewhat limiting in their worldviews and even somewhat intimidating. In fact, the opposite was revealed in my investigation, and the opportunity for education and further research in this genre are vast.

Most of the nuns I have worked with want to be a part of the global world, but they simply do not have the tools to lead them there. Thus, the act of expression, the impulsion as Dewey would call it (1934), is one that is cross-generational, and in many ways, art can be a healing process, a comfort and reprieve from the routines and monotony of certain life paths. Arnheim (1992) writes about the arts as being "an effective means of exploring and caring for the human mind," as "relief" can be found in "some artistic activity" (p. 164). The creation of a community in this case was based around an ongoing art practice and supported by the digital world providing the students, the nuns and the researcher with a sense of connected affiliation (Mentor, 2018).

As this work centered around the concepts of first building a personal aesthetic and then learning how to express such, this idea of a unique aesthetic perspective became a source or avenue for the participants to communicate part of their story to a broader audience while allowing them the breadth to actively learn and employ a whole new set of skills. Video making is a craft that can be somewhat easily taught or be completely overwhelming. The initial lesson plan was constructed with the aim of teaching the girls how to live alongside their cameras, and to rely on their small contraptions to record their experiences, their feelings, and the strides toward their end goals. The videos and pictures then functioned as their journal entries that they would subsequently share with one another on a consistent basis. This experience became the crux of the community which we collectively forged during our 16 months together.

LITERATURE ON ADOLESENCE, SELF-EXPRESSION AND TECHNOLOGY

Adolescence is a time when children are developing a sense of "I" and "self definition" (Kroger, 2009). It is a tender period when young people are on the brink of entering into early adulthood, and "interpsychic restructuring during adolescence brings identity questions to the surface" (p. 7). Essentially, adolescents are in the midst of forming new kinds of relationships while also testing boundaries, where then an "optimal sense" of identity becomes the hope for all young people, and a "sense of well-being" is achieved (Kroger, 2009, p. 23). The development of identity can be described as a "series of interrelated developments that involve changes in the way we view ourselves in relation to others and in relation to the broader society in which we live". (Steinberg, 2016). As relationships shift in a young person's life, so must their sense of self, thus this critical time period is when one is attempting to finally forge a perspective that is centered upon one's future.

Thus it can be easily argued that adolescence is a fundamental time in a person's life when the process of growing into themselves and establishing their beliefs and opinions on a multitude of subjects is strikingly apparent (Burton, 2011; Kroger, 2009; Steinberg,2016), it follows that adolescents, particularly young girls, are on a voyage of sorts to discover how to interpret and relate to the world around them. Their inner lives are saddled with the day-to-day tribulations of assimilating their consistently evolving beliefs with the environment in which they are being raised (Kagan, 1971; Kroger, 2009, Mentor, 2018). Yet, adolescence, despite its onslaught of vulnerabilities, volatilities, and other such associations, is also where we can find the last vestiges of childhood. Personal expression is an integral part of the human experience. When circling back to the initial research questions regarding the implementation of a collaborative art educative project with a population that is aging out of their institutionalized care homes, one can see a clear need for young people to have an outlet of expression.

A fairly recent article in the New York Times titled, "Why Teenagers Act Crazy," the author, Richard Freidman (2014), speaks about the "social and emotional challenges" many teenagers face and the connections scientists are finding in the "quirk of brain development." The author explains why adolescents "on average, experience more anxiety and fear and have a harder time not to be afraid than either children or adults" (p. 1). The science behind the brain development of adolescents is one very large piece to the puzzle of why so many teenagers behave the way they do and waver between feeling afraid and uncertain while also being considered impulsive and compulsive. The fact that the brain matures at different rates and that the region responsible for reasoning and fear processing, called the amygdala, is one of the last regions to develop (p. 1) is yet one more dimension that must be carefully considered when working with this age group. Ironically, it is usually the adolescent age group that begins to be routinely left out of continuing art skill development in their educative trajectories, which seems so strangely counterintuitive.

Judith Burton (2011) speaks about the "advent of puberty" and the "distinctive and often difficult time in the journey towards adulthood" (p. 6). In her work titled, Creative and Mental Growth Revisited, she points out the "inequalities in art education ... that many school art programs do not address adolescent needs, interests, and questions" (p. 9), consequently leaving out a huge demographic that could greatly benefit from acquiring more proficient artistic abilities. The onset and ubiquities of technology in today's times may have somewhat eased the thirst most adolescents seem to have to create and to make their individuated marks upon the world and society. Burton goes on to address the many advantages technology offers in way of virtual museums, social media platforms, computer art skills (drawing, 3-d sculpture, etc.), and photo editing and video editing programs. The list is endless in what technology can provide in these types of artistic skill sets, but there are some concerns about technology and its impact upon our youth.

Young people are becoming increasingly dependent upon their computers, phones, and devices, whereas it has in some instances become problematic. Clearly, technology is beneficial to teenagers in that it opens up new avenues of ways to create art, to design graphically and learn about the world with just a few clicks of a button on a computer. Technology and media can help bring important social causes to the forefront that were once hidden by providing access to online groups catering to marginalized populations or young people in need of support. Social media sites have reunited long-lost friends and family members and created an easier forum for those hoping to keep in contact with their foreign friends and family from afar. The list of benefits is truly endless, if such internet and social media use is not abused.

The growth of technology and its influence upon our collective youth can also be depleting and stunting on other levels, creating isolation for some, while opening up potential dangers for teens from outside influences and exposure that may not be so healthy. This ties into Burton's (2011) point that there is "a debate at present about the social consequences of spending considerable time alone with a computer" (p. 13), which in fact is sharply contrasted to the circumstances of young people in studies similar to my investigation who have had very little access to technology. In fact, according to Lawrence Steinberg (2016) "studies find that more frequent online communication brings friends closer perhaps because online communication facilitates self-disclosure". Yet there must be a balance here because too much time spent online can have an adverse effect on relationships (p.204). Dominic Mentor's publications offers similar balanced sentiments (2016, 2018). Although the participants in Peru must seriously work hard to source a sustainable internet connection and also find the time to actually spend to post pictures. Their leisure time is severely limited due to work and/or study demands, and that may be the fundamental difference between this particular population and more privilege teens/young adults who possess less time constraints.

This was an initial worry I had when I began the project, although I also recognized immediately that this would be the first time the participants would actually learn how to use technology to create photo- narratives and employ social media in personal ways to share their collective stories when we were not together as a group. However, they may have intellectually and emotionally assimilated their time in institutionalized care, young girls living apart from their family are at a particularly complex touchstone in their lives when they face various pressing adult decisions as they leave their place of residence. Accordingly, this study aimed to employ avenues of creativity, technology and the growth of an artistic community around such goals to address their needs going forward.

Simon Sinek (2016), a present-day speaker on technology and the effects of such on our youth, speaks about the consequences of technology, especially upon millennials (born 1994 and after) and the addiction "high" of responsivity one receives when engaging in social media and texting. "The trauma" a kid feels "when they are unfriended" or their "likes" on any other social media platforms, is a sobering reality for younger generations. The constant need for "approval," especially during the onset of adolescence, is particularly delicate as the formation of "deep, meaningful relationships" is decreasing due to devices, and social media to obtain "temporary relief" to cope with adverse events within one's life (Sinek, 2016).

Sherry Turkle (2011) points out both numerous concerns and praise concerning the way people interact with technology and its potential interference with social interactions and growth of meaningful and "authentic relationships" (p. 4). Her largest concern revolves around connectivity and how technology affects society in an increasingly higher level. She writes, "People love their new technologies of connection. They have made parents and children feel more secure and have revolutionized businesses, education, scholarship, and medicine. The global reach of connectivity can make the most isolated output into a center of learning and economic activity" (p. 152). She points out that one of the greatest advantages here about technology is in fact the advent of educational initiatives that possess the ability to reach impoverished and underserved communities as long as access to internet, computers, and devices is readily available. Turkle goes on to say, "Connectivity offers new possibilities for experimenting with identity and, particularly in adolescence, the sense of free space, what Erik Erikson called the moratorium. This is a time, relatively consequence free, for doing what adolescents need to do: fall in and out of love with people and ideas. Real life does not always provide this kind of space, but the Internet does". (p. 152)

At the orphanage cited in the study, there was but one chair in a corner where the residents can go if they wanted to access tnternet wirelessly. This option was only just provided to them a few weeks before my arrival by my initial contact at the Hogar (the orphanage name). The nuns personally possess two or three computers, which they share, and the girls have about nine computers that they share in a library that has nointernet. Thus, the chances of the participants in this study actually spending long periods alone with their devices were increasingly unlikely, as their days were dictated by chores, their school schedules, and caring for the younger children, which all left very little free time.

The fact that the participants of this study did not have exposure of any type to real internet usage or social media was an interesting notion insomuch that preliminary studies have revealed some rather negative effects of "screen time," especially for young children. Steiner-Adair (2013) makes claims that the onslaught of tech culture can have a hugely negative impact upon young and developing minds. Especially if it is not properly monitored, it may cause tremendous disconnections, particularly during adolescence, which she describes as: "associated with the eruption of serious psychological issues, a time where underlying anxiety and depression creep to the surface, where emotions and intensity become unmanageable, where early childhood traumas resurface". Yet she points out that "in some instances media and social networking have been the means by which troubled teens seek help or are identified by others who reach out to them." (p. 219). In this manner, social media can be helpful, but on the other side, like everything else in life, there are addictive components to technology—and especially the Internet.

Even if websites such as Facebook and Instagram can in fact be helpful to society in its global connectivity, the "compulsive" oversharing of photos and videos can potentially wreak havoc on young people's lives and minds. They become so thoroughly entrenched in keeping up their appearances on social medial sites that their real-life peer and familial relationships suffer. The indiscriminate invasion and monetizing of our privacy are still running unchecked while ethics, and laws are still trying to catch up. Yet, most would argue that this seems to be a largely Westernized problem, as all the gadgets and smartphones are quite expensive and supposedly not so available in impoverished communities. There are consistent costs associated with internet plans, and one must have access to WiFi on an ongoing basis in order to be constantly using their computers, tablets, or smart phones. Thus, the nature of the family unit in underserved communities is quite different in that respect from that of richer societies. There is a healthier dynamic on a humanistic level according to my experience; because people are forced to consistently relate to one another, there simply are far fewer distractions. Steiner-Adair (2013) writes about the notion of a "sustainable family," one that is an "ecosystem, hardy, diverse, resilient, fragile" where members create and share an environment that is uniquely theirs; they are interconnected" (p. 261). The idea of connectedness and connectivity is at the forefront of this chapter's aims, and it has stirred up contrasting notions of how technology can serve one population and divide another, this all being based on the idea of privilege.

Davies and Eynon (2013) corroborate such reasoning that teenagers and technology inevitably become inextricably linked and suggest that it is "during the teenage years that young people most intensively develop, share, and establish their own repertoire of technology-enabled activity, forming a set of practices that—though always to some extent are in a state of flux—will remain with them through their studies and on into their working lives" (p. 1). The authors go on to point out that such technology can be particularly helpful when teenagers are in the midst of dramatic change in their lives and seeking that "autonomy and self-determination of adulthood" (p. 1). Again, this opinion ties in with a somewhat universal notion that most teenagers have the privilege of access. Increasingly, this may be so, but if one examines the statistics on poverty levels all around the world, it is clear that access is limited for a majority of young people, and as much as technology can be an aid in a growing individual's life, it can also be a hindrance. In other words, there must be some balance between "real life" and life with a device. At the same time, like this research study showcases, people find ways to own mobile devices, get connected to the internet and grow their physical communities of support, to include and grow those communities beyond their physical geographical limitations.

MOBILE LEARNING

Technology can enhance the experience of co-learning in the sector of sustainability, especially when it entails students and teachers that are not in the same locale or cross-cultural teaching that involves teachers who must travel to foreign communities and establish a lesson plan or educative initiatives and then keep it going from afar. There are some key features to consider here in mobile learning; flexibility, informality (on some levels), and most of all, mobility. Learners are less subjugated to a scheduled time to meet; although assignments and work are most often kept to a timeline of sorts, this is all dependent upon the teacher or facilitator of learning. The expectations are usually laid out from the start and undertaken through group activity and ongoing dialogue (Naismith, 2005, p. 27).

The dialectic nature of mobile learning is perhaps one of the more intriguing aspects of this theory, as learning becomes a constant flow that is not fixed in one space for a set amount of time. "Dynamic" (Naismith, 2005, p. 7) is a useful adjective to describe what might result in an experience that is rewarding for the learner, as long as she/he is actively engaged and held accountable. By "accountable," I mean that the student must be gauged on some level by the educator and maybe even by his/her peers through their form of participation and contribution to the overall experience.

In order to fully define what mobile learning is, we must first define "formal learning" according to the definition provided by the researchers (Curinga & Saravanos, 2016; Mentor, 2018) under the terms of Traxler (2009), where "formal learning is characterized by groups of learners, curricula, courses, semesters, assessments, and physical campuses" (p. 24). Mobile learning could be argued by some (such as Traxler) that it falls under the domain of "informal learning," which is learning that takes place outside the typical classroom environment, yet he sees it at "odds with" formal learning (p. 24). Curinga and Saravanos (2016), as well as Mentor (2018) point out that there are burgeoning drives for mobile first approaches. And that numerous multi-faceted learning management platforms that now support more formalized mobile learning, such as Blackboard, Canvas, Moodle, Schoology and many others, and though their position is one that is not exactly opposed to Traxler's, their research expounds upon the ways that mobile first approaches and mlearning is continuing to expand avenues in which we can acquire new knowledge on-the-go and integrate experiential quality of formal and informal learning.

There is a design aspect to consider for mobile learning, which is key to creating applications and platforms that are mobile first, user friendly and focus on the "useful experience of the user" (Curinga & Saravanos, 2016, p. 25). The principles of such an experience revolve around social interaction of the participants, engagement with the material presented by the teacher, and assessment of such engagement and contributions of learners to the overall agenda (p. 26). This allows mobile learning initiatives to mirror more formal learning criteria and supports the idea of a "structured" approach that is streamlined, inclusive, and innovative. The researchers conclude that design is one of the most important aspects of mobile learning because of its capacity to aid educators in their abilities "to reflect and reevaluate their practices and commitments" (p. 34).

The largest hiccup in mobile learning for my study was the actual access to Wi-Fi and updating the mobile devices. The refurbished tablet devices were somewhat outdated and could not support large video uploading with such little internet capacity. Thus, the eight participants were not able to fully participate all of the time, and I had to essentially downgrade the study from a "video-making" project to a "photo-art making" endeavor. Also, the issue of safety was, of course, a very large consideration. The girls I worked with lived in an unsafe and impoverished community where devices were a rarity, as were flip-top phones (one girl even had her flip-top stolen at knifepoint while commuting from the orphanage to her first-year college).

The idea of collaborative learning that is technology assisted was one that was forwarded in the art educative experience. The sharing of the photo-art making practice and the imprint of such proved to be an effectual teaching/learning pedagogy that did in fact endure for most of the participants. In this specified case, technology enhanced the experience of co-learning in the sector of sustainability, especially because it involved the principal researcher and participants in a dynamic that was tailored to their specific needs. The collaborative learner model that most supports this study work is that of experiential collaborative learning. The act of sharing an experience and the imprint of such is one of the most effectual teaching/learning pedagogies that have endured the test of time. Collaborative mobile learning again would result in a platform that is group-oriented, yet it may lack intimacy on some level. Thus, as with most learning theories, it is one that is highly situational and worked well for the purposes of this particular study and learning initiative.

The participants to this study created a wide range of autobiographical material which most of them very routinely shared on the social media platforms that we collectively used. Below are some examples of the more recent pictures that some of the girls have posted, as one of the most rewarding part of the project is that most of the participants continued sharing their experiences and kept up with one another through this mode of communication. The first picture is from my very first participant, Anna (pseudonym). Her photos have drastically improved in terms of content, imagination, and perspective. This was always a primary aim of our work together but since the end of the project, she has consistently posted pictures that are thoughtful and with purpose whereas at the beginning, like most of the other participants, her photos were somewhat hurried.

The second photo is from Belinda, another participant that has continued to post, mostly about her new life at the university outside of the orphanage. She is not as prolific a social media poster as Anna, but when she does post, they are usually cheerful and lighthearted glimpses of her life.

Figure three is from Celina, a participant that posted infrequently but then posted a number of months later.

The last photo, figure four, is from Madre Alelia, the real champion of this ongoing work, and who has embraced the opportunity to share vignettes from her life at the orphanage with enthusiasm and vigor.

Figure 1. Photo Anna posted to her social media feed

Currently, Madre Alelia is continuing work at the orphanage but there have been some recent shifts in administration and they are taking in fewer girls because of financial issues. Out of the 8 participants to the study, two quit the project (but both are still on social media), one fell very ill and was hospitalized for a severe eating disorder and the other five are finishing up their studies at the university where they received scholarships.

FUTURE RESEARCH IMPLICATIONS

This study offers some recommended actions for further research and educational opportunities in the field of cross-cultural art education and the role of social media as a means to bring connectedness and connectivity in underserved communities. This study argues that art can be utilized as a tool of support, and that if given a

Figure 2. One of the cheerful photos Belinda posted to her social media feed

chance, most young people wish to have their voices heard and their stories told. Over the 16-month period of time of this study, the intention was to create a sense of community based around art making using technology and social media to enable eight young women to tell their stories about their lives and hopes and dreams both singularly and together. In the fashioning of the stories it was anticipated that new identities would emerge, and supportive collaborations formed.

The future of projects similar to these are reliant heavily upon sustainability efforts from the teaching artists that help to facilitate the ongoing work at institutions such as the Hogar. Due to circumstances too numerous to name here, the viability of continuous and reliable internet access and the obtaining of up-to- date devices for the children and young girls that make up this population unfortunately looks very grim. There continues to be some similar initiatives where teaching artists visit and run temporary art workshops on site and certainly Skyping, emailing and social media has facilitated a large portion of these efforts. The next promising step would be to establish a program that would be a blended day-to-day and distance

Figure 3. Participant Celina posted initially, and then only a couple of months later

learning program (Stead, 2006) which seems to be the best way to support such an endeavor. Yet these ideas all come down to funding and leadership roles both of which are fundamentally lacking in this environment.

CONCLUSION

This 16-month semi-longitudinal study began with a vision of creating a collaborative photo art making and video art practice with just eight participants. This vision broadened as the teachers and nuns of the orphanage began to become more involved with the process of photo art making and narrative storytelling via their unique personal practice. Because of the length of the study and the location of the project site, many unexpected events took place in the lives of the participants, most of which were documented through their photos, videos, interviews, personal conversations, Skyping, emailing, and observational journaling. Essentially, as a participatory

Figure 4. Madre Alelia, a social media participant on the project and champion of the art therapy work and supporter of the post orphanage engagement

observer in this study, I was able to witness many transformations over the course of time, while teaching the girls how to build personal aesthetics and utilize social media as a platform so that we could all share our experiences from afar. They also grew in confidence, and developed skills that were transferable to work contexts.

Restated here, the problem is based on how these eight young women in a secluded society constructed renewed agency, and additional identities as they formed new lives outside of the Hogar, whether it was within the university or in some other form. Art then became a sort of formative tool in which they were able to self-reflect and construct either new or additional identities.

Naturally the question arises of "Why art?" in such an instance, why was this route chosen to reach the aims of this study. Simply put art provides an expressive medium through which young people can reflect upon themselves and find an expression for their own identities. Through the giving and exchanging of the artwork an interchange is achieved of a collective identity. These are tools for thinking about something that is critical in the lives of adolescents on the verge of forging new chapters in their lives.

E-&-m-Learning has emerged as an effective and efficient tool for distance learning (Yousuf, 2007) and reaching populations that were once too far away to logistically participate. For my particular project work, there were significant advantages to using this learning pedagogy and hopefully a future wider scope can be achieved through more careful study and application of e-m-learning. Although there is still much work to be done in terms of opening up a broader based dialogue around the concepts of e-learning within my own research and collaborative projects, the resulting data clearly show that e-&-m-Learning offers a great deal of access to people residing within underserved community or populations with little means to afford higher education.

REFERENCES

Arnheim, R. (1991). *Thoughts on art education*. Los Angeles, CA: Getty Center for Education in the Arts.

Arnheim, R. (1992). *To the rescue of art*. Berkeley, CA: University of California Press.

Burton, J. M. (2011). *Creative and mental growth revisited: Parts 3 & 4*. New York, NY: Teachers College, Columbia University.

Chananchida, Y. (2018). *Sindhuphak Apisak, and Krissana Kiddee*. Bangkok, Thailand: King Mongkut's Institute of Technology Ladkrabang.

Curinga, M. X., & Saravanos, A. (2016). *Mobile first e-learning*. Hershey, PA: IGI Learning. doi:10.4018/978-1-5225-0251-7.ch002

Davies, C., & Eynon, R. (2013). *Teenagers and technology*. New York, NY: Routledge. doi:10.4324/9780203079270

Dewey, J. (1934). The need for a philosophy of education. *New Era in Home and School*, 211-214.

Dewey, J. (1934). *Art and experience*. New York, NY: Capricorn Books.

Dewey, J. (1938). *Experience and education*. New York, NY: Touchstone.

Dewey, J. (1958). *Experience and nature*. New York, NY: Dover.

Friedman, R. A. (2014, June 29). Why teenagers act crazy. *New York Times*, pp. SR 1, SR 6.

Garfinkel, H. (2002). *Ethnomethodology's program: Working out Durkheim's aphorism*. New York, NY: Rowman & Littlefield.

Goeser, C. (2013). *Teaching with the iPad: Adding a new dimension to the museum experience*. Retrieved from https://artmuseumteaching.files.wordpress.com/2013/04image005.jpg

Graber, J. A., Brooks-Gunn, J., & Petersen, A. C. (1996). *Transitions through adolescence*. Mahwah, NJ: Erlbaum.

Grierson, E. (2007). Difference: A critical investigation of the creative arts with attention to art as a site of knowledge. *Educational Philosophy and Theory, 39*(2), 531–542. doi:10.1111/j.1469-5812.2007.00246.x

Groark, C. J., McCall, R. B., & Fish, L. (2011). Characteristics of environments, caregivers, and children in three Central American orphanages. *Infant Mental Health Journal, 32*(2), 232–250. doi:10.1002/imhj.20292 PMID:28543023

Izu, M. R. (2006). Quality and equity in early childcare in Peru. *International Review of Education*, 185–199.

Jason, L. A., Keys, B., Suarez-Balcazr, C., Taylar, R. R., Davis, M. I., Durlak, J. A., & Isenberg, D. H. (Eds.). (2002). *Participatory community research*. Washington, DC: American Psychological Association.

Jorgensen, D. L. (1989). *Participatory observation: A methodology for human studies*. Thousand Oaks, CA: Sage. doi:10.4135/9781412985376

Kagan, J. (1971). A conception of early adolescence. *Daedalus, 1*, 12–16.

Karlsen, J. (1991). *Participatory action research*. Newbury Park, CA: Sage.

Kroger, J. (2009). *Identity in adolescence: The balance between the self and the other*. New York, NY: Psychology Press.

Masten, A. S., Obradovic, J., & Keith, B. B. (2006). *Resilience in emerging adulthood: Developmental perspectives on continuity and transformation*. Washington, DC: American Psychological Association.

McCall, R. B., Groark, C. J., Fish, L., Harkins, D., Serrano, G., & Gordon, K. (2010). A socioemotional intervention in a Latin American orphanage. *Infant Mental Health Journal, 31*(5), 521–542. doi:10.1002/imhj.20270 PMID:28543838

McCann, S. (2017). Mobile Learning in Workforce Development: Cultivating Creativity on Action Learning Teams through Higher-Order mLearning. In D. Mentor (Ed.), *Handbook of Research on Mobile Learning in Contemporary Classrooms* (pp. 139–159). Hershey, PA: IGI Global; doi:10.4018/978-1-5225-0251-7.ch007

Mentor, D. (2015). From app attack to goal-oriented tablet use. In Tablets in K-12 Education: Integrated Experiences and Implications (pp. 1-21). IGI Global. doi:10.4018/978-1-4666-6300-8.ch001

Mentor, D. (2016a). EMxC3= e&mLearning Cultivating Connected Communities: Sustainable Workforce Talent Development. In *Handbook of Research on Mobile Learning in Contemporary Classrooms* (pp. 240–259). IGI Global. doi:10.4018/978-1-5225-0251-7.ch012

Mentor, D. (Ed.). (2016b). *Handbook of research on mobile learning in contemporary classrooms*. IGI Global. doi:10.4018/978-1-5225-0251-7

Mentor, D. (2018a). Micro to macro social connectedness through mobile phone engagement. In *Encyclopedia of Information Science and Technology* (4th ed.; pp. 6184–6194). IGI Global. doi:10.4018/978-1-5225-2255-3.ch537

Mentor, D. (2018b). TLC for MOOCS: Teaching and Learning Communities for Computer Programming. In Computer-Mediated Learning for Workforce Development (pp. 93-110). IGI Global.

MoLeNET. (2011). Retrieved from http://www.molenet.org.uk

Naismith, L., Lonsdale, P., Vavoula, G., & Sharples, M. (2004). *Literature review in mobile technologies and learning*. Birmingham, UK: University of Birmingham.

Ntloedibe-Kuswani, S. G. (2007). *Exploring the use of electronic mobile technologies among distance learners in rural communities for safe and disruptive learning*. Syracuse University.

Şahin, F., & Mentor, D. (2016). Using Mobile Phones for Assessment in Contemporary Classrooms. In *Handbook of Research on Mobile Learning in Contemporary Classrooms* (pp. 116–138). IGI Global. doi:10.4018/978-1-5225-0251-7.ch006

Shippee, M., & Keengwe, J. (2014). mLearning: Anytime, anywhere learning transcending the boundaries of the educational box. Academic Press.

SinekS. (2016). Retrieved from: https://m.youtube.com/watch?v=hER0Qp6QJNU

Stienberg, L. (2016). *Adolescence* (11th ed.). New York: McGraw-Hill Education.

Stiener-Adair, C. (2013). *The big disconnect: Protecting childhood and family relationships in the digital age*. New York, NY: Harper.

Tervalon, M., & Murray-Garcia, J. (1998). Cultural humility versus cultural competence: A critical distinction in defining physician training outcomes in multi-cultural education. *Journal of Health Care for the Poor and Underserved*, *9*(2), 117–125. doi:10.1353/hpu.2010.0233 PMID:10073197

Traxler, J. (2009). Mobile learning. A handbook for educators and trainers. *British Journal of Educational Studies*, *57*(3), 337–339. doi:10.1111/j.1467-8527.2009.437_6.x

Turkle, S. (2011). *Alone together: Why we expect more from technology and less from each other*. New York, NY: Basic Books.

Whyte, W. F. (1991). *Participatory action research*. Newbury Park, CA: Sage.

Wiersma, W., & Jurs, S. G. (2009). *Research methods in education: An introduction*. Boston, MA: Pearson Education.

Yousuf, I. M. (2007). Effectiveness of Mobile Learning in Distance Education. *Turkish Online Journal of Distance Education.*, *8*(4), 9.

KEY TERMS AND DEFINITIONS

Action Research: Research that takes the aim of theorizing alongside the end goal of social action taking place after the culmination of the research project.

Aging Out: When children reach a certain age, they usually must leave their orphanage, place of education, and home and either get jobs, go to university or figure out some means of survival. Some orphanages help these young people navigate these options but most do not.

Cross-Cultural Art Education: Art education that is taught to students by a teacher or teaching artist from a foreign culture and usually in a foreign language.

Double Orphan: Child has lost both parents or been abandoned by both parents.

Francisca Daughters of Misericordia: Third regular order of St. Francis (Española: Hijas de la Misericordia de la Tor de San Francisco).

Instagram: Photograph and video sharing social media site owned by Facebook.

Narrative Research: The participant's story as recorded collaboratively by the researcher.

Orphanage: A place where children who are either abandoned, taken from their familial homes or whose parents have passed away, come to live usually in a community with many other children, usually run by the church if not state-funded.

Single Orphan: Child has lost one parent or been abandoned by one parent.

Tablets: Devices such as iPads or Samsung galaxies that are highly portable and allow users internet access, videoing and photographing capabilities and a plethora of applications, many that would be offered on a regular laptop but all consolidated into one small, lightweight, and highly portable device.

Tumblr: Photo and video sharing site that is used for blogging.

Chapter 12
Mobile Apps, Universal Design, and Accessibility in Schools:
Creating an Inclusive Classroom Experience

Tseday Alehegn
Marymount Manhattan College, USA

Richard Keller
Columbia University, USA

ABSTRACT

How can mobile applications may be used to enhance accessibility for individuals with disabilities in educational settings as well as to encourage the development of apps that embrace universal design principles? Smartphones and tablets, often by default, provide alternative accessible features addressing the needs of individuals with visual, auditory, and/or physical disabilities. As schools make the leap into the mobile age, it is imperative to explore mobile applications developed with the universal design learning framework in mind and intent to provide equal access in learning environments. Increased use of mobile apps in the classroom goes hand in hand with the growing popularity of these tools in everyday life outside of the school context, and this chapter is a call for greater engagement with mobile applications that have been designed to be accessible for individuals with disabilities, and to promote a more collaborative and equitable learning environment.

DOI: 10.4018/978-1-5225-9351-5.ch012

INTRODUCTION

Since the introduction of mobile applications for Apple Inc.'s iPhone in 2007 (Strain, 2015), their use has diffused into various sectors of society including in classrooms and institutions of learning. Prensky (2007) refers to the youth of today, whose lives are inevitably intertwined with the use of the Internet and social media, as 'digital natives' while Gardner & Davis (2013) dub these youth as the 'app generation.' Acknowledging the popular use of mobile apps among the student population Antonucci (2014) cites several reasons why such technologies should continue to be promoted in classrooms for purposes including: providing alternative texts in e-book format, the retrieval of supplemental learning resources, and to assist with fieldwork or research. Within classrooms teachers of primary, secondary and college-level students have used mobile phones and apps in various creative ways ranging from the design of assignments incorporating QR codes for additional information on a subject to the production of videos or podcasts, as well as a medium for crowdsourcing and sharing learning materials (Onlineuniversities.com, 2012).

Teachers, like their student population, have increasingly embraced the utilization of mobile technologies in classrooms. A survey of pre-service teachers by O'Bannon & Thomas (2015) has shown that while 25% of pre-service teachers supported the use of mobile phones in the classroom in 2013 that number had increased to almost 50% by 2015. O'Bannon & Thomas (2015) further note that as use of mobile phones in the classroom increased so too did teachers' perceptions of its potential benefits for schoolwork.

Despite the obvious enthusiasm among both teachers and students to integrate mobile phone and apps into the classroom experience, teachers express apprehension about cyber-bullying, cheating, and disruptive behaviors including texting that impede student engagement and performance (O'Bannon & Thomas, 2015). As a result a majority of secondary schools across the nation have continued to prohibit the use of mobile phones in their classrooms (O'Bannon & Thomas, 2015). Contrary to this trend, in March 2015 the Department of Education in New York City ceased its ban of mobile phones in classrooms that had affected approximately 1.1 million students (NY Daily News, 2015).

More importantly, a significant barrier to using mobile phones for instruction in classrooms includes an app's lack of accessibility or universal design, which in turn affects the equal participation of students with disabilities. This chapter focuses on the issue of mobile app accessibility and the inclusion of the needs of individuals with disabilities in the formal learning setting. In addition to featuring existing examples of accessible mobile technologies used by students with disabilities, this chapter delves deeper to inquire about the overall design of mobile apps in use in classrooms today. Are these mobile apps being developed with the principles of universal design in mind? Are we as educators shifting our attitudes from using assistive technology to advocating instead for accessible mobiles that help foster inclusive classroom environments?

This chapter provides an overview of U.S. disability laws that govern equal access to instructional tools and technology including electronic information and use of mobiles, relevant qualities of universal design, and the synergy between the use of accessible mobile apps and the enhanced classroom experience.

BACKGROUND: DISABILITY LAW OVERVIEW AND ASSISTIVE TECHNOLOGY

According to the World Health Organization's disability and health factsheet approximately 1 billion individuals, comprising about 15% of the global population, are living with a disability (World Health Organization, 2018). The World Health Survey likewise reported that 15.6% of individuals aged 15 and older were considered disabled. In the United States the Centers for Disease Control and Prevention reported in 2015 that an estimated 22% of adults nationwide were individuals with disabilities (Courtney-Long et al., 2015). Individuals with disabilities often face barriers in various sectors of society, which impede equal access to services. The World Health Organization (2011) describes the nature and types of barriers including factors such as inadequate regulations and standards; discrimination and negative attitudes towards persons with disabilities that impede their participation; lack of accessible services; inadequate funding for programs that could make resource allocation more inclusive; lack of data and evidence from research to promote accessibility; and lack of direct consultation and feedback from individuals with disabilities regarding their views and needs. The World Health Organization (2011) further notes that in nations that are considered high-income, and where resources are otherwise potentially more available, the number of individuals with disabilities not receiving assistance for everyday tasks may be as high as 40%. Worldwide, 31 out of 93 countries surveyed by the World Health Organization did not have adequate sign language interpreting services for individuals who were deaf or hard-of-hearing, and overall persons with disabilities worldwide had access to significantly less information and services than those without disabilities (World Health Organization, 2011).

The UN's *Convention on the Rights of Persons with Disabilities* was ratified in 2008 drawing attention to the need for increased global accessibility. By November 2015 the convention had been signed by 178 states, yet despite the fact that the United States is a signatory it has not yet ratified the convention (United Nations, n.d.). Preamble r of the convention emphasizes that children with and without disabilities should enjoy the same levels of freedom and protection; while Article 24 calls for equity in education and inclusion in design at all levels of schooling (United Nations, 2006).

Disability laws in the United States were first introduced in 1973 as Section 504 of the Rehabilitation Act, followed by the Individuals with Disabilities Education Act (IDEA) in 1975, and culminating with the passage of the American with Disabilities Act (ADA) in July 1990; all of which required that all organizations serving the public in the United States, including schools, provide accommodations that promote equal access for individuals with disabilities.

Section 504

Section 504 of the Rehabilitation Act of 1973 was one of the earliest U.S. national laws that addressed the rights of individuals with disabilities. It prohibits discrimination on the basis of disability by any entity receiving federal funding. Under Section 504, persons with physical or mental impairments that limit activities of daily living are defined as persons with disabilities. Section 504 seeks to ensure equal access to services through the provision of reasonable accommodations, resulting in a more accessible environment. Discrimination against an individual with a disability or the denial of his/her benefits is strictly prohibited by any organization that receives any form of federal funding (USCode.house.gov, 1973).

Individuals With Disabilities Education Act (IDEA)

Originally passed as the Education of Handicapped Children Act in 1975, the Individuals with Disabilities Education Act (IDEA) as amended in 1990 provided for equal access to public schools for children with disabilities (University of Washington, 2015). This federal regulation primarily deals with national, state and local government initiatives that focus on early intervention and special education services for children and youth ranging in age from birth to 21 (US Dept. of Education, n.d.). The most recent reauthorization of IDEA in 2014 structures this law into four parts: Part A - General Provisions; Part B - Assistance for Education of All Children with Disabilities; Part C - Infants and Toddlers with Disabilities; and Part D - National Activities to Improve Education of Children with Disabilities (IDEA, 2004). The U.S. Department of Education (n.d.) estimates that there are approximately 6.5 million children and youth with disabilities nationwide whose equal access to education was addressed by IDEA, and in 2010 a report released by the U.S Department of Education showed the significant impact of IDEA legislation on protecting the civil rights of students with disabilities including decreased drop out rates from high school and growing enrollment in colleges (APA.org, n.d.).

American With Disabilities Act (ADA)

In 1990 the Americans with Disabilities Act (ADA) was signed into law by President Bush. The intention of this law was to extend protections offered to individuals with visual, hearing, physical, learning, medical or psychiatric disabilities set forth by previous legislation, and for the first time to ensure their equality in all sectors of life as a federal civil rights law. The law's various titles cover specific areas including: participation in schools, places of employment and access to public transportation (ADA.gov, n.d.). The provisions under Title II and III specify obligations most applicable within institutions of learning. In addition to their obligations under the ADA, educational institutions continue their parallel obligations under Section 504 as recipients of federal funding. Although most people remember the passage of the ADA in 1990, all of its Titles actually went into effect in January of 1992 with the adoption of the associated regulations. Enforcement of all Titles of the ADA is overseen by the U.S. Department of Justice (ADA National Network, n.d). Educational entities are predominantly governed by Title II and III depending on whether the institution is part of a state or local government or a place of public accommodation. Title II outlines requirements that state and local governments need to adhere to. Specifically, Title II mandates that government entities must have programs that ensure accessibility in place and where need be provide reasonable accommodations to eliminate discriminatory practices (ADA National Network, n.d). These obligations apply to the removal of physical barriers as well as to the provision of additional resources such as texts in alternative, accessible formats. The mandates governing entities under Title III of the ADA are similar to the mandates associated with Title II, except that these are applicable to places of public accommodation rather than bodies connected to state and local governments. Places of public accommodation include retail stores, hospitals, and non-public schools (ADA National Network, n.d). The American with Disabilities Act was amended in 2008 (ADAA), which refined the definition of disability (ADA.gov, n.d). Essentially, this amendment sought to reclaim disability territory, which had been eliminated by the courts. In addition to redefining what disability is, the 2008 ADAA revised Titles II and III and Standards for Accessible Design, all of which came into effect in March 2011 (ADA.gov, n.d.).

Section 508

The Rehabilitation Act of 1973 was amended in 1998 to promote accessibility of electronic information for persons with disabilities. The amendment required that all federal agencies must make electronic information accessible to both employees and members of the public using government systems. The United States Access Board was tasked to develop accessibility standards for electronic information as outlined by Section 508 (Section508.gov, n.d.). In 2001 the regulations were further amended whereby federal agencies as well as any organization receiving federal funds was obligated to provide accessible electronic information according to those federal standards (Hornor, 2013). The United States Department of Justice defines electronic and information technology (EIT) as any item that is capable of storing, displaying or transmitting information (United States Department of Justice, 2015) and the United States Access Board (n.d.) notes that the Section 508 standards apply to electronic systems such as computers and their networks, materials for telecommunication, web-based applications, and any other multimedia technologies. Horner (2013) notes the three main components outlined in Section 508 regulations including: ensuring the coding of electronic sites and systems work well with assistive technologies used by persons with disabilities; that the electronic system has an overall positive usability by all persons; and that resources and information supplied about the technologies are equally accessible to individuals with disabilities. Horner (2013) also emphasizes that these three criteria, and Section 508 regulations in general, are simply the least acceptable for making electronic information and technology user-friendly and accessible for all individuals. The website on 21st century information fluency (O'Connor, 2003) moreover states that institutions such as public schools, that may receive federal funds, must also be compliant with Section 508 regulations (O'Connor, 2003).

Mobile Applications and Accessibility Standards

The proliferation of mobile applications makes one ponder what standards of accessibility are outlined for the use of this innovation in institutions such as public schools, which need to remain compliant with Section 508 and the ADA. According to Avila (2013) Section 508's regulations regarding web-based content accessibility as well as criteria for overall functional use of electronic information and technology can encompass mobile accessibility requirements. Furthermore, institutions that are required to be ADA compliant must make sure that websites viewed through mobile applications also meet the Web Content Accessibility Guidelines (WCAG) as well as follow the recommendations in the *Mobile Web Best Practices for Mobile*

Web Apps that was developed by the World Wide Web Consortium (Avila, 2013). In addition, President Barack Obama's 21st Century Communications and Video Accessibility Act of 2010 (CVAA) – which outlines performance and functional-related accessibility objectives for electronic communication systems and devices in order to remain compatible with Section 508 regulations – may be applied to mobile applications as well (Avila, 2013).

The Data on Student Population With Disabilities

According to the US Census report titled *Americans with Disabilities:2014* approximately 85 million people – that is 27.2% of the total population in the United States -- lives with a disability (Taylor, 2018). An estimated 58.5% of this population were 65 years and older while 30.3% were between ages of 18 and 64, and 17% were less than 18 years old (Taylor, 2018). Out of the total number of individuals with disabilities recorded by the Census in 2010 approximately 16.7% were in the age range between 21 and 64, 8% were children with disabilities under the age of 15, and an estimated 692,000 children were diagnosed with learning disabilities (Brault, 2012). In addition, data from the National Center for Education Statistics (2015a), shows that the number of students who received disability-related services rose from 8.3% in the mid 1970s to 14% by 2005 with a large portion of this increase stemming from children identified has having learning disabilities.

The Centers for Disease Control and Prevention (2009) rank visual impairment as one of the top 10 disabilities among individuals aged 18 and older, while the 2014 annual report of the American Printing House for the Blind cites the number of visually impaired students in American elementary and high schools who were eligible for alternative texts in braille, large print or audio format as totaling approximately 60 thousand, with 8.5% being braille readers, 29.2% print readers, and 9.2% as auditory readers (National Federation for the Blind, 2014). Approximately 13% of the U.S. population aged 12 years and up has been diagnosed with hearing loss according to the National Institute on Deafness and Other Communication Disorders (2015).

In regards to students with learning disabilities an estimated 11% of youth aged 4-17 were diagnosed with Attention Deficit/Hyperactivity Disorder (ADHD) in 2011 with the percentage continuing to rise (CDC, 2016). In the past decade alone the CDC notes that ADHD rates increased by 53% -- approximately 6 million children diagnosed with the condition -- and with prevalence more evident among boys than girls and in the older teen population than among children (National Center for Learning Disabilities, 2014). It is estimated that 5% of students enrolled in public schools in the United States have been diagnosed with a learning disability, making them the largest group to receive disability-related accommodations among the 5.7

million students categorized as needing special education services (National Center for Learning Disabilities, 2014).

In the 2012-13 academic year the National Center for Education Statistics (2015b) announced that the population of children and youth identified as receiving disability-related accommodations through IDEA totaled 6.4 million or approximately 13% of all students enrolled in U.S. public schools. In the same year, an estimated 87% of students with speech or language impairments, 67% with specific learning disabilities, 64% with visual disabilities, and 62% with developmental delays were recorded to spend the majority of their school day in a regular or general classroom (National Center for Education Statistics, 2015b). Given this fact disability regulations such as IDEA require that technologies that are assistive in nature be utilized to help develop equal access classrooms for all students.

Assistive Technologies for Students With Disabilities

The United States Access Board (2000) defines assistive technology as any item or system that is designed specifically to assist a person with a disability either to increase their functionality or provide greater access. In other words it is any technology geared to enhance the participation of a person with a disability more fully in society. Miller et al. (2013) note that assistive technology ranges from simple items such as pencil grips to mobile devices including iPads. Assistive technologies, such as talking computers for students with visual disabilities, have been cited as effective by various researchers in addition to other services and accommodations provided by institutions of learning mandated to follow federal disability laws (Bryant et al., 1998). Assistive technologies may be categorized according to the functions each assists including: listening devices such as voice recorders; devices related to writing such as spell checkers; reading devices such books on tape; devices that assist with organization and memory; and items for completion of mathematical tasks such as talking calculators (Bryant et al., 1998). In accordance with IDEA students with disabilities in K-12 education settings are provided with an Individualized Education Program (IEP), which further specifies whether the use of assistive technologies is appropriate as part of accommodation recommendations (Bryant et al., 1998).

The use of assistive technologies to enhance the participation of students with disabilities who spend majority of their time in regular classrooms was also supported by Rodriguez et al., (2013) who applied the recommendations to mobile devices citing their high prevalence, which makes them less likely to attract unwanted peer attention from non-disabled students, as well as their open-ended design that makes it easier to tailor its functionality per individual basis.

One of the main barriers faced when considering the use of assistive technologies in a regular classroom is that most of these items were designed primarily as a technology for a specific sub-group – persons who are disabled. Newton & Dell (2011) describe how many students with disabilities don't want to be singled out or made to feel 'different' among their peers by virtue of using 'special' devices. According to Newton & Dell (2011) the use of ubiquitous mobile devices such as cell phones or iPads as assistive devices in classrooms can help reduce the stigma faced by students with disabilities by virtue of the technology not being directed at a single group but rather being inclusive, where all students can use the mobile devices at the same time. New & Dell (2011) also point out the relatively low cost of such mobile technologies in contrast to adaptive software and subsequent licenses that are sometimes too expensive to implement widely in public institutions of learning receiving limited funding. The training of staff and students needed to use assistive technologies can likewise be another barrier to implementation, which mobile phone and tablet utilization may circumvent given their more efficient user-friendly design.

As mobile devices and applications continue to rapidly permeate our society and have seeped into the fabric of educational institutions it is easy to assume that technologies, assistive or otherwise, can provide an increased sense of independence for individuals with disabilities. This follows an 'ableist' discourse where people with disabilities are made to feel that they are succeeding in life if they manage to become more independent by relying on assistive technologies. The danger of this narrative is that it equates technology with liberation while in fact the world remains saturated with technologies that are not accessible to individuals with disabilities. Foley & Ferri (2012) therefore call for an examination of our cultural narrative that normalize certain ways of being and interacting with technology as well as paying attention to the design and implementation of devices that may end up ignoring the daily barriers faced by individuals with disabilities in a technology driven world. Foley & Ferri (2012) also draw our attention to the legal implications of introducing ubiquitous mobile technologies that are not accessible into the classroom environment citing the lawsuit filed by the National Federation of the Blind (NFB) and the American Council of the Blind (ACB) against the use of Amazon's kindle e-reader tablets at four universities in the United States who argued that despite the device's text-to-speech feature several of its other interfaces including ability to switch a page or go to the menu page was inaccessible to students with visual or print disabilities. In effect the incorporation of text-to-speech was not incorporated in the overall design of this mobile tablet and thus created unequal access to information in a learning environment. Such issues shed light on the drawbacks of designing technology with the intention to be assistive rather than accessible for all persons. We argue that shifting the focus from 'assistive' to 'accessible' technology design makes for a more inclusive learning environment.

FROM ASSISTIVE TO ACCESSIBLE: UNIVERSAL DESIGN SOLUTIONS FOR LEARNING

The United States Department of Justice (2015) defines a technology system or device as 'accessible' to individuals with disabilities if it designed to be used in a way that does not rely on only one of our senses, and that the technology is compatible to be utilized in conjunction with other adaptive software. While 'assistive' technology is usually tailored to serve people with disabilities 'accessible' systems are more broader and can be used by both disabled and non-disabled individuals as the need arises. For example, Cragun & Todd (2013) provide scenarios where accessible technologies come in handy for non-disabled persons in everyday life including when outside lighting and other environmental factors require font-enhancing features such as zoom-in interfaces on technology devices. The focus on accessible design also leads to the topic of universal design principles for learning, where the goal is to develop user-friendly design that anyone irrespective of disability status can use with ease including in the mobile app arena.

Universal Design Principles

Given the data on the population of students with disabilities in regular classrooms in public schools it is essential to consider the selection of mobile apps that are not only assistive in nature, but also likewise accessible as a result of developers using universal design principles. Developed in 1997 by Ron Mace, Founder & Program Director of the Center for Universal Design at North Carolina State University, the principles of universal design encourage designers to consider a framework that helps create an inclusive environment so that all persons can have equal access to the features of a technology, thereby reducing the need to re-tailor or re-amend interfaces for individuals with disabilities (Duranczyk et al., 2013). The Center for Inclusive Design and Environmental Access (IDeA Center) defines universal design as something that makes items more user-friendly and accessible to everyone (Center for Inclusive Design & Environmental Access, n.d.). An alternate definition of universal design is suggested by the Institute for Human Centered Design as a framework used when designing spaces, materials, and communication information that can be user-friendly for the highest range of individuals without requirement special alterations (Institute for Human Centered Design, n.d).

Universal design principles appear to promote a broader form of user-friendly interfaces that acknowledge the needs of a diverse population of users. This type of design is also known as 'human-centered' (Institute for Human Centered Design, n.d) with the framework developed through collaborations between professionals from various sectors including architecture, engineering, and product design (Burgstahler, 2012). Universal design is embraced in Article 4 of the 2006 United Nations Convention on the Rights of Persons with Disabilities, which calls for the promotion of goods and services that follow universal design principles when developing protocols (Duranczyk et al., 2013).

Mace's universal design proposal incorporates seven principles. The first principle, equitable use, addresses design that is developed with the diverse abilities of individuals in mind. The second principle, flexibility in use, similarly focuses on acknowledging the diverse range of preferences by individual users of the technology. The third principle promotes simple and intuitive use so that any user, regardless of experience or knowledge level, may be able to interact with the technology with ease. Perceptible information, the fourth principle in universal design, emphasizes the need to communicate information effectively while the fifth principle, tolerance for error, takes into account the necessity to minimize the risk of accidents and adverse reactions from use of an item. The sixth principle of universal design seeks to minimize levels of fatigue and increase efficiency and comfort while using the technology. Last but not least, the seventh principle makes sure that an appropriate amount of space is available for users of a diverse range of body sizes when interacting with the technology and that their mobility or posture is not compromised in the process (Burgstahler, 2012).

In addition to the seven principles of universal design proposed by Mace the Center for Inclusive Design and Environmental Access also suggests eight goals of universal design to consider including body fit; comfort with regards to body function during interaction; awareness of information as easily perceivable; promoting understanding of methods and operational procedures that should be clearly outlined; contributing to the wellness of the individual; promoting social integration; personalizing the experience to include ability to makes choices and individual preferences; and respecting diversity in cultural values (Steinfeld, 2012).

The Universal Design for Learning Framework

Inspired by Mace's universal design principles the National Center on Universal Design for Learning (2014) promotes similar principles among educators involved in curricular development so that academic materials used in classrooms as well as methods of teaching give all students an equal opportunity to participate in the learning environment. Universal Design for Learning (UDL) acknowledges the need to provide open frameworks for instruction that validate the diverse abilities of learners, both disabled and non-disabled in regular classrooms. According to the TEAL Center (n.d.) UDL has three primary areas of focus including: multiple means of representation or the 'what' of learning; multiple means of action and expression – the 'how' of learning; as well as multiple means of engagement– the 'why' of learning.

The National Center on Universal Design for Learning (2014) ties UDL's three areas of focus to the stimulation of three primary brain networks. The first is the recognition network, which is used to gather information and identify reading, writing and other stimuli in the learning process. This network is tied to UDL's multiple means of representation framework with the goal of presenting content via various methods. The strategic network focuses on multiple means of action as students learn to express the knowledge they receive in classrooms. The third brain network is affective and promotes the 'why' of learning seeking ways to motivate and engage students in their learning endeavors. Although Izzo & Bauer (2015) point out that the first federal law to define UDL was in the 2008 Higher Education Opportunities Act, the guidelines are nonetheless pertinent to all levels of learning from K-12 grades as well.

Moreover, the TEAL Center (n.d.) notes that in UDL educators commence the curriculum development process with the intention to disseminate the information to students with a diverse range of abilities and backgrounds and encourages flexibility in instructional methods, techniques and materials. As such UDL serves as an overarching framework that is compatible with various forms of classroom teaching ranging from multimedia based instruction to group-centered learning. While making sure that students with disabilities are learning in an inclusive environment universal design principles for learning simultaneously promote learning for all other students in an enhanced and equitable class environment.

Universal Design for Learning With Mobiles

As learning with mobiles applications is embraced by both youth and educators in schools, the conversation about accessibility and universal design for learning with mobiles has become inevitable. Mobile applications come in a range of formats from cellphones to tablets, and with it comes an immense variability in features and functions that may or may not be designed with universal and inclusive principles in mind. Screens with poor resolutions and keyboards that are too small are just a few examples of the accessibility issues of mobile devices (Elias, 2011). However, with the implementation of universal design principles educators may also be able to harness the benefits of mobile learning including the low-cost of the devices in contrast to other adaptive technologies and licenses; the ability to use create and share multimedia educational content using built-in features such as the microphone and camera; and the ability to continue the learning process outside of the classroom as youth use and interact with their mobile phones and tablets outside of the school environment and at home (Elias, 2011). What the UDL framework advocates for is what Izzo & Bauer (2015) call 'student-centered learning environments' where classroom-related materials, from curriculum to technologies used, is evaluated for its ability to be used by each learner without running into issues of access.

Researchers have argued that mobile applications are well aligned to UDL principles (Reid et al., 2013) observing that their features can be individualized -- as in the example of a dyslexic student using spell-checker apps on his/her phone, iPad or other tablet device. The ability to customize and choose apps tailored to one's needs as well as the ease in portability of a mobile device follows by default the principles and goals of universal design including flexibility in use and increasing efficiency while remaining accessible to a diverse population outside of those diagnosed with a disability. In essence, just as excluding students with disabilities from regular classrooms is prohibited by law, it is equally imperative that mobile technologies introduced into regular classrooms are accessible to all students in the learning environment.

Accessibility on Mobile Devices

The current design of mobile devices such as smartphones and iPads already incorporates several built-in accessibility features that make mobiles prime candidates for use in an inclusive classroom. Operating systems including IOS and Android both provide various formats that are accessible for individuals who are deaf or hard-of-hearing, have visual or print disabilities, or have physical or learning disabilities, and Wee (2013) urges developers to consider some key points while releasing new applications, which includes leveraging the power of voice and including talk-back or text-to-speech features that work on all parts of the interface including: menu and button selections; providing alternative descriptions of images such as logos; designing an accessible layout that individuals with disabilities can navigate efficiently.

Prior to such accessible features being built-in to technology devices individuals with disabilities had to purchase and use adaptive software to assist with zooming in on text or getting access to textual information via screen readers, which Runyan (2013) points out can be an extra expense not otherwise incurred by non-disabled individuals. Today the built-in accessibility features of smartphones and tablets allow individuals to retrieve information without necessitating additional assistive software. On the iPhone, for example, the built-in FaceTime app allows individuals who are deaf or hard-of-hearing to communicate using sign language via its video interface. Izzo & Bauer (2015) also note Apple Inc.'s iSight built-in webcam that also allows for virtual face-to-face interactions for a sign language session. In a research study conducted among students ages ranging from 3 to 19 years old teachers using iPads noted the touch and gesture features of the device in conjunction with multi-media presentations that incorporated visuals and sound in learning sessions that enhanced the students' engagement levels while at the same time assisting students with physical disabilities with an accessible one-touch tactile experience (Flewitt et al., 2014). The flexibility of mobile applications is ideal for inclusive education classrooms as it can be easily tailored to individual student needs, while its portability means the learning session can be further extended outside of the regular classroom hours as the student continues using the mobile device on their own.

Accessibility is also critical when accessing information from websites via mobiles. Recently, Google released an announcement that more searches are now being made from mobiles rather than desktop computers in the United States and in 11 other nations (Sterling 2015). As mobile use outpaces computer use it becomes evident that web accessibility from mobile devices is a challenge to address both as a Section 508 concern as well as from the framework of universal design principles. According to Henry (2015) the World Wide Web Consortium (W3C) – that is in charge of developing web standards – has launched a Web Accessibility Initiative (WAI) and provided guidelines that can be applied to mobile-based web accessibility as well. Henry (2015) shares that WAI's mobile accessibility efforts include the release of a report entitled *Mobile Accessibility: How WCAG 2.0 and Other W3C/WAI Guidelines Apply to Mobile* that supports the ability of individuals with disabilities to access web information via phone, tablet or any another portable device. In addition, the W3C's WAI-Accessible Rich Internet Applications (ARIA) technical documents also elaborate on ways that content for websites may be designed to be accessible on devices including mobile technologies (Henry, 2015). Moreover, W3C's August 2015 report entitled *Standards for Web Applications on Mobile: Current State and Roadmap* further define how web applications in the mobile context can follow W3C guidelines. To test web accessibility on mobiles running IOS developers can use Xcode's Accessibility Inspector that comes with Apple's Developer Tool Kit or IBM's mobile browser simulator on its developer platform Worklight to test hybrid and mobile web applications (Cragun & Todd, 2013). Using mobiles in public school classrooms requires not only that educational apps are Section 508 compliant but also that web accessibility on mobiles is equally enhanced as learning tools shift from desktop to mobile apps.

Accessible Apps for Students With Learning Disabilities

Given the plethora of 'educational' apps available at Android and Apple store that educators can choose from and incorporate below are some examples of accessible mobile applications that individuals with disabilities may benefit from in the educational setting.

The learning disabilities category includes a diverse range of individuals that have been diagnosed with conditions such as Dyslexia (difficulty reading and correctly identifying words and letters); Dyscalculia (difficulty with arithmetic calculations as a result of a disability); Dysgraphia (difficulty in writing as a result of a disability); Attention Deficit/Hyperactivity Disorder (ADHD); deficits in auditory or visual processing; deficits in executive functioning; or non-verbal learning disabilities.

For students with learning disabilities Bookshare is a well-known online library tailored towards individuals with disabilities, but is also popular among non-disabled people with interest in reading books in electronic format. Benetech's Read2Go app is a mobile app released in 2013 that allows students to access the Bookshare library via iPhone and iPad with the additional feature to customize font size and reading voice. The organization Common Sense Media rates Read2Go as an app designed for students ages 5 and older (ADK AccessAbility, n.d).

Kurzweil's firefly is a more comprehensive text-to-speech app also available for iPad that emphasizes universal design principles and is used by individuals with Dyslexia, Dysgraphia, and those with visual or print disabilities along with non-disabled populations such as language learners. In addition to its customizable text-to-speech feature available in 18 languages, firefly also has features for note-taking and outlining, as well as visually highlighting words and phrases as one reads texts including in PDF format (Kurzweil Education, n.d.). Research supports the theory that students diagnosed with learning disabilities may absorb academic material more effectively when it is presented in a multi-sensory mode that includes both audio and visual material (Reid et al., 2013) and Kurzweil's firefly app provides this functionality while developed with universal design principles in mind.

GoodReader is another accessible app developed with universal design principles that allows highlighting of text, note-taking, and reading of any format including books, movies and images. GoodReader also incorporates PDF editing tools including extraction of information from pages (Goodreader.com, n.d.).

StoryboardThat is a digital storytelling app that is available on both Android and IOS platforms that includes the use of OpenDyslexic – an open source typeface tailored towards individuals with disabilities (OpenDyslexic.org, n.d.). Developed in accordance with universal design for learning, StoryboardThat features includes a drag and drop interface, and customizable font size.

Dragon Dictation is an accessible speech-to-text app developed by Nuance that is also available on both IOS and Android platforms for smartphones and iPads that is beneficial for producing various text formats including emails, notes, and essay assignments. Individuals with learning disabilities can use it along with the non-disabled population who simply prefer to dictate rather than type. In accordance with universal design principles Dragon provides a training session where individuals read excerpts of texts as the app prepares a customized profile of the reader and is better able to translate their speech to text including punctuation commands. While the desktop software for Dragon Dictation requires a software license it appears that the smartphone app is free, however Wi-Fi access is necessary to conduct a dictation session. As long as Wi-Fi is available dictation can be done on-the-go, which provides flexibility in use.

Accessible Apps for Students With Visual and Print Disabilities

For students with visual disabilities familiar with braille Apple's BrailleTouch app for iPhone and iPad lets individuals type using braille on their touchscreen devices. Holloway (2012) highlighted the 92% accuracy rate of the app during early phase testing, and noted that BrailleTouch is developed as an open-source technology that reads aloud what is typed using six keys that are gesture-based. The principal investigator of the prototype was also quoted by Holloway (2012) as asserting that users who don't know braille can easily learn the six-digit system to type faster as well. A pilot study conducted to evaluate the use of BrailleTouch for literacy among individuals with visual disabilities recruited 4 braille instructors to assess its functionality and received enthusiastic feedback from a participant regarding preferring to use BrailleTouch as a low-cost method in lieu of other electronic devices they traditionally utilized with their embosser (Frey et al., 2012). Another pilot study that recruited 3 teenage students with visual disabilities from the Nashville Public School System studied their experiences using vibratory touchscreen mobile devices in a math education program focusing on graph-related topics in lieu of traditional methods that included the placement learning objects on tactile graph paper. The participants were enthusiastic about the reduction in time lag between instruction and haptic graphical presentations as well as the touchscreen's overall interactivity (Gorlewicz et al., 2014).

For iPad and iPhone users VoiceOver is Apple's built-in accessibility app for individuals with vision or print disabilities as well as for all other users who prefer text-to-speech output. The VoiceOver app allows users to select use of either braille or regular keyboard typing or even use single finger swiping to go across words that the app reads aloud (Runyan, 2013). In 2015 the American Foundation for the Blind

(AFB) recognized Apple's efforts in developing VoiceOver to improve accessibility in accordance with universal design principles (Chapman, 2015). In addition, the optical character recognition (OCR) app, Prizmo, for IOS platform assists users to extract texts and images working in conjunction with VoiceOver's text-to-speech feature (Runyan, 2013). Taking VoiceOver's accessible interface one-step further researchers Sierra & Roca de Togores (2012) have also developed a low vision mobile portal that includes email, calendar and SMS apps among others features.

Runyan (2013) notes that Bookshare's Read2Go app -- previously described as a mobile tool for students with learning disabilities – was developed and launched following IDEA regulations that required educational institutions to provide all instructional material in accessible format. Read2Go is likewise an accessible app that individuals with visual or print disabilities may also use that is compatible with Apple's VoiceOver built-in accessibility tool, and allows low-vision students to customize font size and color while blind students can likewise customize the text-to-speech interface (Runyan, 2013). Apple's built-in Zoom feature, as well as camera magnification feature, is compatible with several apps including iMagnifier, which serve as accessible tools for individuals with low vision (Runyan, 2013).

Accessible Apps for Deaf/Hard of Hearing Students

As previously mentioned, the FaceTime built-in feature on iPhone and iPad mobile devices have been widely used by individuals using sign language for communication along with the built-in iMessage app for texting (Perez et al., 2014).

Kushalnagar & Trager (2011) have published an intriguing study comparing lecture access in classes attended by 18 deaf students and found that where their attention is divided between the teacher and presentation of multi-media material a single camera installation doesn't offer adequate accessibility to learning materials as would the use of smartphone cameras operated by several participants capturing lecture material in various formats, not to mention the possible use of OCR to produce text from captions, thereby creating a more flexible and universally designed learning environment.

Moore (n.d.) also cites the ClearCaptions app as an accessible tool developed by Purple Communications released in 2011 and available for both Android and IOS devices. ClearCaptions has also been designated as a Federal Communications Commission-certified telephone captioning provider, and in 2013 the Ensemble amplified phone was released that included ClearCaptions as a built-in application (ClearCaptions.com, n.d.).

Accessible Apps for Students With Physical and Developmental Disabilities

There are several apps that are accessible for individuals with physical or developmental disabilities that take into account limitation in gesture and mobility. AssistiveTouch is an app that works with single finger tactile responses for iPad and iPhone, and iPod users (Perez et al., 2014). Featured in the New York Times, AssistiveTouch can be turned on from the IOS settings app, whereby the white circle that is activated works with the use of single finger taps instead of using a feature's regular buttons. AssistiveTouch's tapping feature works with screen rotation, locking screen, and adjusting volume (Pogue, 2011).

In addition to its note-taking and reading interface for students with learning disabilities the GoodReader app cited earlier also has easy tactile feature that allows drawing or highlighting with a single finger as well as use of a bluetooth enabled page turner that can be activated with a foot pedal press for individuals with physical or developmental disabilities (Goodreader.com, 2015).

With regards to individuals diagnosed with autism a recent study conducted among 3 individuals aged 8 to 14 year old examined the effects of visual texts provided via iPad along with pictures for participants to view along with the verb or noun labels. The communication intervention supplied via a mobile device was preferred over traditional cards as it was inconspicuous to others and easily portable, without requiring the carrying around of paper and writing gadgets by the student using the mobile intervention and accessible learning tool (Ganz et al., 2014).

In another small-scale research 4 high school participants diagnosed with moderate to severe intellectual disabilities, recruited from a Midwest public high school, were provided with the same content and instructions and provided with either paper and pencil or an iPad. The researchers found that a majority of the students preferred the electronic mobile device (iPad) to the paper and pencil option and used images, text and dictation features to express themselves on the iPad with greater enthusiasm for the multimedia experience (Miller et al., 2013).

In 2013 a review was conducted to evaluate 15 studies regarding the use of iPods, iPads and iPhones in educational programs for individuals with developmental disorders including their use for instruction as speech-generating devices, teaching a new skill, and improving social communication. Like the study conducted by Miller et al, this review found that mobile devices were increasingly embraced for their reduction in the stigma otherwise experienced by students with disabilities who utilize assistive technologies (Kagohara et al., 2013). Such studies promote the use of accessible mobile applications in classrooms with students diagnosed with physical or developmental disabilities to create a more flexible and universally inclusive environment.

FUTURE RESEARCH DIRECTIONS

In looking to the future one can also envision emerging trends such as wearable apps and portable virtual technologies incorporating augmented reality for students in regular classrooms, which may come with its own set of challenges as it relates to accessibility. The utilization of universal design principles when selecting these future technologies may likewise greatly reduce the disparities in access for individuals with disabilities and help educators to create nurturing, interactive, and inclusive learning environments.

Future research opportunities not only include scouting for emerging trends but also researching appropriate use of mobile and wearable technologies. Newton & Dell (2011) remind us that it is not appropriate to use an app simply because a student or a parent is requesting it to be included in the curriculum (Newton & Dell, 2011). Furthermore, More & Travers (2013) urge educators to evaluate the efficacy of the educational apps marketed to instructors focusing on their accessibility, developmental appropriateness, content suitability, and relevance to the students' academic needs. As mandated by IDEA regulations electronic tools such as mobile applications should also be pertinent to a student's IEP recommendations and specifically address the challenges a student faces when working on academic tasks. Furthermore, it is equally important to evaluate how the use of a mobile application impacts a student's learning experience (Newton & Dell, 2011). Another area of future research that can better enhance the mobile or wearable app selection process for use in schools is related to the rating of these technologies. What standards can be used to rate them? Are they universal? Are there randomized controlled studies that can support the efficacy of using accessible mobile apps among diverse populations? These are questions that need to be further addressed.

CONCLUSION

In conclusion, using mobiles for inclusive education is a first step in developing and promoting enabling learning environments. As the World Health Organization (2011) states, moving forward with electronic information and services includes the development of standards and regulations as well as promotion of awareness raising programs that encourage accessibility for all persons. Designing an inclusive curriculum using accessible technologies in classrooms is a positive step that addresses existing barriers in education. More importantly, educators should reach out to communities that have been discriminated against or stigmatized, including populations diagnosed with disabilities, and receive feedback on how the learning environment can be more inclusive from their point of view. In an article aptly titled

Technology for People, Not Disabilities: Ensuring Access and Inclusion, Foley & Ferry (2012) point out that while technology has been used to provide access at some levels it can also isolate certain individuals if not accessible, and they propose that the dialogue on technology be shifted from an ableist view of functionality to that of equity. In the same vein, rather than designing mobile applications as assistive technologies -- geared only towards a certain segment of the classroom -- it is more efficient, cost-effective and ethical to develop them using universal design principles and with a more broader and flexible view that is beneficial and accessible for all learners.

REFERENCES

ADA National Network. (n.d.). *What is the Americans with Disabilities Act (ADA).* Retrieved December 8, 2015, from https://adata.org/learn-about-ada

ADA.Gov. (n.d.). *Information and Technical Assistance on the Americans with Disabilities Act.* Retrieved December 8, 2015, from http://www.ada.gov/2010_regs.htm

American Psychological Association (APA). (n.d.). *Individuals with Disabilities Education Act (IDEA).* Retrieved December 8, 2015, from http://www.apa.org/about/gr/issues/disability/idea.aspx

Antonucci, N. (2014). *5 reasons technology should be allowed in the classroom.* Retrieved December 8, 2015, from http://www.edtechmagazine.com/higher/article/2014/11/5-reasons-technology-should-be-allowed-classroom

Avila, J. (2013). *What accessibility standards apply to mobile applications?* Retrieved December 8, 2015, from http://www.ssbbartgroup.com/blog/what-accessibility-standards-apply-to-mobile-phone-applications/

Brault, M. W. (2012). *Americans with disabilities: 2010.* Retrieved December 8, 2015, from http://www.census.gov/prod/2012pubs/p70-131.pdf

Bryant, D. P., Bryant, B. R., & Raskind, M. H. (1998). Using assistive technology to enhance the skills of students with learning disabilities. *Intervention in School and Clinic, 34*(1), 53–58. doi:10.1177/105345129803400109

Burgstahler, S. (2012). *Universal design in education: principles and applications.* Retrieved December 8, 2015, from http://www.washington.edu/doit/universal-design-education-principles-and-applications

Center for Inclusive Design and Environmental Access. (n.d.). *What is universal design*. Retrieved December 8, 2015, from http://udeworld.com/

Centers for Disease Control and Prevention (CDC). (2009). *The burden of vision loss*. Retrieved December 8, 2015, from http://www.cdc.gov/visionhealth/basic_information/vision_loss_burden.htm

Centers for Disease Control and Prevention (CDC). (2016). *Attention deficit/hyperactivity disorder (ADHD): data and statistics*. Retrieved December 8, 2015, from http://www.cdc.gov/ncbddd/adhd/data.html

Chapman, T. (2015). *American foundation for the blind announces 2015 Helen Keller achievement award honorees*. Retrieved December 8, 2015, from http://www.afb.org/info/about-us/press-room/press-release-archive/afb-announces-2015-helen-keller-achievement-award-honorees-/1245

ClearCaptions.com. (n.d.). *About ClearCaptions*. Retrieved December 8, 2015, from http://www.clearcaptions.com/Content/about

Common Sense Media. (n.d.). *Read2Go Review*. Retrieved December 8, 2015, from https://www.commonsensemedia.org/app-reviews/read2go#

Courtney-Long, E. A., Carroll, D. D., Zhang, Q. C., Stevens, A. C., Griffin-Blake, S., Armour, B. S., & Campbell, V. A. (2015). *Prevalence of disability and disability type among adults – United States 2013*. Retrieved December 8, 2015, from http://www.cdc.gov/mmwr/preview/mmwrhtml/mm6429a2.htm?s_cid=mm6429a2_w

Cragun, B., & Todd, D. (2013). *How do mobile apps measure up to accessibility standards?* Retrieved December 8, 2015, from https://www.03.ibm.com/able/education/downloads/IBM_How_Do_Mobile_Apps_Measure_Up_To_Standards_CSUN13.pdf

Duranczyk, I. M., Myers, K. A., Couillard, E. K., Schoen, S., & Higbee, J. L. (2013). Enacting the spirit of the United Nations convention on the rights of persons with disabilities: The role of postsecondary faculty in ensuring access. *Journal of Diversity Management*, *8*(2), 63–72. doi:10.19030/jdm.v8i2.8232

Elias, T. (2011). Universal instructional design principles for mobile learning. *International Review of Research in Open and Distance Learning*, *12*(2), 143–156. doi:10.19173/irrodl.v12i2.965

Flewitt, R., Kucirkova, N., & Messer, D. (2014). Touching the virtual, touching the real: iPads and enabling literacy for students experiencing disability. *Australian Journal of Language and Literacy*, *37*(2), 107–116.

Foley, A., & Ferri, B. A. (2012). Technology for people, not disabilities: Ensuring access and inclusion. *Journal of Research in Special Educational Needs*, *12*(4), 192–200. doi:10.1111/j.1471-3802.2011.01230.x

Frey, B., Southern, C., Rosier, K., & Romero, M. (2012). *From Texting App to Braille Literacy*. Austin, TX: CHI.

Ganz, J. B., Boles, M. B., Goodwyn, F. D., & Flores, M. M. (2014). Efficacy of handheld electronic visual supports to enhance vocabulary in children with ASD. *Focus on Autism and Other Developmental Disabilities*, *29*(1), 3–12. doi:10.1177/1088357613504991

Gardner, H., & Davis, K. (2013). *The app generation: how today's youth navigate identity, intimacy, and imagination in a digital world*. New Haven, CT: Yale University Press.

Goodreader.com. (n.d.). *GoodReader is the super-robust PDF reader app*. Retrieved December 8, 2015, from http://www.goodreader.com/

Gorlewicz, J. L., Burgner, J., Withrow, T. J., & Webster, R. J. III. (2014). Initial Experiences Using Vibratory Touchscreens to Display Graphical Math Concepts to Students with Visual Impairments. *Journal of Special Education Technology*, *29*(2), 17–25. doi:10.1177/016264341402900202

Henry, S. L. (2015). *Mobile accessibility*. Retrieved December 8, 2015, from https://www.w3.org/WAI/mobile/

Holloway, J. (2012). *Brailletouch app to bring eyes-free texting iPhone and iPad*. Retrieved December 8, 2015, from http://www.gizmag.com/brailletouch/21527/

Hornor, T. (2013). *Designing for disabilities: section 508 and international accessibility compliance for beginners*. Retrieved December 8, 2015, from http://www.sitepoint.com/designing-disabilities-section-508-international-accessibility-compliance-beginners/

Individuals with Disabilities Education Act (IDEA). (2004). *Public law 108-446: 108th congress*. Retrieved December 8, 2015, from http://www.parentcenterhub.org/wp-content/uploads/repo_items/PL108-446.pdf

Institute for Human Centered Design. (n.d.). *What is universal design?* Retrieved December 8, 2015, from http://www.humancentereddesign.org/universal-design

Izzo, M. V., & Bauer, W. M. (2015). Universal design for learning: Enhancing achievement and employment of STEM students with disabilities. *Universal Access in the Information Society*, *14*(1), 17–27. doi:10.100710209-013-0332-1

Kagohara, D. M., Van der Meer, L., Ramdoss, S., O'Reilly, M. F., Lancioni, G. E., Davis, T. N., ... Sigafoos, J. (2013). Using iPods and iPads in teaching programs for individuals with developmental disabilities: A systematic review. *Research in Developmental Disabilities*, *34*(1), 147–156. doi:10.1016/j.ridd.2012.07.027 PMID:22940168

Kurzweil Education. (n.d.). *Text-to-Speech Software*. Retrieved December 8, 2015, from https://www.kurzweiledu.com/about-kurzweil/text-to-speech.html

Kushalnagar, R. S., & Trager, B. P. (2011). *Improving classroom visual accessibility with cooperative smartphone recordings*. Retrieved December 8, 2015, from http://www.researchgate.net/publication/254003873

Miller, B. T., Krockover, G. H., & Doughty, T. (2013). Using iPads to teach inquiry science to students with a moderate to severe intellectual disability: A pilot study. *Journal of Research in Science Teaching*, *50*(8), 887–911. doi:10.1002/tea.21091

Moore, J. (n.d.). *Mobile apps and accessibility*. Retrieved December 8, 2015, from http://tinyurl.com/z39laae

More, C. M., & Travers, J. C. (2013). What's app with that? Selecting educational apps for young children with disabilities. *Young Exceptional Children*, *16*(2), 15–32. doi:10.1177/1096250612464763

National Center for Education Statistics. (2015a). *Fast facts: Students with disabilities*. Retrieved December 8, 2015, from https://nces.ed.gov/fastfacts/display.asp?id=64

National Center for Education Statistics. (2015b). *Children and youth with disabilities*. Retrieved December 8, 2015, from http://nces.ed.gov/programs/coe/indicator_cgg.asp

National Center for Learning Disabilities. (2014). *The state of learning disabilities: facts, trends and emerging issues*. New York, NY: Author.

National Center on Universal Design for Learning. (2014). *What is UDL?* Retrieved December 8, 2015, from http://www.udlcenter.org/aboutudl/whatisudl

National Federation for the Blind (NFB). (2014). *Blindness statistics*. Retrieved December 8, 2015, from https://nfb.org/blindness-statistics

National Institute on Deafness and Other Communication Disorders (NIDCD). (2015). *Quick statistics*. Retrieved December 8, 2015, from http://www.nidcd.nih.gov/health/statistics/Pages/quick.aspx

New York Daily News. (2015). *Department of education lifts ban on cell phones in New York City schools*. Retrieved December 8, 2015, from http://www.nydailynews.com/new-york/dept-education-ends-cell-phone-ban-nyc-schools-article-1.2134970

Newtown, D., & Dell, A. G. (2011). Mobile devices and students with disabilities: What do best practices tell us? *Journal of Special Education Technology*, *26*(3), 47–49.

O'Bannon, B. W., & Thomas, K. M. (2015). Mobile phones in the classroom: Preservice teachers answer the call. *Computers & Education*, *85*, 110–122. doi:10.1016/j.compedu.2015.02.010

O'Connor, D. (2003). *21st century information fluency*. Retrieved December 8, 2015, from http://21cif.com/tutorials/micro/mm/508/page7.htm

OnlineUniversities.com. (2012). *40 creative ways teachers are using cell phones in the classroom*. Retrieved December 8, 2015, from http://www.onlineuniversities.com/blog/2012/10/40-creative-ways-teachers-are-using-cell-phones-classroom/

OpenDyslexic.org. (n.d.). *Available applications that include OpenDyslexic*. Retrieved December 8, 2015, from http://opendyslexic.org/category/products/applications/

Perez, L., Furler, B., & Reat, C. C. (2014). *Mobile devices empower students with special needs*. Retrieved December 8, 2015, from http://www.educationworld.com/a_tech/apps-special-needs-disabilities-assistive-technology-students.shtml

Pogue, D. (2011). *Apple's assistive touch helps the disabled use a smartphone*. Retrieved December 8, 2015, from http://pogue.blogs.nytimes.com/2011/11/10/apples-assistivetouch-helps-the-disabled-use-a-smartphone/

Prensky. (2007). How to teach with technology: keeping both teachers and students comfortable in an era of exponential change. In *Emerging technologies for learning* (vol. 2, pp. 40-46). British Educational Communications and Technology Agents (Becta).

Reid, G., Strnadova, I., & Cumming, T. (2013). Expanding horizons for students with dyslexia in the 21st century: Universal design and mobile technology. *Journal of Research in Special Educational Needs*, *13*(3), 175–181. doi:10.1111/1471-3802.12013

Rodriguez, C. D., Strnadova, I., & Cumming, T. (2013). Using ipads with students with disabilities: Lessons learned from students, teachers, and parents. *Intervention in School and Clinic*, *49*(4), 244–250. doi:10.1177/1053451213509488

Runyan, M. (2013). Seeing is believing! *Learning and Leading with Technology*, 12–17.

Section508.gov. (n.d.). *Section 508 law and related laws and policies*. Retrieved December 8, 2015, from https://www.section508.gov/content/learn/laws-and-policies

Steinfeld, E. (2012). *The goals of universal design*. Retrieved December 8, 2015, from http://udeworld.com/presentations/oslo/Steinfeld.Goals%20of%20UD-Oslo_Final_web.pdf

Sterling, G. (2015). *It's official: Google says more searches now on mobile than on desktop*. Retrieved December 8, 2015, from http://searchengineland.com/its-official-google-says-more-searches-now-on-mobile-than-on-desktop-220369

Storyboardthat.com. (n.d.). *Digital storytelling: powerful visual communication, made easy*. Retrieved December 8, 2015, from http://www.storyboardthat.com/

Strain, M. (2015). *1983 to today: a history of mobile apps*. Retrieved December 8, 2015, from http://www.theguardian.com/media-network/2015/feb/13/history-mobile-apps-future-interactive-timeline

Taylor, D. M. (2018). *Americans with disabilities: 2014*. Retrieved March 30, 2019 from https://www.census.gov/content/dam/Census/library/publications/2018/demo/p70-152.pdf

TEAL Center. (n.d.). *Fact sheet – universal design for learning*. Retrieved December 8, 2015, from https://teal.ed.gov/tealguide/udl

United Nations. (2006). *Convention on the rights of persons with disabilities*. Retrieved December 8, 2015, from http://www.un.org/disabilities/convention/conventionfull.shtml

United Nations. (n.d.). *Convention and optional protocol signature and ratifications*. Retrieved December 8, 2015, from http://www.un.org/disabilities/countries.asp?navid=17&pid=166

United States Access Board. (2000). *Section 508 standards for electronic and information technology*. Retrieved December 8, 2015, from https://www.access-board.gov/guidelines-and-standards/communications-and-it/about-the-section-508-standards/section-508-standards

United States Access Board. (n.d.). *About the section 508 standards.* Retrieved December 8, 2015, from https://www.access-board.gov/guidelines-and-standards/communications-and-it/about-the-section-508-standards

United States Department of Justice. (2015). *United States department of education q&a: title IV rehabilitation act amendments of 1998 - section 508 electronic and information technology.* Retrieved December 8, 2015, from http://www.justice.gov/crt/department-education-q

University of Washington. (2015). *What is the individuals with disabilities education act?* Retrieved December 8, 2015, from http://www.washington.edu/doit/what-individuals-disabilities-education-act

US Department of Education. (n.d.). *Building the legacy: IDEA 2004.* Retrieved December 8, 2015, from http://idea.ed.gov/

USCode.House.Gov. (1973). *Public Law 93-113. Section 504 of the Rehabilitation Act of 1973.* Retrieved December 8, 2015, from http://uscode.house.gov/statviewer.htm?volume=87&page=394

Wee, M. (2013). *Three things to make your app stand out when building for accessibility.* Retrieved December 8, 2015, from http://www.wired.com/insights/2013/03/three-things-to-make-your-app-stand-out-when-building-for-accessibility/

World Health Organization (WHO). (2011). *Summary world report on disability.* Geneva, Switzerland: WHO.

World Health Organization (WHO). (2018). *Disability and health – key facts.* Retrieved March, 30, 2019 from https://www.who.int/en/news-room/fact-sheets/detail/disability-and-health

KEY TERMS AND DEFINITIONS

Accessible Technology: A technology or system that is designed to be used by the widest range of populations that is user-friendly in design.

Assistive Technology: A technology or system that is usually designed specifically to enhance the abilities or assist a person with a disability.

Barrier: Any situation or item that blocks or prevents access or movement without impediments.

Disability: A medical, physical, psychiatric, or cognitive condition that limits or impairs one's abilities or sense(s) when participating in daily society.

Inclusive Classroom: A classroom design that promotes equal access to instruction materials and sessions for students of all backgrounds including those with disabilities.

Mobile Application: Applications that are used on mobile devices such as smartphones, tablets, or wearable and portable technologies.

Universal Design: A design that is accessible for populations across all spectrums of society including individuals with disabilities, non-disabled persons, the elderly, or anyone with any type of limitation.

Compilation of References

Abedi, M., Badragheh, A., & Varamin, I. (2011). Review definitions and principles relating to Andragogy. *The Journal of American Science*, *7*(5).

Abel, R. (2015, May 12). *Find out more about IMS*. Retrieved November 30, 2015, from https://www.imsglobal.org/aboutims.html

About Us. (2019). Retrieved March 5, 2019, from https://www.teacherspayteachers.com/About-Us

Abu-Al-Aish, A. (2014). *Toward mobile learning deployment in higher education* (Doctoral dissertation). Brunel University, School of Information Systems, Computing and Mathematics.

Acedo, F. J., & Casillas, J. C. (2005). Current paradigms in the international management field: An author co-citation analysis. *International Business Review*, *14*(5), 619–639. doi:10.1016/j.ibusrev.2005.05.003

ADA National Network. (n.d.). *What is the Americans with Disabilities Act (ADA)*. Retrieved December 8, 2015, from https://adata.org/learn-about-ada

ADA.Gov. (n.d.). *Information and Technical Assistance on the Americans with Disabilities Act*. Retrieved December 8, 2015, from http://www.ada.gov/2010_regs.htm

Afflerbach, P. (2016). Reading Assessment. *The Reading Teacher*, *69*(4), 413–419. doi:10.1002/trtr.1430

Ahmad, N., & Orion, P. (2010). *Smartphones make IBM smarter, but not as expected*. American Society for Training and Development.

Ainsworth, S. (1999). The functions of multiple representations. *Computers & Education*, *33*(2), 131–152. doi:10.1016/S0360-1315(99)00029-9

Ajegbomogun, F. O., & Oduwole, O. K. (2017, October). Social media trends and collaborative learning for scholarly research among postgraduate students. In *European Conference on e-Learning* (pp. 9-17). Academic Conferences International Limited.

Ajzen, I. (1985). *From intentions to actions: A theory of planned behavior*. Springer Berlin Heidelberg.

Al-Durra, M., Torio, M. B., & Cafazzo, J. A. (2015). The Use of Behavior Change Theory in Internet-Based Asthma Self-Management Interventions: A Systematic Review. *Journal of Medical Internet Research*, *17*(4), e89. doi:10.2196/jmir.4110 PMID:25835564

Alessa, T., Abdi, S., Hawley, M., & de Witte, L. (2018). Mobile Apps to Support the Self-Management of Hypertension: Systematic Review of Effectiveness, Usability, and User Satisfaction. *JMIR mHealth and uHealth*, *6*(7), e10723. doi:10.2196/10723 PMID:30037787

Al-Okaily, R. (2013). Mobile learning and BYOD: Implementations in an intensive English program. *Learning & Teaching in Higher Education: Gulf Perspectives*, *10*(2), 1–17.

Alwashmi, M., Hawboldt, J., Davis, E., Marra, C., Gamble, J. M., & Ashour, W. A. (2016). The Effect of Smartphone Interventions on Patients with Chronic Obstructive Pulmonary Disease Exacerbations: A Systematic Review and Meta-Analysis. *JMIR mHealth and uHealth*, *4*(3), e105. doi:10.2196/mhealth.5921 PMID:27589898

Amanda, C. (2014, June 4). Dr. Panda's Restaurant 2 gets kids cooking [Web log post]. Retrieved from http://www.geekswithjuniors.com/blog/2014/6/4/dr-pandas-restaurant-2-gets-kids-cooking.html

American College of Cardiology. (2017). *New ACC/AHA High Blood Pressure Guidelines Lower Definition of Hypertension.* Retrieved March 3, 2019 from https://www.acc.org/latest-in-cardiology/ articles/2017/11/08/11/47/mon-5pm-bp-guideline-aha-2017

American Heart Association. (2018). *More than 100 million Americans have high blood pressure, AHA says.* Retrieved February 9, 2019 from 2018/05/01/more-than-100-million-americans-have-high-blood-pressure-aha-says

American Psychological Association (APA). (n.d.). *Individuals with Disabilities Education Act (IDEA).* Retrieved December 8, 2015, from http://www.apa.org/about/gr/issues/disability/idea.aspx

American Social History Project. (2018, September 14). *Mission US: Time Snap – Can VR Enhance the Teaching of History?* Retrieved from https://ashp.cuny.edu/news/mission-us-time-snap-can-vr-enhance-teaching-history-0

Anastasiadis, T., Lampropoulos, G., & Siakas, K. (2018). *Digital Game-based Learning and Serious Games in Education.* Academic Press.

Anderson, J. R. (2005). *Cognitive Psychology and Its Implications* (6th ed.). Worth Publishing.

Anderson, L. W., Krathwohl, D. R., & Bloom, B. S. (2001). *A taxonomy for learning, teaching, and assessing: A revision of Bloom's taxonomy of educational objectives.* Allyn & Bacon.

Andrade, J., Ares, J., García, R., Rodríguez, S., Seoane, M., & Suárez, S. (2008). Guidelines for the development of e-learning systems by means of proactive questions. *Computers & Education*, *51*(4), 1510–1522. doi:10.1016/j.compedu.2008.02.002

Annamma, S. A., Connor, D. J., & Ferri, B. A. (2016). Disability critical race studies (DisCrit): Theorizing at the intersections of race and dis/ability. In J. D. Connor, B. A. Ferri, & S. A. Annamma (Eds.), *DisCrit: Disability studies and critical race theory in education*. New York: Teachers College Press.

Antonucci, N. (2014). *5 reasons technology should be allowed in the classroom*. Retrieved December 8, 2015, from http://www.edtechmagazine.com/higher/article/2014/11/5-reasons-technology-should-be-allowed-classroom

Anzai, Y., & Simon, H. A. (1979). The theory of learning by doing. *Psychological Review*, *86*(2), 124–140. doi:10.1037/0033-295X.86.2.124 PMID:493441

AppMatrix, Inc. (2014). Nicolas' Garden (Version 1.1.0) [Mobile application software]. Retrieved from http:// http://www.nicolasgarden.com

Ardito, C., Buono, P., Costabile, M. F., Lanzilotti, R., & Pederson, T. (2007, November). Re-experiencing history in archaeological parks by playing a mobile augmented reality game. In *OTM Confederated International Conferences on the Move to Meaningful Internet Systems* (pp. 357-366). Springer. 10.1007/978-3-540-76888-3_58

Argyris, C. (1976). Single-loop and double-loop models in research on decision making. *Administrative Science Quarterly*, *21*(3), 363–375. doi:10.2307/2391848

Arnesen, K., Sivertsen Korpås, G., Hennissen, J. E., & Birger Stav, J. (2013). Experiences with use of various pedagogical methods utilizing a student response system – Motivation and learning outcome. *Electronic Journal of E-Learning*, *11*, 169–181.

Arnheim, R. (1991). *Thoughts on art education*. Los Angeles, CA: Getty Center for Education in the Arts.

Arnheim, R. (1992). *To the rescue of art*. Berkeley, CA: University of California Press.

Ashby, F. G., Isen, A. M., & Turken, A. U. (1999). A neuropsychological theory of positive affect and its influence on cognition. *Psychological Review*, *106*(3), 529–550. doi:10.1037/0033-295X.106.3.529 PMID:10467897

Attewell, J., & Savill-Smith, C. (2004). Mobile learning and social inclusion: focusing on learners and learning. In J. Attewell & C. Savill-Smith (Eds.), *Learning With Mobile Devices: Research and Development* (pp. 3–11). London: Learning and Skills Development Agency.

Avila, J. (2013). *What accessibility standards apply to mobile applications?* Retrieved December 8, 2015, from http://www.ssbbartgroup.com/blog/what-accessibility-standards-apply-to-mobile-phone-applications/

Bakhsh, S. S. H. (2015). Mobile learning adoption by language instructors in Taibah University. *Education Research International*, *4*(5), 117–150.

Bandura, A. (1986). *Social foundations of thought and action: A social cognitive theory*. Englewood Cliffs, NJ: Prentice Hall.

Barbosa, A. F. S., Pereira, N. M. P., Dias, A. F. F. J., & Silva, G. M. F. (2014). *A New Methodology of Design and Development of Serious Games*. Retrieved from Hindawi Publishing Corporation: http://www.hindawi.com/journals/ijcgt/aip/817167/

Barbour, M., & Rich, P. (2004). Social constructivist e-learning: A case study. In *E-Learn: World Conference on E-Learning in Corporate, Government, Healthcare, and Higher Education* (pp. 1754-1759). Association for the Advancement of Computing in Education (AACE).

Bargerhuff, M. E., Cowan, H., Oliveira, F., Quek, F., & Fang, B. (2010). Haptic glove technology: Skill development through video game play. *Journal of Visual Impairment & Blindness*, *104*(11), 688–699. doi:10.1177/0145482X1010401103

Batagelj, V., & Mrvar, A. (1998). Pajek: A program for large network analysis. *Connections*, *21*, 47–57.

Beach, R. (2015). *CI5472 Teaching Film, Television, and Media*. Retrieved April 19, 2015. Retrieved from http://www.tc.umn.edu/~rbeach/teachingmedia/module7/3.htm

Berila, B. (2016). Mindfulness as a healing, liberatory practice in queer anti-oppression pedagogy. *Social Alternatives*, *35*, 5–10.

Biocca, F. A. (2012). Opposing Conceptions of the Audience: The Active and Passive Hemisphere's of Mass Communication Theory. *Communication Yearbook*, *11*, 51–80.

Blaschke, L. M. (2012). Heutagogy and lifelong learning: A review of heutagogical practice and self-determined learning. *The International Review of Research in Open and Distributed Learning*, *13*(1), 56–71. doi:10.19173/irrodl.v13i1.1076

Bloom, B. S. (1956). *Taxonomy of educational objectives: The classification of educational goals. Handbook 1-2*. Longmans: McKay.

Bobis, J., Sweller, J., & Cooper, J. (1993). Cognitive load effects in primary-school geometry task. *Learning and Instruction*, *3*(1), 1–21. doi:10.1016/S0959-4752(09)80002-9

Bodily, R., Leary, H., & West, R. E. (2019). Research trends in instructional design and technology journals. *British Journal of Educational Technology*, *50*(1), 64–79. doi:10.1111/bjet.12712

Bogdanović, Z., Barać, D., Jovanić, B., Popović, S., & Radenković, B. (2014). Evaluation of mobile assessment in a learning management system (LMS). *British Journal of Educational Technology*, *45*(2), 231–244. doi:10.1111/bjet.12015

Bolsinova, M., Deonovic, B., Attali, M., Settles, B., Hagiwara, M., von Davier, A., & Maris, G. (2018). *Modeling hint use and response accuracy in learning environments*. Paper presented at Maryland Assessment Conference. Retrieved from https://education.umd.edu/file/11330/download?token=MYO8oBKU

Bonoto, B. C., de Aruajo, V. E., Godoi, I. P., Pires de Lemos, L. L., Godman, B., Bennie, M., ... Guerra, A. A. (2017). Efficacy of Mobile Apps to Support the Care of Patients with Diabetes Mellitus: A Systematic Review and Meta-Analysis of Randomized Controlled Trials. *JMIR mHealth and uHealth, 5*(3), e4. doi:10.2196/mhealth.6309 PMID:28249834

Borgatti, S. P., Everett, M. G., & Freeman, L. C. (1999). *UCINET 6.0 Version 1.00*. Natick: Analytic Technologies.

Bow Girls. (2017, May 17). *How to Make a Mystic Mermaid Shake* [Video File]. Retrieved from https://www.youtube.com/watch?v=zXE3lGHCX1U

Boyle, T. (1997). *Design for multimedia learning*. London: Prentice Hall.

Brainster Apps. (2014). Veggie Circus Farm (Version 1.1) [Mobile application software]. Retrieved from http://www.toddlerwords.com

Brandon, B. (2006). *311 Tips for the Successful Management of an LMS or LCMS*. The eLearning Guild.

Bransford, J. D., Franks, J. J., Vye, N. J., & Sherwood, R. D. (1989). New approaches to instruction: Because wisdom can't be told. In S. Vosniadou & A. Ortony (Eds.), *Similarity and analogical reasoning* (pp. 470–497). New York: Cambridge University Press. doi:10.1017/CBO9780511529863.022

Brault, M. W. (2012). *Americans with disabilities: 2010*. Retrieved December 8, 2015, from http://www.census.gov/prod/2012pubs/p70-131.pdf

Brett, P. (2011). Students' experiences and engagement with SMS for learning in higher education. *Innovations in Education and Teaching International, 48*(2), 137–147. doi:10.1080/14703297.2011.564008

British Broadcasting Corporation. (2012). *UN turns players into refugees in a new app*. Retrieved from BBC Technology: http://www.bbc.com/news/technology-18553601

Brookfield, S. D. (2011). *Teaching for critical thinking: Tools and techniques to help students question their assumptions*. John Wiley & Sons.

Brown, E., Börner, D., Sharples, M., Glahn, C., De Jong, T., & Specht, M. (2010). *Location-based and contextual mobile learning*. A STELLAR Small-Scale Study.

Brown, S., Birch, D., Teufel, J., & Kancherla, V. (2006). Overweight in children: The perspectives of 9-13 year olds. *American Journal of Health Education, 37*(5), 296–305. doi:10.1080/19325037.2006.10598918

Bruner, J. (1966). *Toward a Theory of Instruction*. Cambridge, MA: Harvard University Press.

Bryant, D. P., Bryant, B. R., & Raskind, M. H. (1998). Using assistive technology to enhance the skills of students with learning disabilities. *Intervention in School and Clinic, 34*(1), 53–58. doi:10.1177/105345129803400109

Burgstahler, S. (2012). *Universal design in education: principles and applications*. Retrieved December 8, 2015, from http://www.washington.edu/doit/universal-design-education-principles-and-applications

Burston, J. (2014). The reality of MALL: Still on the fringes. *CALICO Journal, 31*(1), 103–125. doi:10.11139/cj.31.1.103-125

Burston, J. (2015). Twenty years of MALL project implementation: A meta-analysis of learning outcomes. *ReCALL, 27*(1), 4–20. doi:10.1017/S0958344014000159

Burton, J. M. (2011). *Creative and mental growth revisited: Parts 3 & 4*. New York, NY: Teachers College, Columbia University.

Caceres, N., Wideberg, J. P., & Benitez, F. G. (2007). Deriving origin–destination data from a mobile phone network. *IET Intelligent Transport Systems, 1*(1), 15–26. doi:10.1049/iet-its:20060020

Carpenter, K., Pagar, D., & Labrecque, R. (2013). *Teachley: Addimal Adventure: Bridging Research and Technology to Help Children Foster Strategy Development*. Conceptual Understanding, and Number Fact Fluency.

Carroll, L. (1865). *Alice in Wonderland*. Academic Press.

Carroll, L. (1871). *Through the Looking Glass*. Academic Press.

Casale, R., & Mentor, D. (2018). Digital Narratives of Immigrant Youth: Cultivating Mobile Activism and Mobile Journalism Skills. In Computer-Mediated Learning for Workforce Development (pp. 132-149). IGI Global.

Casey, P. H., Szeto, K. L., Robbins, J. M., Stuff, J. E., Connell, C., Gossett, J. M., & Simpson, P. M. (2005). Child health-related quality of life and household food security. *Archives of Pediatrics & Adolescent Medicine, 159*(1), 51–56. doi:10.1001/archpedi.159.1.51 PMID:15630058

Center for Inclusive Design and Environmental Access. (n.d.). *What is universal design*. Retrieved December 8, 2015, from http://udeworld.com/

Centers for Disease Control and Prevention (CDC). (2009). *The burden of vision loss*. Retrieved December 8, 2015, from http://www.cdc.gov/visionhealth/basic_information/vision_loss_burden.htm

Centers for Disease Control and Prevention (CDC). (2016). *Attention deficit/hyperactivity disorder (ADHD): data and statistics*. Retrieved December 8, 2015, from http://www.cdc.gov/ncbddd/adhd/data.html

Centers for Disease Control and Prevention. (2018). *Cancer Data and Statistics*. Retrieved February 22, 2019 from https://www.cdc.gov/cancer/dcpc/data/index.htm

Centers for Disease Control and Prevention. (2019). *About Chronic Diseases*. Retrieved February 8, 2019 from https://www.cdc.gov/chronicdisease/about/index.htm

Chaiklin, S. (2003). The zone of proximal development in Vygotsky's analysis of learning and instruction. *Vygotsky's Educational Theory in Cultural Context, 1*, 39-64.

Chananchida, Y. (2018). *Sindhuphak Apisak, and Krissana Kiddee*. Bangkok, Thailand: King Mongkut's Institute of Technology Ladkrabang.

Chao, K. H., Chang, K. E., & Lan, C. H., Kinshuk, & Sung, Y. T. (2016). Integration of Mobile AR Technology in Performance Assessment. *Journal of Educational Technology & Society, 19*(4), 239–251.

Chapman, T. (2015). *American foundation for the blind announces 2015 Helen Keller achievement award honorees*. Retrieved December 8, 2015, from http://www.afb.org/info/about-us/press-room/press-release-archive/afb-announces-2015-helen-keller-achievement-award-honorees-/1245

Charli's Crafty Kitchen. (2014, September 25). *Personalized Healthy Fruit Kebobs- Kababs Skewers – Educational how to DIY* [Video File]. Retrieved from https://www.youtube.com/watch?v=G_IsaJIDbp0

Chase, W. G., & Simon, H. A. (1973). The mind's eye in chess. In W. G. Chase (Ed.), *Visual information processing* (pp. 215–281). New York, NY: Academic Press. doi:10.1016/B978-0-12-170150-5.50011-1

Chayko, M. (2012). *Connecting: How we form social bonds and communities in the Internet age*. Suny Press.

Chen, C., & Teh, C. (2013). Enhancing an instructional design model for virtual reality-based learning. *Australasian Journal of Educational Technology, 29*(5). doi:10.14742/ajet.247

Chen, K. (2017). Examining EFL instructors' and students' perceptions and acceptance toward M-learning in higher education. *International Journal (Toronto, Ont.), 16*(4), 967–976. doi:10.100710209-016-0494-8

Chen, R. C. Y., Lee, M. S., Chang, Y. H., & Wahlqvist, M. L. (2012). Cooking frequency may enhance survival in Taiwanese elderly. *Public Health Nutrition, 15*(7), 1142–1149. doi:10.1017/S136898001200136X PMID:22578892

Chew, F., Palmer, S., & Kim, S. (1995). Sources of information and knowledge about health and nutrition: Can viewing one television programme make a difference? *Public Understanding of Science (Bristol, England), 4*(1), 17–29. doi:10.1088/0963-6625/4/1/002

Chew, F., Palmer, S., & Soohong, K. (1998). Testing the influence of the health belief model and a television program on nutrition behavior. *Health Communication, 10*(3), 227–245. doi:10.120715327027hc1003_3 PMID:16370984

Chocolapps. (2013). *KidECook (Version 1)* [Mobile application software]. Retrieved from http://en.chocolapps.com

Choi, J. I., & Hannafin, M. (1995). Situated cognition and learning environments: Roles, structures, and implications for design. *Educational Technology Research and Development*, *43*(2), 53–69. doi:10.1007/BF02300472

Clark, R. E. (1983). Reconsidering research on learning from media. *Review of Educational Research*, *53*(4), 445–459. doi:10.3102/00346543053004445

Clark, R., & Mayer, R. (2011). *E-learning and the science of instruction proven guidelines for consumers and designers of multimedia learning* (3rd ed.). San Francisco, CA: Pfeiffer. doi:10.1002/9781118255971

ClearCaptions.com. (n.d.). *About ClearCaptions*. Retrieved December 8, 2015, from http://www.clearcaptions.com/Content/about

Clements, D. H., & Sarama, J. (2003). Strip mining for gold: Research and policy in educational technology—A response to "fool's gold." *AACE Journal*, *11*(1), 7–69.

Clifford, D., Anderson, J., Auld, G., & Champ, J. (2009). Good grubbin': Impact of a TV cooking show for students living off campus. *Journal of Nutrition Education and Behavior*, *41*(3), 194–200. doi:10.1016/j.jneb.2008.01.006 PMID:19411053

Cobcroft, R. (2006). *Literature review into mobile learning in the university context* (Unpublished doctoral dissertation). Queensland University of Technology.

Cobcroft, R. S., Towers, S. J., Smith, J. E., & Bruns, A. (2006). Mobile learning in review: Opportunities and challenges for learners, teachers, and institutions. In *Proceedings Online Learning and Teaching (OLT) conference 2006* (pp. 21-30). Brisbane: Queensland University of Technology. Available at http://eprints.qut.edu.au/5399/1/5399.pdf

Cochrane, T., Antonczak, L., & Wagner, D. (2013). Post-web 2.0 pedagogy: From student-generated content to international co-production enabled by mobile socia media. *International Journal of Mobile and Blended Learning*, *5*(4), 1–18. doi:10.4018/ijmbl.2013100101

Collins, A., Neville, P., & Bielaczyc, K. (2000). The role of different media in designing learning environments. *International Journal of Artificial Intelligence in Education*, *11*(1), 144–162.

Common Sense Media. (n.d.). *Read2Go Review*. Retrieved December 8, 2015, from https://www.commonsensemedia.org/app-reviews/read2go#

CondeNet. (2015). Epicurious (Version 4.5.2) [Mobile application software]. Retrieved from http://www.epicurious.com/services/mobile

Contento, I. R., Koch, P. A., Lee, H. W., Sauberli, W., & Calabrese-Barton, A. (2007). Enhancing personal agency and competence in eating and moving: Formative evaluation of a middle school curriculum—Choice, Control, and Change. *Journal of Nutrition Education and Behavior*, *39*(5), S179–S186. doi:10.1016/j.jneb.2007.02.006 PMID:17826699

Cook, J. T., & Frank, D. A. (2008). Food security, poverty, and human development in the United States. *Annals of the New York Academy of Sciences*, *1136*(1), 193–209. doi:10.1196/annals.1425.001 PMID:17954670

Cook, V. E., Ellis, A. K., & Hildebrand, K. J. (2016). Mobile Health Applications in Clinical Practice: Pearls, Pitfalls, and Key Considerations. *Annals of Allergy, Asthma & Immunology*, *117*(2), 143–149. doi:10.1016/j.anai.2016.01.012 PMID:27499542

Courtney-Long, E. A., Carroll, D. D., Zhang, Q. C., Stevens, A. C., Griffin-Blake, S., Armour, B. S., & Campbell, V. A. (2015). *Prevalence of disability and disability type among adults – United States 2013*. Retrieved December 8, 2015, from http://www.cdc.gov/mmwr/preview/mmwrhtml/mm6429a2.htm?s_cid=mm6429a2_w

Cox, K. A. (2011). *Helping to Feed and Educate*. Undergraduate Works, Award-Winners & Notable.

Cragun, B., & Todd, D. (2013). *How do mobile apps measure up to accessibility standards?* Retrieved December 8, 2015, from https://www.03.ibm.com/able/education/downloads/IBM_How_Do_Mobile_Apps_Measure_Up_To_Standards_CSUN13.pdf

Craik, K. (1943). *The nature of explanation*. Cambridge, UK: Cambridge University Press.

Crompton, H. (2013). A historical overview of mobile learning: Toward learner-centered education. In Z. L. Berge & L. Y. Muilenburg (Eds.), *Handbook of mobile learning* (pp. 3–14). Florence, KY: Routledge.

Crompton, H. (2014). A diachronic overview of mobile learning: A shift toward student-centered pedagogies. In M. Alley (Ed.), *Perspectives on Open and Distance Learning: Increasing Access through Mobile Learning*. Athabasca University.

Csikszentmihalyi, M. (1990). *Flow: The psychology of optimal experience*. New York: Harper & Row.

Csikszentmihalyi, M. (1997). *Flow and the Psychology of Discovery and Invention*. Harper Perennial.

Csikszentmihalyi, M. (2014). *Applications of Flow in Human Development and Education*. Springer Netherlands. doi:10.1007/978-94-017-9094-9

Cummings, M. (2015, April 2). *There's a big hole in how teachers build skills, and Pinterest is helping fill it*. Retrieved from https://slate.com/human-interest/2015/04/pinterest-and-teachers-how-the-site-is-filling-a-gap-in-teacher-training.html

Curinga, M. X., & Saravanos, A. (2016). *Mobile first e-learning*. Hershey, PA: IGI Learning. doi:10.4018/978-1-5225-0251-7.ch002

da Silva, S. L., da Silva, R. L., Juniora, G., Goncalves, E., & Viana, E. R. (2014). *Animation with Algodoo: a simple tool for teaching and learning physics*. arXiv preprint arXiv:1409.1621

Dahya, N. (2009). Serious Learning in Playful Roles: Socio-political games for education and social change. Academic Press.

Dalgarno, B., & Lee, M. J. (2010). What are the learning affordances of 3-D virtual environments? *British Journal of Educational Technology*, *41*(1), 10–32. doi:10.1111/j.1467-8535.2009.01038.x

Darlow, S., & Wen, K. Y. (2016). Development Testing of Mobile Health Interventions for Cancer Patient Self-Management. *Health Informatics Journal*, *22*(3), 633–650. doi:10.1177/1460458215577994 PMID:25916831

Dasgupta, P., Tureski, K., Lenzi, R., Bindu, K., & Nanda, G. (2012). Half the Sky Movement Multimedia Communication Initiative: An Evaluation of the 9-Minutes Mobile Game and Video. Washington, DC: C-Change/FHI 360.

Dashtestani, R. (2013). Implementing mobile-assisted language learning (MALL) in an EFL context: Iranian EFL teachers' perspectives on challenges and affordances. *JALT CALL Journal*, *9*(2), 149–168.

Davidson, B. (Director). (2011). King Eddie who loved spaghetti [Television series episode]. In S. Freeman Alpert (Executive producer), Super Why! Toronto, Canada: Out of the Blue Productions & DHX Media.

Davies, C., & Eynon, R. (2013). *Teenagers and technology*. New York, NY: Routledge. doi:10.4324/9780203079270

Davis, F. D. (1989). Perceived usefulness, perceived ease of use, and user acceptance of information technology. *Management Information Systems Quarterly*, *13*(3), 319–340. doi:10.2307/249008

De Lisi, R., & Wolford, J. L. (2002). Improving children's mental rotation accuracy with computer game playing. *The Journal of Genetic Psychology*, *163*(3), 272–282. doi:10.1080/00221320209598683 PMID:12230149

de Solier, I. (2005). TV dinners: Culinary television, education and distinction. *Continuum (Perth)*, *19*(4), 465–481. doi:10.1080/10304310500322727

de Waard, I., Abajian, S., Gallagher, M. S., Hogue, R., Keskin, N., Koutropoulos, A., & Rodriguez, O. C. (2011). Using mLearning and MOOCs to understand chaos, emergence, and complexity in education. *The International Review of Research in Open and Distributed Learning*, *12*(7), 94–115. doi:10.19173/irrodl.v12i7.1046

Dearnley, C. A., Taylor, J. D., Laxton, J. C., Rinomhota, S., & Nkosana-Nyawata, I. (2013). The student experience of piloting multi-modal performance feedback tools in health and social care practice (work)-based settings. *Assessment & Evaluation in Higher Education*, *38*(4), 436–450. doi:10.1080/02602938.2011.645014

Delbridge, A., & Helman, L. (2015). *Evidence-Based Strategies for Fostering Biliteracy in Any Classroom*. New York: Springer Science & Business Media.

de-Marcos, L., Hilera, J. R., Barchino, R., Jimenez, L., Martinez, J. J., Gutierrez, J. A., ... Oton, S. (2010). An experiment for improving students' performance in secondary and tertiary education by means of m-learning auto-assessment. *Computers & Education, 55*(3), 1069–1079. doi:10.1016/j.compedu.2010.05.003

Dewey, J. (1934). The need for a philosophy of education. *New Era in Home and School*, 211-214.

Dewey, J. (1934). *Art and experience*. New York, NY: Capricorn Books.

Dewey, J. (1938). *Experience and education*. New York, NY: Touchstone.

Dewey, J. (1958). *Experience and nature*. New York, NY: Dover.

Dillon, P., & Åhlberg, M. (2006). Integrativism as a theoretical and organizational framework for e-learning and practitioner research. *Technology, Pedagogy and Education, 15*(1), 7–30. doi:10.1080/14759390500435747

Diodato, V. (1994). *Dictionary of Bibliometrics*. Binghamton, NY: Haworth Press.

Dogan, A., & Akbarov, A. (2016). Teachers' attitudes toward the usage of mobile devices in EFL classroom. *European Journal of Educational Research, 5*(1), 11–17. doi:10.12973/eu-jer.5.1.11

Dohn, N. B. (2010). Teaching with wikis and blogs: Potentials and pitfalls. In *Proceedings of the 7th International conference on networked learning* (pp. 142-150). Lancaster University.

Domagk, S., Schwartz, R. N., & Plass, J. L. (2010). Interactivity in multimedia learning: An integrated model. *Computers in Human Behavior, 26*(5), 1024–1033. doi:10.1016/j.chb.2010.03.003

Donaldson, L. F. (2016). *Temporality and Film Analysis* by Matilda Mroz (review). *Cinema Journal, 55*(2), 159-163.

Dr. Panda Limited. (2014). Dr. Panda's Restaurant 2 (Version 1.2) [Mobile application software]. Retrieved from http://drpandagames.com/app/dr-panda-restaurant2/

Duman, G., Orhon, G., & Gedik, N. (2015). Research trends in Mobile Assisted Language Learning from 2000 to 2012. *ReCALL, 27*(2), 197–216. doi:10.1017/S0958344014000287

Duranczyk, I. M., Myers, K. A., Couillard, E. K., Schoen, S., & Higbee, J. L. (2013). Enacting the spirit of the United Nations convention on the rights of persons with disabilities: The role of postsecondary faculty in ensuring access. *Journal of Diversity Management, 8*(2), 63–72. doi:10.19030/jdm.v8i2.8232

Eat2Explore. (2019). Retrieved from https://eat2explore.com/

Edge, D., Searle, E., Chiu, K., Zhao, J., & Landay, J. A. (2011). *CHI '11 Proceedings of the SIGCHI Conference on Human Factors in Computing Systems*. ACM.

Elias, T. (2011). Universal instructional design principles for mobile learning. *International Review of Research in Open and Distance Learning, 12*(2), 143–156. doi:10.19173/irrodl.v12i2.965

Ellis, R., & Goodyear, P. (2010). Students' Experiences of e-Learning in Higher Education: The Ecology of Sustainable Innovation (Open and Flexible Learning Series). Sydney, Australia: Routledge.

Ellis, A. E. (2003). Personality Type and Participation in Networked Learning Environments. *Educational Media International*, *40*(1/2), 101–114. doi:10.1080/0952398032000092152

Epstein, E. (2013). *Get Water! A Mobile Game that Advocates for Change*. Retrieved from Girls Globe: http://girlsglobe.org/2013/01/24/get-water-a-mobile-game-that-advocates-for-change/

ERIC. (2019). *ERIC thesaurus*. Retrieved from https://eric.ed.gov/?ti=M

Erlich, R., Yngve, A., & Wahlqvist, M. L. (2012). Cooking as a healthy behaviour. *Public Health Nutrition*, *15*(7), 1139–1140. doi:10.1017/S1368980012002662 PMID:23062443

Ervin, J., & Slife, N. (2019). *Double-Loop Learning: An Approach to Critical Thinking*. Academic Press.

Facione, P. A. (1990). *Critical thinking: A statement of expert consensus for purposes of educational assessment and instruction*. Newark, DE: American Philosophical Association.

Fairlady Media. (2015). Grandma's Kitchen (Version 2.1) [Mobile application software]. Retrieved from http://fairladymedia.com

Farzandipour, M., Nabovati, E., Sharif, R., Arani, M. H., & Anvari, S. (2017). Patient Self-Management of Asthma Using Mobile Health Applications: A Systematic Review of the Functionalities and Effects. *Applied Clinical Informatics*, *8*(4), 1068–1081. doi:10.4338/ACI-2017-07-R-0116 PMID:29241254

Felder, R. M., & Silverman, L. K. (1988). Learning and teaching styles in engineering education. *Engineering education*, *78*(7), 674–681.

Ferenchick, G. S., Solomon, D., Foreback, J., Towfiq, B., Kavanaugh, K., Warbasse, L., ... Mohmand, A. (2013). Mobile technology for the facilitation of direct observation and assessment of student performance. *Teaching and Learning in Medicine : An International Journal*, *25*(4), 291–299. doi:10.1080/10401334.2013.827972 PMID:24112197

Ferri, G., & Fusaroli, R. (2009). *Which Narrations for Persuasive Technologies? Habits and Procedures in Ayiti: The Cost of Life*. Academic Press.

Fildes, A., van Jaarsveld, C. H., Wardle, J., & Cooke, L. (2014). Parent-administered exposure to increase children's vegetable acceptance: A randomized controlled trial. *Journal of the Academy of Nutrition and Dietetics*, *114*(6), 881–888. doi:10.1016/j.jand.2013.07.040 PMID:24091061

Fiol, C. M., & Lyles, M. A. (1985). Organizational learning. *Academy of Management Review*, *10*(4), 803–813. doi:10.5465/amr.1985.4279103

Fisher, M., & Baird, D. E. (2006). Making mLearning work: Utilizing mobile technology for active exploration, collaboration, assessment, and reflection in higher education. *Journal of Educational Technology Systems*, *35*(1), 3–30. doi:10.2190/4T10-RX04-113N-8858

Flewitt, R., Kucirkova, N., & Messer, D. (2014). Touching the virtual, touching the real: iPads and enabling literacy for students experiencing disability. *Australian Journal of Language and Literacy*, *37*(2), 107–116.

Foley, A., & Ferri, B. A. (2012). Technology for people, not disabilities: Ensuring access and inclusion. *Journal of Research in Special Educational Needs*, *12*(4), 192–200. doi:10.1111/j.1471-3802.2011.01230.x

Food and Drug Administration. (n.d.). *Mobile Medical Applications.* Retrieved February 9, 2019 from https://www.fda.gov/medicaldevices/digitalhealth/mobilemedicalapplications/default.htm

Food for Life Partnership. (2015). *Cooking in school: Your guide to planning and teaching cooking in primary and early years settings*. Retrieved from http://www.foodforlife.org.uk/~/media/files/resources/fof%20resources/fflp-cooking-in-school-checklist.pdf

Food Network's Star Kids. (2016). [Television Program]. Retrieved from https://www.foodnetwork.com/shows/food-network-star-kids

Food Span. (2011). Retrieved from http://www.foodspanlearning.org/

Food Truth Online. (2015). Food Truth (Version 2.0) [Mobile application software]. Retrieved from http://www.foodtruthonline.com/app

FoodStirs. (2019). Retrieved from https://foodstirs.com/

Forma Animada. (2013). Super Abby vs. The Enchanted Sweet Stall (Version 1.0) [Mobile application software]. Retrieved from http://www.superabbyworld.com

Freire, P. (1970). *Pedagogy of the Oppressed.* Academic Press.

Frey, B., Southern, C., Rosier, K., & Romero, M. (2012). *From Texting App to Braille Literacy.* Austin, TX: CHI.

Friedman, B. (Writer), & Kapridov, V. (Director). (2012). Be a vegetable taster! / Daniel tries a new food [Television series episode]. In W. Harris (Supervising producer), & A. Santomero (Executive producer), Daniel Tiger's neighborhood. Pittsburgh, PA: The Fred Rogers Company.

Friedman, R. A. (2014, June 29). Why teenagers act crazy. *New York Times*, pp. SR 1, SR 6.

Furió, D., Juan, M. C., Seguí, I., & Vivó, R. (2015). Mobile learning vs. traditional classroom lessons: A comparative study. *Journal of Computer Assisted Learning*, *31*(3), 189–201. doi:10.1111/jcal.12071

Futurelab, N., Naismith, L., Lonsdale, P., Vavoula, G., Sharples, M., & Series, N. F. (2004). *Literature review in mobile technologies and learning*. Academic Press.

Gagné, R. M., & Merrill, M. D. (1990). Integrative goals for instructional design. *Educational Technology Research and Development*, *38*(1), 23–30. doi:10.1007/BF02298245

Games, S. (2018, October 22). *'HoloLAB Champions' Now Available on Oculus Rift*. Retrieved from https://www.schellgames.com/blog/schell-games-adds-oculus-rift-support-to-its-award-winning-virtual-reality-chemistry-lab-game-hololab-champions

Ganz, J. B., Boles, M. B., Goodwyn, F. D., & Flores, M. M. (2014). Efficacy of handheld electronic visual supports to enhance vocabulary in children with ASD. *Focus on Autism and Other Developmental Disabilities*, *29*(1), 3–12. doi:10.1177/1088357613504991

Garabedian, L. F., Ross-Degnan, D., & Wharam, J. F. (2015). Mobile Phone and Smartphone Technologies for Diabetes Care and Self-Management. *Current Diabetes Reports*, *15*(109). PMID:26458380

Gardner, H., & Davis, K. (2013). *The app generation: how today's youth navigate identity, intimacy, and imagination in a digital world*. New Haven, CT: Yale University Press.

Garfield, E., Malin, M. V., & Small, H. (1975). A system for automatic classification of scientific literature. *Indian Institute of Science*, *57*(2), 61–74.

Garfinkel, H. (2002). *Ethnomethodology's program: Working out Durkheim's aphorism*. New York, NY: Rowman & Littlefield.

Garris, R., Ahlers, R., & Driskell, J. E. (2002). Games, motivation, and learning: A research and practice model. *Simulation & Gaming*, *33*(4), 441–467. doi:10.1177/1046878102238607

Gellersen, H. W., Schmidt, A., & Beigl, M. (2002). Multi-sensor context-awareness in mobile devices and smart artifacts. *Mobile Networks and Applications*, *7*(5), 341–351. doi:10.1023/A:1016587515822

Georgiev, T., Smrikarov, A., & Georgieva, E. (2004). *m-learning: a new stage of e-learning*. doi:10.1145/1050330.1050437

Ghrieb, E. B. (2015). *Teachers' and students' attitudes towards the use of mobile assisted language learning* (Master's thesis). University of Biskra, Biskra, Algeria.

Gibson, D. (2013). Assessing teaching skills with a mobile simulation. *Journal of Digital Learning in Teacher Education*, *30*(1), 4–10. doi:10.1080/21532974.2013.10784720

Gibson, J. J., & Gibson, E. J. (1955). Perceptual learning: Differentiation or enrichment. *Psychological Review*, *62*(1), 32–51. doi:10.1037/h0048826 PMID:14357525

Giles, J. (2010, April 13). Barcodes help objects tell their stories. *New Scientist*. Retrieved from http://www.newscientist.com/article/dn18766-barcodes-help-objects-tell-their-stories.html

Ginsburg, H. P. (2014). My entirely plausible fantasy: Early mathematics education in the age of the touchscreen computer. *Journal of Mathematics Education at Teachers College*, *5*(1), 9–17.

Ginsburg, H. P., Jamalian, A., & Creighan, S. (2013). Cognitive guidelines for the design and evaluation of early mathematics software: The example of MathemAntics. In L. D. English & J. T. Mulligan (Eds.), *Reconceptualizing early mathematics learning* (pp. 83–120). Springer Netherlands. doi:10.1007/978-94-007-6440-8_6

Ginwright, S. (2016). *Hope and Healing in Urban Education: How Urban Activist and Teachers are Reclaiming Matters of the Heart*. New York, NY: Routledge.

Glavinic, V., Ljubic, S., & Kukec, M. (2008, September). A Holistic approach to enhance universal usability in m-learning. In *Mobile Ubiquitous Computing, Systems, Services and Technologies, 2008. UBICOMM'08. The Second International Conference on* (pp. 305-310). IEEE. 10.1109/UBICOMM.2008.54

Glossary: Learning Activity, New Learning - Transformational Designs for Pedagogy and Assessment. (2015, January 1). Retrieved from http://newlearningonline.com/learning-by-design/glossary/learning-activity

Goeser, C. (2013). *Teaching with the iPad: Adding a new dimension to the museum experience*. Retrieved from https://artmuseumteaching.files.wordpress.com/2013/04image005.jpg

Goodreader.com. (n.d.). *GoodReader is the super-robust PDF reader app*. Retrieved December 8, 2015, from http://www.goodreader.com/

Gordon Ramsy's Master Chef Junior. (2019). Retrieved from https://www.fox.com/masterchef-junior/article/about-the-show-597bbb42ef528f0026dc0305/

Gorlewicz, J. L., Burgner, J., Withrow, T. J., & Webster, R. J. III. (2014). Initial Experiences Using Vibratory Touchscreens to Display Graphical Math Concepts to Students with Visual Impairments. *Journal of Special Education Technology*, *29*(2), 17–25. doi:10.1177/016264341402900202

Graber, J. A., Brooks-Gunn, J., & Petersen, A. C. (1996). *Transitions through adolescence*. Mahwah, NJ: Erlbaum.

Greene, H., & Crespi, C. (2012). The value of student created videos in the college classroom–an exploratory study in marketing and accounting. *The International Journal of the Arts in Society*, *5*(1), 273–283.

Greenwood, D. A., Gee, P. M., Fatkin, K. J., & Peeples, M. (2017). A Systematic Review of Reviews Evaluating Technology-Enabled Diabetes Self-Management Education and Support. *Journal of Diabetes Science and Technology*, *11*(5), 1015–1027. doi:10.1177/1932296817713506 PMID:28560898

Grierson, E. (2007). Difference: A critical investigation of the creative arts with attention to art as a site of knowledge. *Educational Philosophy and Theory*, *39*(2), 531–542. doi:10.1111/j.1469-5812.2007.00246.x

Grimshaw, J., Cardoso, W., & Collins, L. (2017). Teacher perspectives on the integration of Mobile-Assisted Language Learning. In K. Borthwick, L. Bradley, & S. Thouësny (Eds.), CALL in a climate of change: A dapting to turbulent global conditions – short papers from EUROCALL 2017 (pp. 135-139). Research-publishing.net.

Groark, C. J., McCall, R. B., & Fish, L. (2011). Characteristics of environments, caregivers, and children in three Central American orphanages. *Infant Mental Health Journal, 32*(2), 232–250. doi:10.1002/imhj.20292 PMID:28543023

Grow Garden. (2019). Grow Play (Version 1) [Mobile application software]. Retrieved from https://www.groplay.com/apps/grow-garden/

Hachey, A. C. (2009). I hate math: What we want young children NOT to learn. *Texas Child Care Quarterly*, 2-7.

Halliday-Wynes, S., & Beddie, F. (2009). *Informal learning: At a glance*. Adelaide: National Centre for Vocational Education Research.

Halupa, C. (2015). *Pedagogy, Andragogy, and Heutagogy*. Retrieved from https://www.researchgate.net/publication/297767648_Pedagogy_Andragogy_and_Heutagogy

Halupa, C. M. (2015). Pedagogy, Andragogy, and Heutagogy. In *Transformative Curriculum Design in Health Sciences Education* (pp. 143–158). IGI Global. doi:10.4018/978-1-4666-8571-0.ch005

Hanson, K., & Shelton, B. (2008). *Design and Development of Virtual Reality: Analysis of Challenges Faced by Educators*. ITLS Faculty Publications. Retrieved from https://www.researchgate.net/publication/43952612_Design_and_Development_of_Virtual_Reality_Analysis_of_Challenges_Faced_by_Educators

Hase, S., & Kenyon, C. (2000). *From andragogy to heutagogy*. Ulti-BASE In-Site.

Hedberg, J., & Alexander, S. (1994). Virtual reality in education: Defining researchable issues. *Educational Media International, 31*(4), 214–220. doi:10.1080/0952398940310402

Hegarty, M. (1992). Mental animation: Inferring motion from static diagrams of mechanical systems. *Journal of Experimental Psychology. Learning, Memory, and Cognition, 18*(5), 1084–1102. doi:10.1037/0278-7393.18.5.1084 PMID:1402712

Henry, S. L. (2015). *Mobile accessibility*. Retrieved December 8, 2015, from https://www.w3.org/WAI/mobile/

Heric, M., & Carter, J. (2011). Cognitive readiness assessment and reporting: An open source mobile framework for operational decision support and performance improvement. *Performance Improvement, 50*(7), 5–17. doi:10.1002/pfi.20227

Holloway, J. (2012). *Brailletouch app to bring eyes-free texting iPhone and iPad*. Retrieved December 8, 2015, from http://www.gizmag.com/brailletouch/21527/

Holmberg, B., Moore, G. M., & Peters, O. (2007). *The Theories and the Theorists: Why Theory is Important for Research.* Presented at the 4 th EDEN Research Workshop on "Research into Online Distance Education and E-Learning: Making the Difference," Castelldefels, Spain.

Holton, E. F., Swanson, R. A., & Naquin, S. S. (2001). Andragogy in practice: Clarifying the andragogical model of adult learning. *Performance Improvement Quarterly, 14*(1), 118–143. doi:10.1111/j.1937-8327.2001.tb00204.x

Holtzblatt, L. J., Damianos, L. E., & Weiss, D. (2010, April). Factors impeding Wiki use in the enterprise: a case study. In CHI' 10 Extended Abstracts on Human Factors in Computing Systems (pp. 4661-4676). ACM. doi:10.1145/1753846.1754208

Hornor, T. (2013). *Designing for disabilities: section 508 and international accessibility compliance for beginners.* Retrieved December 8, 2015, from http://www.sitepoint.com/designing-disabilities-section-508-international-accessibility-compliance-beginners/

Hsu, C. C., & Sandford, B. A. (2007). The Delphi technique: Making sense of consensus. *Practical Assessment, Research & Evaluation, 12*(10), 1–8.

Huang, Y. M., Lin, Y. T., & Cheng, S. C. (2009). An adaptive testing system for supporting versatile educational assessment. *Computers & Education, 52*(1), 53–67. doi:10.1016/j.compedu.2008.06.007

Hui, C. Y., Walton, R., McKinstry, B., Jackson, T., Parker, R., & Pinnock, H. (2016). The Use of Mobile Applications to Support Self-Management for People with Asthma: A Systematic Review of Controlled Studies to Identify Features Associated with Clinical Effectiveness and Adherence. *Journal of the American Medical Informatics Association, 24*(3), 619–632. PMID:27694279

Huizenga, J., Admiraal, W., Akkerman, S., & Dam, G. T. (2009). Mobile game-based learning in secondary education- engagement, motivation and learning in a mobile city game. *Journal of Computer Assisted Learning, 25*(4), 332–344. doi:10.1111/j.1365-2729.2009.00316.x

Huminah Huminah Interactive. (2013). Perfect Picnic (Version 1.2.3) [Mobile application software]. Retrieved from http://fightbac.org/kids

Hung, J. L., & Zhang, K. (2012). Examining mobile learning trends 2003–2008: A categorical meta-trend analysis using text mining techniques. *Journal of Computing in Higher Education, 24*(1), 1–17. doi:10.100712528-011-9044-9

Hung, P. H., Hwang, G. J., Lin, Y. F., Wu, T. H., & Su, I. H. (2013). Seamless connection between learning and assessment – applying progressive learning tasks in mobile ecology inquiry. *Journal of Educational Technology & Society, 16*, 194–205.

Hung, P. H., Hwang, G. J., Su, I., & Lin, I. H. (2012). A concept-map integrated dynamic assessment system for improving ecology observation competences in mobile learning activities. *Turkish Online Journal of Educational Technology-TOJET, 11*, 10–19.

Hunicke, R., LeBlanc, M., & Zubek, R. (2004, July). MDA: A formal approach to game design and game research. *Proceedings of the AAAI Workshop on Challenges in Game AI*, 4.

Hwang, G. J., & Tsai, C. C. (2011). Research trends in mobile and ubiquitous learning: A review of publications in selected journals from 2001 to 2010. *British Journal of Educational Technology*, *42*(4), E65–E70. doi:10.1111/j.1467-8535.2011.01183.x

Hwang, G. J., & Wu, P. H. (2014). Applications, impacts and trends of mobile technology-enhanced learning: A review of 2008–2012 publications in selected SSCI journals. *International Journal of Mobile Learning and Organisation*, *8*(2), 83–95. doi:10.1504/IJMLO.2014.062346

Hwang, G.-J., & Chang, H.-F. (2011). A formative assessment-based mobile learning approach to improving the learning attitudes and achievements of students. *Computers & Education*, *56*(4), 1023–1031. doi:10.1016/j.compedu.2010.12.002

Immerse. (2019a, February 1). *Immerse Emerges from Stealth to Launch First Ever Virtual Reality Language Tutoring Platform* [Press release]. Retrieved from https://www.immerse.online/press-kit

Immerse. (2019b, February 1). *The Immersive Method: A New Era of Language Learning* (White paper). Retrieved from https://app.hubspot.com/documents/4379388/view/34399933?accessId=919a7a

Individuals with Disabilities Education Act (IDEA). (2004). *Public law 108-446: 108th congress.* Retrieved December 8, 2015, from http://www.parentcenterhub.org/wp-content/uploads/repo_items/PL108-446.pdf

Institute for Human Centered Design. (n.d.). *What is universal design?* Retrieved December 8, 2015, from http://www.humancentereddesign.org/universal-design

Islas, A., Contento, I. R., Koch, P. A., Hagiwara, S., & Barton, A. C. (2000). Food choice criteria in elementary school age children incorporating ecological concerns. *Appetite*, *34*, 125.

Issa's Edible Adventures. (2017). ALK Ventures (Version 1) [Mobile application software]. Retrieved from https://issasedibleadventures.com/

Izu, M. R. (2006). Quality and equity in early childcare in Peru. *International Review of Education*, 185–199.

Izzo, M. V., & Bauer, W. M. (2015). Universal design for learning: Enhancing achievement and employment of STEM students with disabilities. *Universal Access in the Information Society*, *14*(1), 17–27. doi:10.100710209-013-0332-1

Jacobs, C. (2013). *Film Theory and Approaches to Criticism, or, What did that movie mean?* Retrieved from http://www.und.edu/instruct/cjacobs/FilmTheory&Analysis.htm

James, A., & Brookfield, S. D. (2014). *Engaging imagination: helping students become creative and reflective thinkers*. John Wiley & Sons.

Jason, L. A., Keys, B., Suarez-Balcazr, C., Taylar, R. R., Davis, M. I., Durlak, J. A., & Isenberg, D. H. (Eds.). (2002). *Participatory community research*. Washington, DC: American Psychological Association.

Johnson-Laird, P. N., & Wason, P. C. (1970). A theoretical analysis of insight into a reasoning task. *Cognitive Psychology, 1*(2), 134–148.Jump App (2015). Cute Food (Version 1.5) [Mobile application software]. Retrieved from http://jumpapp.eu/?post_type=portfolio&p=2532

Johnson-Laird, P. N. (1994). Mental models and probabilistic thinking. *Cognition, 50*(1), 189–209. doi:10.1016/0010-0277(94)90028-0 PMID:8039361

Johnson-Laird, P. N. (2005). Mental models and thoughts. In K. Holyoak & R. Morrison (Eds.), *The Cambridge handbook of thinking and reasoning* (pp. 185–208). New York, NY: Cambridge University Press.

Johnson-Laird, P. N., & Byrne, R. M. (1991). *Deduction*. Lawrence Erlbaum Associates, Inc.

Jorgensen, D. L. (1989). *Participatory observation: A methodology for human studies*. Thousand Oaks, CA: Sage. doi:10.4135/9781412985376

Joseph, B., Deepika, R., Pal, A., Tunks, J., & Mehta, G. (2019). Intrinsic vs. Extrinsic Motivation in an Interactive Engineering Game. *Journal of Advances in Computer Engineering and Technology*.

Jung, H.-J. (2015). Fostering an English teaching environment: Factors influencing English as a foreign language teachers' doption of mobile learning. *Informatics in Education, 14*(2), 219–241. doi:10.15388/infedu.2015.13

Kagan, J. (1971). A conception of early adolescence. *Daedalus, 1*, 12–16.

Kagohara, D. M., Van der Meer, L., Ramdoss, S., O'Reilly, M. F., Lancioni, G. E., Davis, T. N., ... Sigafoos, J. (2013). Using iPods and iPads in teaching programs for individuals with developmental disabilities: A systematic review. *Research in Developmental Disabilities, 34*(1), 147–156. doi:10.1016/j.ridd.2012.07.027 PMID:22940168

Kalyuga, S., Chandler, P., & Sweller, J. (1999). Managing split-attention and redundancy in multimedia instruction. *Applied Cognitive Psychology, 13*(4), 351–371. doi:10.1002/(SICI)1099-0720(199908)13:4<351::AID-ACP589>3.0.CO;2-6

Karlsen, J. (1991). *Participatory action research*. Newbury Park, CA: Sage.

Karnjanapun, S. (2015). *Nakhon Ratchasima teachers' using of smartphones in English Languageteaching*. Paper Presented at the EdMedia: *World Conference on Educational Media and Technology*, Montreal, Quebec, Canada.

Kay, M., Santos, J., & Takane, M. (2011). Barriers to mHealth implementation. In mHealth: New horizons for health through mobile technologies (Vol. 3, p. 112). Geneva: World Health Organization.

Kelsey, T., & Cavendish, W. (2014, November 13). *Personalised health and care 2020: A framework for action.* Retrieved December 14, 2015, from https://www.gov.uk/government/publications/personalised-health-and-care-2020/using-data-and-

Kidstir. (2018). Retrieved from https://kidstir.com/

Kihm, H., & Singleton, K. (2012). Creative cooking with preschoolers: A service learning experience for FCS students. *Journal of Family and Consumer Sciences Education, 30,* 1.

Kiili, K. (2005). Digital game-based learning: Towards an experiential gaming model. *The Internet and Higher Education, 8*(1), 13–24. doi:10.1016/j.iheduc.2004.12.001

Kim (2015). *Introduction to the Study of Theory and Film.* Retrieved from http://arts.columbia.edu/summer/film/course/study-and-theory

Kim, N. (2015, January 1). *Introduction to the Study and Theory of Film.* Retrieved January 1, 2015, from http://arts.columbia.edu/summer/film/course/study-and-theory

Kitsiou, S., Pare, G., Janna, M., & Gerber, B. (2017). Effectiveness of mHealth Interventions for Patients with diabetes: An Overview of Systematic Reviews. *PLoS One, 12*(3), e0173160. doi:10.1371/journal.pone.0173160 PMID:28249025

Kittler, P. G., Sucher, K., & Nelms, M. (2012). *Food and culture.* Belmont, CA: Cengage Learning.

Kizito, R. (2011). Pretesting mathematical concepts with the mobile phone: Implications for curriculum design. *International Review of Research in Open and Distance Learning, 13*(1), 38–55. doi:10.19173/irrodl.v13i1.1065

Klibanoff, R. S., Levine, S. C., Huttenlocher, J., Vasilyeva, M., & Hedges, L. V. (2006). Preschool children's mathematical knowledge: The effect of teacher "math talk." *Developmental Psychology, 42*(1), 59–69. doi:10.1037/0012-1649.42.1.59 PMID:16420118

Klopfer, E., & Woodruff, E. (2002, January). The impact of distributed and ubiquitous computational devices on the collaborative learning environment. In *Proceedings of the Conference on Computer Support for Collaborative Learning: Foundations for a CSCL Community* (pp. 702-703). International Society of the Learning Sciences. 10.3115/1658616.1658795

Knight, A., O'Connell, R., & Brannen, J. (2014). The temporality of food practices: Intergenerational relations, childhood memories and mothers' food practices in working families with young children. *Families, Relationships and Societies, 3*(2), 303–318. doi:10.1332/204674313X669720

Knowles, M. S. (1970). *The modern practice of adult education* (Vol. 41). New York: New York Association Press.

Knowles, M. S. (1978). *The adult learner: A neglected species* (2nd ed.). Houston, TX: Gulf Publishing Company.

Koch, P., Calabrese Barton, A., Contento, I., & Crabtree, M. (2008). Farm to table and beyond: Helping students make sense of the global food system. *Science Scope, 31*(9), 36–39.

KoolAppz. (2011). Film Making – Learn Basics (1.0) [Mobile application software]. Retrieved from https://play.google.com

Koole, M. (2010, May). The web of identity: Selfhood and belonging in online learning networks. *Seventh International Conference on Networked Learning 2010.*

Korpershoek, Y. J. G., Vervoort, S. C. J. M., Trappenburg, J. C. A., & Schuurmans, M. J. (2018). Perceptions of Patients with Chronic Obstructive Pulmonary Disease and Their Health Care Providers Towards Using mHealth for Self-Management of Exacerbations: A Qualitative Study. *BMC Health Services Research, 18*(757). PMID:30286761

Kozma, R. B. (1991). Learning with media. *Review of Educational Research, 61*(2), 179–211. doi:10.3102/00346543061002179

Krenn, J. & Hachey, A. C. (2015). Cooking with attitude: How preschool children and early primary educators can promote positive lifelong skills in the kitchen. *Texas Childcare, 38*(4).

Kristof, N. D., & WuDunn, S. (2012). *Mobile Games: Reaching the Hardest to Reach.* Retrieved from Half the Sky Movement: http://www.halftheskymovement.org/pages/mobile-games.html

Kroger, J. (2009). *Identity in adolescence: The balance between the self and the other.* New York, NY: Psychology Press.

Kruger, J., Blanck, H. M., & Gillespie, C. (2006). Dietary and physical activity behaviors among adults successful at weight loss maintenance. *The International Journal of Behavioral Nutrition and Physical Activity, 3*(3), 1–10. Retrieved from http://www.ijbnpa.org/content/pdf/1479-5868-3-17.pdf PMID:16854220

Kukulska-Hulme, A. (2007). Mobile usability in educational contexts: What have we learnt? *The International Review of Research in Open and Distributed Learning, 8*(2). doi:10.19173/irrodl.v8i2.356

Kukulska-Hulme, A. (2009). Will mobile learning change language learning? *ReCALL, 21*(02), 157–165. doi:10.1017/S0958344009000202

Kukulska-Hulme, A., Sharples, M., Milrad, M., Arnedillo-Sánchez, I., & Vavoula, G. (2009). Innovation in mobile learning: A European perspective. *International Journal of Mobile and Blended Learning, 1*(1), 13–35. doi:10.4018/jmbl.2009010102

Kurzweil Education. (n.d.). *Text-to-Speech Software.* Retrieved December 8, 2015, from https://www.kurzweiledu.com/about-kurzweil/text-to-speech.html

Kushalnagar, R. S., & Trager, B. P. (2011). *Improving classroom visual accessibility with cooperative smartphone recordings.* Retrieved December 8, 2015, from http://www.researchgate.net/publication/254003873

Larson, N., Perry, C., Story, M., & Neumark-Sztainer, D. (2006). Food preparation by young adults is associated with better diet quality. *Journal of the American Dietetic Association, 106*(12), 2001–2007. doi:10.1016/j.jada.2006.09.008 PMID:17126631

Lave, J., & Wenger, E. (1991). *Situated Learning: Legitimate Peripheral Participation.* Cambridge, UK: Cambridge University Press. doi:10.1017/CBO9780511815355

Lee, H., Contento, I. R., Sauberli, W., Koch, P. A., & Calabrese Barton, A. (2007). Using science education to move middle schoolers toward more healthful food and activity choices: An outcome evaluation of Choice, Control, & Change (C3). *Journal of Nutrition Education and Behavior, 39*(4S), S96. doi:10.1016/j.jneb.2007.04.270

Lee, L. (2015a). Digital media and young children's learning: A case study of using iPads in American preschools. *International Journal of Information and Education Technology (IJIET), 5*(12), 947–951. doi:10.7763/IJIET.2015.V5.643

Lee, L. (2015b). Technology play and learning. In D. P. Fromberg & D. Bergen (Eds.), *Play from birth to twelve: Contexts, perspectives, and meanings* (pp. 217–224). New York, NY: Routledge.

Lee, M. R., & Chen, T. T. (2010). Visualising intellectual structure of ubiquitous computing. In *Knowledge Management and Acquisition for Smart Systems and Services* (pp. 261–272). Springer Berlin Heidelberg. doi:10.1007/978-3-642-15037-1_22

Li, M., Sareh, S., Xu, G., Ridzuan, M. B., Luo, S., Xie, J., ... Althoefer, K. (2016). Evaluation of pseudo-haptic interactions with soft objects in virtual environments. *PLoS One, 11*(6), e0157681. doi:10.1371/journal.pone.0157681 PMID:27352234

Lin, J., Hsu, H.-Y., & Lee, G. (2015). *Inducing language teachers to use tablets in the classroom: Lessons learned.* Paper Presented at the E-Learn: World Conference on E-Learning in Corporate, Government, Healthcare, and Higher Education 2015, Kona, HI.

Lin, J., & Rivera-Sánchez, M. (2012). Testing the information technology continuance model on a mandatory SMS-based student response system. *Communication Education, 61*(2), 89–110. doi:10.1080/03634523.2011.654231

Linstone, H. A., & Turoff, M. (2002). The Delphi Method. *Techniques and Applications, 53.*

Liquori, T., Koch, P. D., Contento, I. R., & Castle, J. (1998). The Cookshop Program: Outcome evaluation of a nutrition education program linking lunchroom food experiences with classroom cooking experiences. *Journal of Nutrition Education, 30*(5), 302–313. doi:10.1016/S0022-3182(98)70339-5

Liu, T. C., Wang, H. Y., Liang, J. K., Chan, T. W., Ko, H. W., & Yang, J. C. (2003). Wireless and mobile technologies to enhance teaching and learning. *Journal of Computer Assisted Learning, 19*(3), 371–382. doi:10.1046/j.0266-4909.2003.00038.x

Liu, Y., Han, S., & Li, H. (2010). Understanding the factors driving m-learning adoption: A literature review. *Campus-Wide Information Systems, 27*(4), 210–226. doi:10.1108/10650741011073761

Looi, C. K., Lim, K. F., Pang, J., Koh, A. L. H., Seow, P., Sun, D., & Soloway, E. (2016). Bridging Formal and Informal Learning with the Use of Mobile Technology. In *Future Learning in Primary Schools* (pp. 79–96). Springer Singapore. doi:10.1007/978-981-287-579-2_6

Lou, Y., Abrami, P. C., & d'Apollonia, S. (2001). Small group and individual learning with technology: A meta-analysis. *Review of Educational Research, 71*(3), 449–521. doi:10.3102/00346543071003449

Mahmud, N., Rodriguez, J., & Nesbit, J. (2010). A text message-based intervention to bridge the healthcare communication gap in the rural developing world. *Technology and Health Care, 18,* 137–144. doi:10.323 d3/THC-2010-0576

Mango Tree. (2015). WeCookit (Version 1.0.2) [Mobile application software]. Retrieved from http://www.we-cookit.com

Ma, R., Dai, Q., Ni, C., & Li, X. (2009). An author co-citation analysis of information science in China with Chinese Google Scholar search engine, 2004–2006. *Scientometrics, 81*(1), 33–46. doi:10.100711192-009-2063-x

Marquardt, M. J. (2000). Action Learning and Leadership. *The Learning Organization, 7*(5), 233–241. doi:10.1108/09696470010352990

Marsick, V., & O'Neil, J. (2007). *Understanding Action Learning (AMA Innovations in Adult Learning).* New York: AMACOM.

Martin, A., & Chamberlain, L. J. (2015). Technology-Driven Service Learning: The Instructor's Role and the Student's Experience. *EDUCAUSE Review*. Retrieved from http://www.educause.edu/ero/article/technology-driven-service-learning-instructors-roleand-students-experience

Martín, E., Andueza, N., Carro, R. M., & Rodríguez, P. (2006). Recommending activities in collaborative m-learning. *Lecture Notes in Learning and Teaching*, 197-204.

Martínez, M., & Hilera, J. R. (2010). A new m-learning application designed for self-assessment. *Pulso: Revista de Educación, 33*, 207–215.

Martinez, S. M., Rhee, K., Blanco, E., & Boutelle, K. (2014). Maternal attitudes and behaviors regarding feeding practices in elementary school–aged Latino children: A pilot qualitative study on the impact of the cultural role of mothers in the U.S.–Mexican border region of San Diego, California. *Journal of the Academy of Nutrition and Dietetics, 114*(2), 230–237. doi:10.1016/j.jand.2013.09.028 PMID:24315129

Martiz, G. (2015). *A qualitative case study on cell phone appropriation for language learning purposes in a Dominican context* (M. Recker, K. DeJonge-Kannan, A. Diekema, V. Lee, & J. Rogers, Eds.). ProQuest Dissertations Publishing.

Masten, A. S., Obradovic, J., & Keith, B. B. (2006). *Resilience in emerging adulthood: Developmental perspectives on continuity and transformation.* Washington, DC: American Psychological Association.

Mayer, R. (2009). *The promise of multimedia learning, Multimedia Learning* (2nd ed.). New York: Cambridge University Press. doi:10.1017/CBO9780511811678

Mayer, R. E. (2002). *Multimedia learning*. Cambridge, UK: Cambridge University Press.

Mayer, R. E. (2005). Cognitive theory of multimedia learning. In R. E. Mayer (Ed.), *The Cambridge handbook of multimedia learning* (pp. 31–48). New York: Cambridge University Press. doi:10.1017/CBO9780511816819.004

Mayer, R. E., Heiser, J., & Lonn, S. (2001). Cognitive constraints on multimedia learning: When presenting more material results in less understanding. *Journal of Educational Psychology*, *93*(1), 187–198. doi:10.1037/0022-0663.93.1.187

Mayer, R. E., & Moreno, R. (2003). Nine ways to reduce cognitive load in multimedia learning. *Educational Psychologist*, *38*(1), 43–52. doi:10.1207/S15326985EP3801_6

McCabe, C., McCann, M., & Brady, A. M. (2017). Computer and Mobile Technology Interventions for Self-Management in Chronic Obstructive Pulmonary Disease [Review]. *Cochrane Database of Systematic Reviews*, *5*(CD-11425). PMID:28535331

McCain, K. W. (1990). Mapping authors in intellectual space: A technical overview. *Journal of the American Society for Information Science*, *41*(6), 433–443. doi:10.1002/(SICI)1097-4571(199009)41:6<433::AID-ASI11>3.0.CO;2-Q

McCall, R. B., Groark, C. J., Fish, L., Harkins, D., Serrano, G., & Gordon, K. (2010). A socioemotional intervention in a Latin American orphanage. *Infant Mental Health Journal*, *31*(5), 521–542. doi:10.1002/imhj.20270 PMID:28543838

McCann, S. (2017). Mobile Learning in Workforce Development: Cultivating Creativity on Action Learning Teams through Higher-Order mLearning. In D. Mentor (Ed.), *Handbook of Research on Mobile Learning in Contemporary Classrooms* (pp. 139–159). Hershey, PA: IGI Global. doi:10.4018/978-1-5225-0251-7.ch007

McClean, G., Murray, E., Band, R., Moffatt, K. R., Hanlon, P., Bruton, A., ... Mair, F. S. (2016b). Interactive Digital Interventions to Promote Self-Management in Adults with Asthma: Systematic Review and Meta-Analysis. *BMC Pulmonary Medicine*, *16*(83). PMID:27215329

McFarlane, A., Sparrowhawk, A., & Heald, Y. (2002). Report on the educational use of games: An exploration by TEEM of the contribution which games can make to the education process. Cambridgeshire, UK: Academic Press.

McLean, G., Band, R., Saunderson, K., Hanlon, P., Murray, E., Little, P., ... Mair, F. (2016a). Digital interventions to promote self-management in adults with hypertension systematic review and meta-analysis. *Journal of Hypertension*, *34*(4), 600–612. doi:10.1097/HJH.0000000000000859 PMID:26845284

McPherson, M., & Baptista Nunes, M. (2006). Organizational issues for e-Learning: Critical success factors as identified by HE practitioners. *International Journal of Educational Management*, *20*(7), 542–558. doi:10.1108/09513540610704645

McTigue, E. M., & Uppstad, P. H. (2019). Getting Serious About Serious Games: Best Practices for Computer Games in Reading Classrooms. *The Reading Teacher*, *72*(4), 453–461. doi:10.1002/trtr.1737

Mehdipour, Y., & Zerehkafi, H. (2013). Mobile learning for education: Benefits and challenges. *International Journal of Computational Engineering Research*, *3*(6), 93–100. Retrieved from http://www.ijceronline.com/papers/Vol3_issue6/part%203/P03630930100.pdf

Melnick, K. (2018, July 5). *VR Chemistry Game 'Tablecraft' Makes Learning The Periodic Table Fun*. Retrieved from https://vrscout.com/news/vr-chemistry-game-tablecraft/

Mentor, D. (2011, October). Digit-all texting thumbs as a social driving force: A texting service study catering to the socio-emotional challenges facing the Higher Education student. In *World Conference on E-Learning in Corporate, Government, Healthcare, and Higher Education* (*Vol. 2011*, No. 1, pp. 1493-1498). Academic Press.

Mentor, D. (2011b). *Supporting students' wellbeing via a texting service*. Invited guest speaker iClicker Conference, Houston, TX.

Mentor, D. (2014a). *Showcasing a Criteria Taxonomy for selecting a 21st Century Learning Management System to support blended teaching and mLearning*. Poster Presentation. The 21st Century Academic Forum Conference at Harvard, Cambridge, MA. Retrieved from http://www.21caf.org/uploads/1/3/5/2/13527682/21st_century_academic_forum_conference_book.pdf

Mentor, D. (2015). From App Attack to Goal-Oriented Tablet Use. In Tablets in K-12 Education: Integrated Experiences and Implications (pp. 1-21). IGI Global. doi:10.4018/978-1-4666-6300-8.ch001

Mentor, D. (2018). mClass Planet of the Apps: The Rise of Mobile Learning. In Computer-Mediated Learning for Workforce Development (pp. 196-215). IGI Global.

Mentor, D. (2018). TLC for MOOCS: Teaching and Learning Communities for Computer Programming. In Computer-Mediated Learning for Workforce Development (pp. 93-110). IGI Global.

Mentor, D. (2018, March). *TLC for MOOCs - Teaching and Learning Communities for MOOCs*. Paper presented at UNESCO's Mobile Learning Week Sympoisum at UNESCO headquarters, Paris, France.

Mentor, D. (2018a). mClass Planet of the Apps: The Rise of Mobile Learning. In Computer-Mediated Learning for Workforce Development (pp. 196-215). IGI Global.

Mentor, D. (2018a). Teachers College, Columbia University, Department of Biobehavioral Science. *BBS 5199 002: mHealth - Mobile Health App Development CRN 32100*. Retrieved from tc.edu

Mentor, D. (2018b). TLC for MOOCS: Teaching and Learning Communities for Computer Programming. In Computer-Mediated Learning for Workforce Development (pp. 93-110). IGI Global.

Mentor, D. (2018c). mClass Planet of the Apps: The Rise of Mobile Learning. In Computer-Mediated Learning for Workforce Development (pp. 196-215). IGI Global.

Mentor, D. J. (2011). *Exploring social connectedness via mobile phone texting* (Doctoral dissertation). Teachers College, Columbia University.

Mentor, D. J. (2011a). *Exploring social connectedness via mobile phone texting* (Ph.D. thesis). Teachers College, Columbia University. Retrieved November 25, 2018 from https://www.learntechlib.org/p/11

Mentor, D. (2010). *Book Review of Students' experiences of E-learning in higher education; the ecology of sustainable innovation by Robert Ellis and Peter Goodyear*. New York: TC Record.

Mentor, D. (2014b). From app attack to goal oriented tablet use. In H. An, S. Alon, & D. Fuentes (Eds.), *Tablets in K-12 education: Integrated experiences and implications*. Hershey, PA: IGI Global. doi:10.4018/978-1-4666-6300-8

Mentor, D. (2016). EMxC3= e&mLearning Cultivating Connected Communities: Sustainable Workforce Talent Development. In *Handbook of Research on Mobile Learning in Contemporary Classrooms* (pp. 240–259). IGI Global. doi:10.4018/978-1-5225-0251-7.ch012

Mentor, D. (2018). Micro to macro social connectedness through mobile phone engagement. In *Encyclopedia of Information Science and Technology* (4th ed.; pp. 6184–6194). IGI Global. doi:10.4018/978-1-5225-2255-3.ch537

Mentor, D. (Ed.). (2016b). *Handbook of research on mobile learning in contemporary classrooms*. IGI Global. doi:10.4018/978-1-5225-0251-7

Mentor, D. (March, 2014). *Showcasing a Criteria Taxonomy for selecting a 21st Century Learning Management System to support blended teaching and mLearning*. Poster Presentation delivered at the *21st Century Academic Forum Conference*, Harvard University, Cambridge, MA.

Merton, R. K. (1973). *The sociology of science: Theoretical and empirical investigations*. University of Chicago Press.

Mezirow, J. (1990). Fostering critical reflection in adulthood. San Francisco: Academic Press.

Mezirow, J. (1998). On critical reflection. *Adult Education Quarterly*, *48*(3), 185–198. doi:10.1177/074171369804800305

Miller, B. T., Krockover, G. H., & Doughty, T. (2013). Using iPads to teach inquiry science to students with a moderate to severe intellectual disability: A pilot study. *Journal of Research in Science Teaching*, *50*(8), 887–911. doi:10.1002/tea.21091

Mobile Learning International Conference. (2019). *Previous Editions: From 2005-2018 | ML 2019*. Retrieved from https://www.mlearning-conf.org/previous-editions-2005-2015/

MoLeNET. (2011). Retrieved from http://www.molenet.org.uk

Moore, J. (n.d.). *Mobile apps and accessibility*. Retrieved December 8, 2015, from http://tinyurl.com/z39laae

Moore, M. (2015, February 2). 6 great apps to teach kids about nutrition [Web log post]. Retrieved from http://www.eatright.org/resource/food/nutrition/healthy-eating/6-great-apps-to-teach-kids-about-nutrition

More, C. M., & Travers, J. C. (2013). What's app with that? Selecting educational apps for young children with disabilities. *Young Exceptional Children*, *16*(2), 15–32. doi:10.1177/1096250612464763

MoreChinese. (2015). Dodo China Pro: the trip of experiencing Chinese culture, food and characters (Version 3.22) [Mobile application software]. Retrieved from https://itunes.apple.com/us/app/dodo-china-pro-trip-experiencing/id598851707?mt=8

Moreno, R., & Mayer, R. (2000). A coherence effect in multimedia learning: The case for minimizing irrelevant sounds in the design of multimedia instructional messages. *Journal of Educational Psychology*, *92*(1), 117–125. doi:10.1037/0022-0663.92.1.117

Morgan, A., Pendergast, D., Brown, R., & Heck, D. (2015). Relational ways of being an educator: Trauma-informed practice supporting disenfranchised young people. *International Journal of Inclusive Education*, *19*(10), 1037–1051. doi:10.1080/13603116.2015.1035344

Morgan, P. J., Collins, C. E., Plotnikoff, R. C., Callister, R., Burrows, T., Fletcher, R., & Lubans, D. R. (2014). The 'Healthy Dads, Healthy Kids' community randomized controlled trial: A community-based healthy lifestyle program for fathers and their children. *Preventive Medicine*, *61*, 90–99. doi:10.1016/j.ypmed.2013.12.019 PMID:24380796

Morrison, D., Mair, F. S., Yardley, L., Kirby, S., & Thomas, M. (2016). Living with Asthma and Chronic Obstructive Airway Disease: Using Technology to Support Self-Management—An Overview. *Chronic Respiratory Disease*, *14*(4), 407–419. doi:10.1177/1479972316660977 PMID:27512084

Mosa, A. S. M., Yoo, I., & Sheets, L. (2012). A systematic review of healthcare applications for smartphones. *BMC Medical Informatics and Decision Making*, *12*(1), 67. doi:10.1186/1472-6947-12-67 PMID:22781312

Motiwalla, L. F. (2007). Mobile learning: A framework and evaluation. *Computers & Education*, *49*(3), 581–596. doi:10.1016/j.compedu.2005.10.011

Mroz, M. (2013). *Temporality and Film Analysis*. Edinburgh, UK: Edinburgh University Press.

Mulrennan, D. (2018). Mobile social media and the news: Where heutagogy enables journalism education. *Journalism & Mass Communication Educator*, *73*(3), 322–333. doi:10.1177/1077695817720762

Naismith, L. Lonsadale, P., Vavoula, G., & Sharples, M. (2006). *Literature Review in Mobile Technologies and Learning*. Academic Press.

Naismith, L., Lonsdale, P., Vavoula, G. N., & Sharples, M. (2004). *Mobile technologies and learning*. Academic Press.

Naismith, L., Lonsdale, P., Vavoula, G., & Sharples, M. (2004). Literature review in mobile technologies and learning. Futurelab. *Literature review in mobile technologies and learning: Futurelab.*

Naismith, L., Lonsdale, P., Vavoula, G., & Sharples, M. (2004). *Literature review in mobile technologies and learning*. Birmingham, UK: University of Birmingham.

Naremore, J. (1996). American Film Noir: The History of an Idea. *Film Quarterly*, *49*(2), 12–28. doi:10.2307/1213310

National Cancer Institute. (2015). *What is Cancer?* Retrieved February 22, 2019 from https://www.cancer.gov/about-cancer/understanding/what-is-cancer

National Center for Education Statistics. (2015a). *Fast facts: Students with disabilities*. Retrieved December 8, 2015, from https://nces.ed.gov/fastfacts/display.asp?id=64

National Center for Education Statistics. (2015b). *Children and youth with disabilities*. Retrieved December 8, 2015, from http://nces.ed.gov/programs/coe/indicator_cgg.asp

National Center for Learning Disabilities. (2014). *The state of learning disabilities: facts, trends and emerging issues*. New York, NY: Author.

National Center on Universal Design for Learning. (2014). *What is UDL?* Retrieved December 8, 2015, from http://www.udlcenter.org/aboutudl/whatisudl

National Federation for the Blind (NFB). (2014). *Blindness statistics.* Retrieved December 8, 2015, from https://nfb.org/blindness-statistics

National Institute on Deafness and Other Communication Disorders (NIDCD). (2015). *Quick statistics.* Retrieved December 8, 2015, from http://www.nidcd.nih.gov/health/statistics/Pages/quick.aspx

Neale, S. (2000). Questions of Genre. In *Film and Theory: An Anthology* (pp. 157–178). Malden, MA: Blackwell Publishing.

Neisser, U., & Becklen, R. (1975). Selective looking: Attending to visually specified events. *Cognitive Psychology*, *7*(4), 480–494. doi:10.1016/0010-0285(75)90019-5

Nelson, L. A., Coston, T. D., Cherrington, A. L., & Osborn, C. Y. (2016). Patterns of User Engagement with Mobile- and Web-Delivered Self-Care Interventions for Adults with T2DM: A Review of the Literature. *Current Diabetes Reports*, *16*(66). PMID:27255269

New Schools. (n.d.). *Wireless Generation.* Retrieved from http://www.newschools.org/venture/wireless-generation

New York Daily News. (2015). *Department of education lifts ban on cell phones in New York City schools*. Retrieved December 8, 2015, from http://www.nydailynews.com/new-york/dept-education-ends-cell-phone-ban-nyc-schools-article-1.2134970

Newtown, D., & Dell, A. G. (2011). Mobile devices and students with disabilities: What do best practices tell us? *Journal of Special Education Technology*, *26*(3), 47–49.

Ng, M., Fleming, T., Robinson, M., Thomson, B., Graetz, N., Margono, C., ... Gupta, R. (2014). Global, regional, and national prevalence of overweight and obesity in children and adults during 1980–2013: A systematic analysis for the Global Burden of Disease Study 2013. *Lancet*, *384*(9945), 766–781. doi:10.1016/S0140-6736(14)60460-8 PMID:24880830

Niccolai, J., & Gohring, N. (2010, April). A brief history of palm II. *Network World*.

Night & Day Studios. (2015). Yummiloo Rainbow Power (Version 1.1.2) [Mobile application software]. Retrieved from http://www.nightanddaystudios.com

NMSU Board of Regents. (2015). Eat & Move-O-Matic (Version 2.2) [Mobile application software]. Retrieved from http://mediaproductions.nmsu.edu/eatmove/

Nomster Chef. (2019). Retrieved from http://nomsterchef.com

Norman, Å., Berlin, A., Sundblom, E., Elinder, L. S., & Nyberg, G. (2015). Stuck in a vicious circle of stress: Parental concerns and barriers to changing children's dietary and physical activity habits. *Appetite*, *87*, 137–142.

Norris, C., & Soloway, E. (2004). Envisioning the handheld-centric classroom. *Journal of Educational Computing Research*, *30*(4), 281–294. doi:10.2190/MBPJ-L35D-C4K6-AQB8

Ntloedibe-Kuswani, S. G. (2007). *Exploring the use of electronic mobile technologies among distance learners in rural communities for safe and disruptive learning*. Syracuse University.

O'Bannon, B. W., & Thomas, K. M. (2015). Mobile phones in the classroom: Preservice teachers answer the call. *Computers & Education*, *85*, 110–122. doi:10.1016/j.compedu.2015.02.010

O'Connor, D. (2003). *21st century information fluency*. Retrieved December 8, 2015, from http://21cif.com/tutorials/micro/mm/508/page7.htm

Oberle, E., Domitrovich, C., Meyers, D., & Weissberg, R. (2016). Establishing systemic social and emotional learning approaches in schools: A framework for schoolwide implementation. *Cambridge Journal of Education*, *46*(3), 277–297. doi:10.1080/0305764X.2015.1125450

Octave Media International. (2012). Smash Your Food (Version 1.3) [Mobile application software]. Retrieved from http://smashyourfood.foodnme.com

Ogata, H., & Uosaki, N. (2012). A new trend of mobile and ubiquitous learning research: Towards enhancing ubiquitous learning experiences. *International Journal of Mobile Learning and Organisation*, *6*(1), 64–78. doi:10.1504/IJMLO.2012.046883

Oh, J., Lee, H.-K., Park, M., & Cho, Y. (2014). Exploring the effects of tablet PC-based English learning. *English Teaching*, *69*(3), 151–176. doi:10.15858/engtea.69.3.201409.151

Okoli, C., & Pawlowski, S. D. (2004). The Delphi method as a research tool: An example, design considerations and applications. *Information & Management*, *42*(1), 15–29. doi:10.1016/j.im.2003.11.002

Olvera, N., Knox, B., Scherer, R., Maldonado, G., Sharma, S., Alastuey, L., & Bush, J. (2008). A healthy lifestyle program for Latino daughters and mothers: The BOUNCE overview and process evaluation. *American Journal of Health Education*, *39*(5), 283–295. doi:10.1080/19325037.2008.10599052

OnlineUniversities.com. (2012). *40 creative ways teachers are using cell phones in the classroom.* Retrieved December 8, 2015, from http://www.onlineuniversities.com/blog/2012/10/40-creative-ways-teachers-are-using-cell-phones-classroom/

OpenDyslexic.org. (n.d.). *Available applications that include OpenDyslexic*. Retrieved December 8, 2015, from http://opendyslexic.org/category/products/applications/

Osherson, D. N. (1976). *Logical abilities in children: IV. Reasoning and concepts.* New York: Lawrence Erlbaum.

Özçınar, H. (2015). Mapping teacher education domain: A document co-citation analysis from 1992 to 2012. *Teaching and Teacher Education*, *47*, 42–61. doi:10.1016/j.tate.2014.12.006

Özhan, Ş. Ç., & Kocadere, S. A. (2019). The Effects of Flow, Emotional Engagement, and Motivation on Success in a Gamified Online Learning Environment. *Journal of Educational Computing Research*, 0735633118823159.

Pagel, J., Lambacher, S., Kikuchi, H., & Mori, S. (2018). *Student and instructor attitudes toward CALL and MALL in the L2 classroom.* Editorial Universitat Politècnica de València. doi:10.4995/HEAD18.2018.8108

Pappas, C. (2015, April 21). How to develop mobile learning assessments [Web Blog Post]. Retrieved from http://elearningindustry.com/5-tips-develop-mobile-learning-assessments

Parati, G., Torlasco, C., Omboni, S., & Pellegrini, D. (2017). Smartphone Applications for Hypertension Management: A Potential Game-Changer That Needs More Control. *Current Hypertension Reports*, *19*(48). PMID:28501982

Paris, D. (2012). Culturally sustaining pedagogy: A needed change in stance, terminology, and practice. *Educational Researcher*, *41*(3), 93–97. doi:10.3102/0013189X12441244

Parker, K. R., & Chao, J. T. (2007). Wiki as a teaching tool. *Interdisciplinary Journal of Knowledge and Learning Objects*, *3*(1), 57–72.

Park, M., & Slater, T. (2014). A typology of tasks for mobile-assisted language learning: Recommendations from a small-scale needs analysis. *TESL Canada Journal*, *31*, 93–115. doi:10.18806/tesl.v31i0.1188

Pask, A. G. S. (1976). *Conversation Theory: Applications in Education and Epistemology*. Amsterdam: Elsevier.

Paul, R., & Elder, L. (2003, Spring). Critical thinking: Teaching students how to study and learn (Part III). *Journal of Developmental Education*, *26*(3), 36–37.

Pedro, L. F. M. G., & de Oliveira Barbosa, C. M. M. (2018). A critical review ofmobile learning integration in formal educational contexts. *International Journal of EducationalTechnology in Higher Education*, *15*(1), 10. doi:10.118641239-018-0091-4

Pegrum, M., Oakley, G., & Faulkner, R. (2013). Schools going mobile: A study of the adoption of mobile handheld Technologies in Western Australian independent schools. *Australasian Journal of Educational Technology*, *29*(1), 66–81. doi:10.14742/ajet.64

Peirce, C. S. (1903). Abduction and induction. In J. Buchler (Ed.), *Philosophical writings of Peirce* (pp. 150–156). New York, NY: Dover.

Peña, J. J. S., Tello, J. C., & de la Cámara, S. P. (2011). *A comparative study between free e-learning platforms*. Academic Press.

Perez, L., Furler, B., & Reat, C. C. (2014). *Mobile devices empower students with special needs*. Retrieved December 8, 2015, from http://www.educationworld.com/a_tech/apps-special-needs-disabilities-assistive-technology-students.shtml

Pérez-Paredes, P., Ordoñana Guillamón, C., & Aguado Jiménez, P. (2018). Language teachers' perceptions on the use of OER language processing technologies in MALL. *Computer Assisted Language Learning, 31*(5-6), 522-526. doi:10.1080/09588221.2017.1418754

Persson, O., Danell, R., & Schneider, J. W. (2009). How to use Bibexcel for various types of bibliometric analysis. *Celebrating scholarly communication studies: A Festschrift for Olle Persson at his 60th Birthday*, 9-24.

Persson, O. (2001). All author citations versus first author citations. *Scientometrics*, *50*(2), 339–344. doi:10.1023/A:1010534009428

Peters, J., Parletta, N., Lynch, J., & Campbell, K. (2014). A comparison of parental views of their pre-school children's 'healthy' versus 'unhealthy' diets. A qualitative study. *Appetite*, *76*, 129–136. doi:10.1016/j.appet.2014.02.001 PMID:24524974

Pew Research Center. (2014, February 13). *Emerging nations embrace Internet, mobile technology - Cell Phones Nearly Ubiquitous in Many Countries*. Retrieved from http://www.pewglobal.org/2014/02/13/emerging-nations-embrace-internet-mobile-technology/

Pew Research Center. (2015). *Device ownership over time*. Retrieved from http://www.pewinternet.org/data-trend/mobile/device-ownership/

Pew Research Center. (2015). Retrieved from http://www.pewglobal.org/2018/06/19/2-smartphone-ownership-on-the-rise-in-emerging-economies/#smartphone-ownership-growing

Pew Research Center. (2017). *Tech Adoption Climbs Among Older Adults.* Retrieved February 9, 2019 from http://www.pewinternet.org/2017/05/17/technology-use-among-seniors/

Pew Research Center. (2018). *Mobile Fact Sheet.* Retrieved February 8, 2019 from http://www.pewinternet.org/fact-sheet/mobile/

Playdek. (2012). Food Fight (Version 1.0.1) [Mobile application software]. Retrieved from http://www.playdekgames.com

Pogue, D. (2011). *Apple's assistive touch helps the disabled use a smartphone.* Retrieved December 8, 2015, from http://pogue.blogs.nytimes.com/2011/11/10/apples-assistivetouch-helps-the-disabled-use-a-smartphone/

Poll Everywhere. (n.d.). *Live audience participation.* Retrieved from http://www.polleverywhere.com

Prensky. (2007). How to teach with technology: keeping both teachers and students comfortable in an era of exponential change. In *Emerging technologies for learning* (vol. 2, pp. 40-46). British Educational Communications and Technology Agents (Becta).

Prochaska, J., Coughlin, S. S., & Lyons, E. (2017). Social Media and Mobile Technology for Cancer Prevention and Treatment. *American Society of Clinical Oncology Educational Book, 37*, 128–137. doi:10.14694/EDBK_173841 PMID:28561647

Public Broadcasting Service. (n.d.). *Julia Child: PBS icon and cooking pioneer.* Retrieved from http://www.pbs.org/food/julia-child

Purple Carrot Books. (2013). The Prisoner of Carrot Castle (Version 1.1.2) [Mobile application software]. Retrieved from http://purplecarrotbooks.com

Puska, P., & McAlister, A. (1981). Television in health promotion: Evaluation of a national programme in Finland. *International Journal of Health Education, 24*, 238–250. PMID:7344372

Qualtrics. (2015). *Qualtrics Mobile.* Retrieved from http://www.qualtrics.com/research-suite/mobile-surveys

Radish Kids. (2019). Retrieved from https://www.raddishkids.com

Raghupathi, W., & Raghupathi, V. (2018). An Empirical Study of Chronic Diseases in the United States: A Visual Analytics Approach to Public Health. *International Journal of Environmental Research and Public Health, 15*(431).

Rashid, N. A. M., Salleh, S. M., & Noor, N. M. (2018). The Role of Game Elements in Improving Jawi Skills through a Mobile Game 'G-Jawi'. *International Journal of Interactive Mobile Technologies, 12*(7), 20–30. doi:10.3991/ijim.v12i7.9636

Rashid, S., Cunningham, U., & Watson, K. (2017). Task-based language teaching with smartphones: A case study in Pakistan. *Teachers and Curriculum, 17*(2), 33–40. doi:10.15663/tandc.v17i2.167

Reid, G., Strnadova, I., & Cumming, T. (2013). Expanding horizons for students with dyslexia in the 21st century: Universal design and mobile technology. *Journal of Research in Special Educational Needs*, *13*(3), 175–181. doi:10.1111/1471-3802.12013

Richards, R., Kinnersley, P., Brain, K., McCutchan, G., Staffurth, J., & Wood, F. (2018). Use of Mobile Devices to Help Cancer Patients Meet Their Information Needs in Non-Inpatient Settings: Systematic Review. *JMIR mHealth and uHealth*, *6*(12), e10026. doi:10.2196/10026 PMID:30552082

Riconscente, M. M. (2013). Results from a controlled study of the iPad fractions game Motion Math. *Games and Culture*, *8*(4), 186–214. doi:10.1177/1555412013496894

Robertson, M. (2009). Young "netizens" Creating Public Citizenship In Cyberspace. *International Research in Geographical and Environmental Education*, *18*(4), 287–293. doi:10.1080/10382040903251158

Rodriguez, C. D., Strnadova, I., & Cumming, T. (2013). Using ipads with students with disabilities: Lessons learned from students, teachers, and parents. *Intervention in School and Clinic*, *49*(4), 244–250. doi:10.1177/1053451213509488

Romero, C., Ventura, S., & de Bra, P. (2009). Using mobile and web-based computerized tests to evaluate university students. *Computer Applications in Engineering Education*, *17*(4), 435–447. doi:10.1002/cae.20242

Ronkowitz, K. (2013, February 1). *Andragogy and pedagogy*. Retrieved December 20, 2015, from https://web.njit.edu/~ronkowit/teaching/andragogy.htm

Rosas, R., Nussbaum, M., Cumsille, P., Marianov, V., Correa, M., Flores, P., ... Salinas, M. (2003). Beyond Nintendo: Design and assessment of educational video games for first and second grade students. *Computers & Education*, *40*(1), 71–94. doi:10.1016/S0360-1315(02)00099-4

Roschelle, J. (2003). Keynote paper: Unlocking the learning value of wireless mobile devices. *Journal of Computer Assisted Learning*, *19*(3), 260–272. doi:10.1046/j.0266-4909.2003.00028.x

Rosenkranz, R. R. (2008). *The impact of family mealtime on human nutrition*. Paper presented at the Food & Youth: Finding the Tipping Point for Wellness, Kansas Nutrition Conference, Junction City, KS.

Ross, C., Gray, S., Warwick, C., Hudson-Smith, A., & Terras, M. (2012, July). *Engaging the museum space: Mobilising visitor engagement with digital content creation*. Paper presented at Twenty-third Digital Humanities Conference, Hamburg, Germany. Retrieved from http://www.dh2012.uni-hamburg.de/conference/programme/abstracts/engaging-the-museum-space-mobilising-visitor-engagement-with-digital-content-creation.1.html

Runyan, M. (2013). Seeing is believing! *Learning and Leading with Technology*, 12–17.

Russell, C. G., Worsley, A., & Liem, D. G. (2015). Parents' food choice motives and their associations with children's food preferences. *Public Health Nutrition*, *18*(06), 1018–1027. doi:10.1017/S1368980014001128 PMID:24887485

Ryu, H., & Parsons, D. (2012). Risky business or sharing the load? Social flow in collaborative mobile learning. *Computers & Education, 58*(2), 707–720. doi:10.1016/j.compedu.2011.09.019

SageMilk. (2011). Film, TV and Video Production (1.0) [Mobile application software]. Retrieved from https://play.google.com

Şahin, F., & Mentor, D. (2016). Using Mobile Phones for Assessment in Contemporary Classrooms. In *Handbook of Research on Mobile Learning in Contemporary Classrooms* (pp. 116–138). IGI Global. doi:10.4018/978-1-5225-0251-7.ch006

Şahin, F., & Yan, Z. (2013). Mobile phones in data collection: A systematic review. *International Journal of Cyber Behavior, Psychology and Learning, 3*(3), 67–87. doi:10.4018/ijcbpl.2013070106

Saidouni, K., & Bahloul, A. (2016). Teachers and students' attitudes towards using mobile-assisted language learning in higher education. *Arab World English Journal, 3*, 123–140.

Samaie, M., Mansouri Nejad, A., & Qaracholloo, M. (2018). An inquiry into the efficiency of WhatsApp for self-and peer-assessments of oral language proficiency. *British Journal of Educational Technology, 49*(1), 111–126. doi:10.1111/bjet.12519

Sangrà, A., Vlachopoulos, D., & Cabrera, N. (2012). Building an inclusive definition of e-learning: An approach to the conceptual framework. *The International Review of Research in Open and Distributed Learning, 13*(2), 145–159. doi:10.19173/irrodl.v13i2.1161

Santos, P., Pérez-Sanagustín, M., Hernández-Leo, D., & Blat, J. (2011). QuesTInSitu: From tests to routes for assessment in situ activities. *Computers & Education, 57*(4), 2517–2534. doi:10.1016/j.compedu.2011.06.020

Saric, L. (Writer), Bakalian, P. (Writer), Rey, H.A. (Writer) & Malkasian, C. (Director). (2009). Juicy George/The Big Picture [Television series episode]. In B. Howard, B. Grazer, D. Kirschner, & J. Shapiro (Executive producers), Curious George. Image Entertainment & Universal Animation Studios WGBH-TV.

Sariola, J. (2003). The boundaries of university teaching: Mobile learning as a strategic choice for the virtual university. *Mobile Learning*, 71-78.

Satter, E. (2008). *Secrets of feeding a healthy family: How to eat, how to raise good eaters, how to cook*. Kelcy Press.

Savas, P. (2014). Tablet PCs as instructional tools in English as a foreign language education. *Turkish Online Journal of Educational Technology, 13*(1), 217–222.

Schaeken, W., Johnson-Laird, P. N., & d'Ydewalle, G. (1996). Mental models and temporal reasoning. *Cognition, 60*(3), 205–234. doi:10.1016/0010-0277(96)00708-1 PMID:8870513

Schreiner, K. (2008). Digital Games Target Social Change. *IEEE Computer Graphics and Applications, 28*(1), 12–17. doi:10.1109/MCG.2008.4 PMID:18240782

Schroeder, R. (2006). Being there together and the future of connected presence. *Presence (Cambridge, Mass.)*, *15*(4), 438–454. doi:10.1162/pres.15.4.438

Schuler, C. (2009). *Pockets of potential: Using mobile technologies to promote children's learning*. New York: The Joan Ganz Cooney Center at the Sesame Workshop.

Schwabe, G., & Göth, C. (2005). Mobile learning with a mobile game: Design and motivational effects. *Journal of Computer Assisted Learning*, *21*(3), 204–216. doi:10.1111/j.1365-2729.2005.00128.x

Schwartz, D. L., & Black, J. B. (1996). Shuttling between depictive models and abstract rules: Induction and fallback. *Cognitive Science*, *20*(4), 457–497. doi:10.120715516709cog2004_1

Schwartz, D., Martin, T., & Nasir, N. (2005). Design for knowledge evolution: Towards a prescriptive theory for integrating first- and second-hand knowledge. In P. Gardenfors, P. Johanssson, & N. J. Mahwah (Eds.), *Cognition, education, and communication technology* (pp. 21–54). Mahwah, NJ: Lawrence Erlbaum Associate, Inc.

Section508.gov. (n.d.). *Section 508 law and related laws and policies*. Retrieved December 8, 2015, from https://www.section508.gov/content/learn/laws-and-policies

Seo, M., & Creed, D. (2002). Institutional Contradictions, Praxis, and Institutional Change: A Dialectical Perspective. *Academy of Management Review*, *27*(2), 222–247. doi:10.5465/amr.2002.6588004

Seppälä, P., Sariola, J., & Kynäslahti, H. (2002). Mobile learning in personnel training of university teachers. In *Wireless and Mobile Technologies in Education, 2002. Proceedings. IEEE International Workshop on* (pp. 136-139). IEEE 10.1109/WMTE.2002.1039236

Seppälä, P., & Alamäki, H. (2003). Mobile learning in teacher training. *Journal of Computer Assisted Learning*, *19*(3), 330–335. doi:10.1046/j.0266-4909.2003.00034.x

Serf-Walls, S. (2013, December 7). Super Abby vs. the enchanted sweet stall [Web log post]. Retrieved from http://www.bestappsforkids.com/2013/super-abby-vs-the-enchanted-sweet-stall

Shah, S. A., Velardo, C., Farmer, A., & Tarassenko, L. (2017). Exacerbations in Chronic Obstructive Pulmonary Disease: Identification and Production Using a Digital Health System. *Journal of Medical Internet Research*, *19*(3), e69. doi:10.2196/jmir.7207 PMID:28270380

Shannon, B., Thurman, G., & Schiff, W. (1979). Foodsense: A pilot TV show on nutrition issues. *Journal of Nutrition Education*, *11*(1), 15–18. doi:10.1016/S0022-3182(79)80057-6

Sharples, M. (2005) *Learning As Conversation: Transforming Education in the Mobile Age*. Paper to presented at Conference on Seeing, Understanding, Learning in the Mobile Age, Budapest, Hungary.

Sharples, M., Taylor, J., & Vavoula, G. (2005). Towards a theory of mobile learning. *Proceedings of mLearn 2005, 1*(1), 1-9.

Sharples, M. (2000). The design of personal mobile technologies for lifelong learning. *Computers & Education, 34*(3), 177–193. doi:10.1016/S0360-1315(99)00044-5

Shaw, E. (Writer), & Martin, H. (Director). (2008). My ice pops [Television series episode]. In B. Henson, L. Henson, H. Stanford, & B. Zweig (Executive producers), Sid the science kid. Hollywood, CA: The Jim Henson Company.

Shih, Y. E. (2007). Setting the new standard with mobile computing in online learning. *The International Review of Research in Open and Distributed Learning, 8*(2). doi:10.19173/irrodl.v8i2.361

Shippee, M. & Keengwe, J. (2014). *mLearning: Anytime, anywhere learning transcending the boundaries of the educational box.* Academic Press.

Shippee, M., & Keengwe, J. (2014). mLearning: Anytime, anywhere learning transcending the boundaries of the educational box. Academic Press.

Shoe the Goose. (2014). Cake Doodle (Version 1.15) [Mobile application software]. Retrieved from http://www.shoethegoose.com

Shuler, C. (2012). iLearn; An Analysis of the Education Category of the iTunes AppStore. New York: The Joan Ganz Cooney Center at Sesame Workshop.

Silva, E. H., Lawler, S., & Langbecker, D. (2019). The Effectiveness of mHealth for Self-Management in Improving Pain, Psychological Distress, Fatigue, and Sleep in Cancer Survivors: A Systematic Review. *Journal of Cancer Survivorship: Research and Practice, 13*(1), 97–107. doi:10.100711764-018-0730-8 PMID:30635865

Simpson, A. J., Honkoop, P. J., Kennington, E., Snoeck-Strobrand, J. B., Smith, I., East, J., ... Fowler, S. J. (2017). Perspectives of Patients and Healthcare Professionals of mHealth for Asthma Self-Management. *The European Respiratory Journal, 49*(5), 1601966. doi:10.1183/13993003.01966-2016 PMID:28461291

SinekS. (2016). Retrieved from: https://m.youtube.com/watch?v=hER0Qp6QJNU

Sites, J. D., & Potter, R. F. (2018). Everything Merges with the Game: A Generative Music System Embedded in a Videogame Increases Flow. *Game Studies, 18*(2).

Skinner, B. F. (1968). The Technology of Teaching. New York: Appleton-CenturyCrofts.

Smith, L. T., Tuck, E., & Yang, K. W. (2019). Indigenous and decolonizing studies in education: Mapping the long view. New York, NY: Routledge, an imprint of the Taylor & Francis Group.

Smith, J., George, V., & Easton, P. S. (2001). Home-grown television: A way to promote better nutrition. *Journal of Nutrition Education, 33*(1), 59–60. doi:10.1016/S1499-4046(06)60013-4 PMID:12031209

Sobnath, D. D., Philip, N., Kayyali, R., Nabhani-Gebara, S., Pierscionek, B., Vaes, A. W., ... Kaimakamis, E. (2017). Features of a Mobile Support App for Patients With Chronic Obstructive Pulmonary Disease: Literature Review and Current Applications. *JMIR mHealth and uHealth*, *5*(2), e17. doi:10.2196/mhealth.4951 PMID:28219878

Specht, M. (2014). Design of contextualised mobile learning applications. *Increasing Access, 61*.

Statista. (2019). *Mobile Gaming Industry - Statistics & Facts*. Retrieved from https://www.statista.com/topics/1906/mobile-gaming/

Statista. (2019). *Mobile phone gaming penetration in the United States from 2011 to 2020*. Retrieved from https://www.statista.com/statistics/234649/percentage-of-us-population-that-play-mobile-games/

Statista. (2019). *Number of mobile phone gamers in the United States from 2011 to 2020 (in millions)*. Retrieved from https://www.statista.com/statistics/234635/number-of-mobile-gamers-forecast/

Stauffer, K. (2014). *Technology use in post secondary language education* (M. L. Thomas-Ruzic, K. Hall, & P. Mosele, Eds.). ProQuest Dissertations Publishing.

Stav, J., Nielsen, K., Hansen-nygård, G., & Thorseth, T. (2010). Experiences obtained with integration of student response systems for iPod touch and iPhone into e- learning environments. *Electronic Journal of E-Learning*, *8*, 179–190.

Steinfeld, E. (2012). *The goals of universal design*. Retrieved December 8, 2015, from http://udeworld.com/presentations/oslo/Steinfeld.Goals%20of%20UD-Oslo_Final_web.pdf

Sterling, G. (2015). *It's official: Google says more searches now on mobile than on desktop*. Retrieved December 8, 2015, from http://searchengineland.com/its-official-google-says-more-searches-now-on-mobile-than-on-desktop-220369

Stienberg, L. (2016). *Adolescence* (11th ed.). New York: McGraw-Hill Education.

Stiener-Adair, C. (2013). *The big disconnect: Protecting childhood and family relationships in the digital age*. New York, NY: Harper.

Storyboardthat.com. (n.d.). *Digital storytelling: powerful visual communication, made easy*. Retrieved December 8, 2015, from http://www.storyboardthat.com/

Strain, M. (2015). *1983 to today: a history of mobile apps*. Retrieved December 8, 2015, from http://www.theguardian.com/media-network/2015/feb/13/history-mobile-apps-future-interactive-timeline

SurveyMonkey. (n.d.). *Mobile Surveys*. Retrieved from http://www.surveymonkey.com/mp/mobile-surveys

Swain, C. (2007). *Designing Games to Effect Social Change*. DiGRA.

Sweller, J. (1988). Cognitive load during problem solving: Effects on learning. *Cognitive Science*, *12*(2), 257–285. doi:10.120715516709cog1202_4

Sweller, J., Van Merriënboer, J. J. G., & Paas, F. G. W. C. (1998). Cognitive architecture and instructional design. *Educational Psychology Review*, *10*(3), 251–297. doi:10.1023/A:1022193728205

Takahashi, D. (2018). *Smartphone users will top 3 billion in 2018, hit 3.8 billion by 2021.* Retrieved Feb, 2019 from https://venturebeat.com/2018/09/11/newzoo-smartphone-users-will-top-3-billion-in-2018- hit-3-8-billion-by-2021/

Tayan, B. M. (2017). Students and teachers' perceptions into the viability of mobile technology implementation to support language learning for first year business students in a Middle Eastern university. *International Journal of Education and Literacy Studies*, *5*(2), 74–83. doi:10.7575/aiac.ijels.v.5n.2p.74

Taylor, D. M. (2018). *Americans with disabilities: 2014.* Retrieved March 30, 2019 from https://www.census.gov/content/dam/Census/library/publications/2018/demo/p70-152.pdf

Taylor, J., & Sharples, M. (2006). *The Pedagogical Perspectives of Mobile Learning.* Available at: http://www2.le.ac.uk/projects/impala/presentations/Berlin/The%20Pedagogical%20 Perspectives%20of%20Mobile %20Learning/view

Taylor, M. (1987). Self-directed learning: More than meets the observer's eye. In Appreciating adults learning: From the learner's perspective (pp. 179-96). London: Kogan Page.

Taylor, J. D., Dearnley, C. A., Laxton, J. C., Coates, C. A., Treasure-Jones, T., Campell, R., & Hall, I. (2010). Developing a mobile learning solution for health and social care practice. *Distance Education*, *31*(2), 175–192. doi:10.1080/01587919.2010.503343

TEAL Center. (n.d.). *Fact sheet – universal design for learning.* Retrieved December 8, 2015, from https://teal.ed.gov/tealguide/udl

Television Food Network. (2015). Food Network In the Kitchen (Version 4.8.1) [Mobile application software]. Retrieved from http://www.foodnetwork.com/site/mobile.html

Tervalon, M., & Murray-Garcia, J. (1998). Cultural humility versus cultural competence: A critical distinction in defining physician training outcomes in multi-cultural education. *Journal of Health Care for the Poor and Underserved*, *9*(2), 117–125. doi:10.1353/hpu.2010.0233 PMID:10073197

The Office of the United Nations High Commissioner for Refugees. (2012). *Could you survive?* Retrieved from My Life as a Refugee: http://mylifeasarefugee.org/index.html

The Statistics Portal. (2019). *Number of Smartphone Users Worldwide from 2014 to 2020 (in billions).* Retrieved February 8, 2019 from number-of-smartphone-users-worldwide/

The Statistics Portal. (2019). *Smartphone and feature phone penetration in the United States in 2009 and 2011.* Retrieved February 8, 2019 https://www.statista.com/statistics/216340/us-smartphone-and- feature-phone-penetration-since-2009/

Thompson, S. (2014). Encyclopedia of Diversity and Social Justice. Academic Press.

Thornton, P., & Houser, C. (2005). Using mobile phones in English education in Japan. *Journal of Computer Assisted Learning*, *21*(3), 217–228. doi:10.1111/j.1365-2729.2005.00129.x

Tiggly. (2015). Tiggy Chef (Version 1.1) [Mobile application software]. Retrieved from https://www.tiggly.com

Tinschert, P., Jakob, R., Barata, F., Kramer, J. N., & Kowatsch, T. (2017). The Potential of Mobile Apps for Improving Asthma Self-Management: A Review of Publicly Available and Well-Adopted Asthma Apps. *JMIR mHealth and uHealth*, *5*(8), e113. doi:10.2196/mhealth.7177 PMID:28768606

Tisdell, E. (2008). Critical Media Literacy And Transformative Learning: Drawing On Pop Culture And Entertainment Media In Teaching For Diversity In Adult Higher Education. *Journal of Transformative Education*, *6*(1), 48–67. doi:10.1177/1541344608318970

Toca Boca. (2015). Toca Kitchen 2 (Version 1.2) [Mobile application software]. Retrieved from http://tocaboca.com/app/toca-kitchen-2

Torbert, B., & ... (2004). *Action inquiry: The secret to timely and transforming leadership*. San Francisco: Barrett-Koehler.

Traxler, J., & Kukulska-Hulme, A. (2005). *Evaluating mobile learning: Reflections on current practice*. Paper presented at the mLearn 2005: Mobile technology: The future of learning in your hands. Retrieved from http://oro.open.ac.uk/12819/1/

Traxler, J. (2007). Defining, Discussing and Evaluating Mobile Learning: The moving finger writes and having writ..... *The International Review of Research in Open and Distributed Learning*, *8*(2). doi:10.19173/irrodl.v8i2.346

Traxler, J. (2009). Mobile learning. A handbook for educators and trainers. *British Journal of Educational Studies*, *57*(3), 337–339. doi:10.1111/j.1467-8527.2009.437_6.x

Traxler, J. (2012). Mobile learning-The future already behind us. *Proceedings of 2012 International Conference on Interactive Mobile and Computer Aided Learning (IMCL)*. 10.1109/IMCL.2012.6396442

Triantafillou, E., Georgiadou, E., & Economides, A. A. (2008). The design and evaluation of a computerized adaptive test on mobile devices. *Computers & Education*, *50*(4), 1319–1330. doi:10.1016/j.compedu.2006.12.005

Trilling, B., & Fadel, C. (2009a). *Career And Life Skills: Work Ready, Prepared for Life. In 21st century skills: Learning for life in our times*. San Francisco, CA: Jossey-Bass.

Trilling, B., & Fadel, C. (2009b). *21st Century Learning and Teaching. In 21st century skills: Learning for life in our times*. San Francisco, CA: Jossey-Bass.

Trowbridge Filippone, P. (2010). *Tomato history: The history of tomatoes as food*. Retrieved from http://homecooking.about.com/od/foodhistory/a/tomatohistory.htm

Tuckman, B. W., & Jensen, M. A. C. (1977). Stages of small-group development revisited. *Group & Organization Management*, *2*(4), 419–427.

Turchinsky, M. (2011). Smartphone Apps: The Hype, the Reality and Everything in Between. In *Proceedings of The American Alliance of Museums*. Houston, TX: PROLibraries.

Turkle, S. (2011). *Alone together: Why we expect more from technology and less from each other.* New York, NY: Basic Books.

Tüzün, H., Yılmaz-Soylu, M., Karakuş, T., İnal, Y., & Kızılkaya, G. (2009). The effects of computer games on primary school students' achievement and motivation in geography learning. *Computers & Education*, *52*(1), 68–77. doi:10.1016/j.compedu.2008.06.008

Uden, L. (2006). Activity theory for designing mobile learning. *International Journal of Mobile Learning and Organization*, *1*(1), 81–102. doi:10.1504/IJMLO.2007.011190

United Nations. (2006). *Convention on the rights of persons with disabilities*. Retrieved December 8, 2015, from http://www.un.org/disabilities/convention/conventionfull.shtml

United Nations. (n.d.). *Convention and optional protocol signature and ratifications.* Retrieved December 8, 2015, from http://www.un.org/disabilities/countries.asp?navid=17&pid=166

United States Access Board. (2000). *Section 508 standards for electronic and information technology*. Retrieved December 8, 2015, from https://www.access-board.gov/guidelines-and-standards/communications-and-it/about-the-section-508-standards/section-508-standards

United States Access Board. (n.d.). *About the section 508 standards.* Retrieved December 8, 2015, from https://www.access-board.gov/guidelines-and-standards/communications-and-it/about-the-section-508-standards

United States Department of Justice. (2015). *United States department of education q&a: title IV rehabilitation act amendments of 1998 - section 508 electronic and information technology.* Retrieved December 8, 2015, from http://www.justice.gov/crt/department-education-q

University College London Centre for Digital Humanities, Centre for Advanced Spatial Analysis, & University College London Museums. (2011). *QRator*. Retrieved from http://www.qrator.org

University of Washington. (2015). *What is the individuals with disabilities education act?* Retrieved December 8, 2015, from http://www.washington.edu/doit/what-individuals-disabilities-education-act

US Department of Education. (n.d.). *Building the legacy: IDEA 2004.* Retrieved December 8, 2015, from http://idea.ed.gov/

USCode.House.Gov. (1973). *Public Law 93-113. Section 504 of the Rehabilitation Act of 1973.* Retrieved December 8, 2015, from http://uscode.house.gov/statviewer.htm?volume=87&page=394

Uther, M. (2019). Mobile Learning—Trends and Practices. *Education in Science*, *9*(1), 33. doi:10.3390/educsci9010033

Uy, C., Lopez, J., Trinh-Shevrin, C., Kwon, S. C., Sherman, S., & Liang, P. S. (2017). Text Messaging Interventions on Cancer Screening Rates: A Systematic Review. *Journal of Medical Internet Research*, *19*(8), e296. doi:10.2196/jmir.7893 PMID:28838885

van der Linden, W. J., & Pashley, P. J. (2010). Item selection and ability estimation in adaptive testing. In Elements of adaptive testing. New York: Springer.

Veazie, S., Winchell, K., Paynter, R., Ivlev, I., Eden, K., Nussbaum, K., . . . Hefland, M. (2018). Mobile Applications for Self-Management of Diabetes. Technical Brief No. 31. Agency for Healthcare Research and Quality. No. 18-EHC010-EF. doi:10.23970/AHRQEPCTB31

Venkatesh, V. (2000). Determinants of perceived ease of use: Integrating control, intrinsic motivation, and emotion into the technology acceptance model. *Information Systems Research*, *11*(4), 342–365. doi:10.1287/isre.11.4.342.11872

Venkatesh, V., & Davis, F. D. (2000). A theoretical extension of the technology acceptance model: Four longitudinal field studies. *Management Science*, *46*(2), 186–204. doi:10.1287/mnsc.46.2.186.11926

Venkatesh, V., Morris, M. G., Davis, G. B., & Davis, F. D. (2003). User acceptance of information technology: Toward a unified view. *Management Information Systems Quarterly*, *27*(3), 425–478. doi:10.2307/30036540

Voronov, M., & Yorks, L. (2015). "Did you notice that?" Theorizing differences in the capacity to apprehend institutional contradictions. *Academy of Management Review*, *40*(4), 563–586. doi:10.5465/amr.2013.0152

Vosniadou, S., & Brewer, W. F. (1989). *The concept of the earth's shape: A study of conceptual change in childhood* (Technical Report No. 467). Retrieved from Illinois Digital Environment for Access to Learning and Scholarship website: https://www.ideals.illinois.edu/bitstream/handle/2142/17996 /ctrstreadtechrepv01989i00467_opt.pdf

Waller, T., Balanyk, J., & Thornquist, E. (2015). *Three perspectives of mobile learning integration at a tertiary institution in the UAE.* Paper presented at the 9th International Technology, Education and Development Conference, Madrid, Spain.

Wankel, L. A., & Blessinger, P. (2013). *Increasing Student Engagement and Retention Using Mobile Technologies: Smartphones.* Skype and Texting Technologies. doi:10.1108/S2044-9968(2013)6_Part_D

Wartella, E., Beaudoin-Ryan, L., Blackwell, C. K., Cingel, D. P., Hurwitz, L. B., & Lauricella, A. R. (2016). What kind of adults will our children become? The impact of growing up in a media-saturated world. *Journal of Children and Media*, *10*(1), 13–20. doi:10.1080/17482798.2015.1124796

Webster, J., Trevino, L. K., & Ryan, L. (1994). The dimensionality and correlates of flow in human-computer interactions. *Computers in Human Behavior*, *9*(4), 411–426. doi:10.1016/0747-5632(93)90032-N

Wee, M. (2013). *Three things to make your app stand out when building for accessibility*. Retrieved December 8, 2015, from http://www.wired.com/insights/2013/03/three-things-to-make-your-app-stand-out-when-building-for-accessibility/

Wege des Lernens, E. V. N. (2014). Close-Up Film Language (2.6.1) [Mobile application software]. Retrieved from https://itunes.apple.com

Wei, R. (2013). Mobile media: Coming of age with a big splash. *Mobile Media & Communication*, *1*(1), 50–56. doi:10.1177/2050157912459494

Westera, W. (2019). Why and How Serious Games can Become Far More Effective: Accommodating Productive Learning Experiences, Learner Motivation and the Monitoring of Learning Gains. *Journal of Educational Technology & Society*, *22*(1), 59–69.

White, L. T. (2019). *Video games could teach spatial skills lost to a society dependent on devices*. Academic Press.

White, H. D., & McCain, K. W. (1998). Visualizing a discipline: An author co-citation analysis of information science, 1972-1995. *Journal of the American Society for Information Science*, *49*(4), 327–355.

Whitelock, D., Brna, P., & Holland, S. (1996). What is the value of virtual reality for conceptual learning? Towards a theoretical framework. In P. Brna, A. Paiva & J. A. Self (Eds), *Proceedings of the European Conference on artificial intelligence in education (EuroAIE)* (pp. 136–141). Lisbon: Edicões Colibri.

Whole Foods Market. (2015). Whole Foods Market (Version 5.2.0) [Mobile application software]. Retrieved from http://www.wholefoodsmarketcom/apps

Whole Kids Foundation. (2013). Awesome Eats (Version 1.3) [Mobile application software]. Retrieved from http://www.fun-machine.com

Wiersma, W., & Jurs, S. G. (2009). *Research methods in education: An introduction*. Boston, MA: Pearson Education.

Wiggins, S. (2014). Adult and child use of love, like, don't like and hate during family mealtimes: Subjective category assessments as food preference talk. *Appetite*, *80*, 7–15. doi:10.1016/j.appet.2014.04.024 PMID:24793063

Williams van Rooij, S. (2011). Open-source learning management systems: A predictive model for higher education. *Journal of Computer Assisted Learning*, *28*(2), 114–125. doi:10.1111/j.1365-2729.2011.00422.x

Williams, V. (2003). Designing simulations for learning. *E-journal of Instructional Science and Technology*, *6*(1), 50–71.

Winters, N. (2007). What is mobile learning. *Big Issues in Mobile Learning*, 7–11.

Wong, L.-H., & Looi, C.-K. (2011). What seams do we remove in mobile assisted seamless learning? A critical review of the literature. *Computers & Education, 57*(4), 2364–2381. doi:10.1016/j.compedu.2011.06.007

Woodruff, S. J., & Kirby, A. R. (2013). The associations among family meal frequency, food preparation frequency, self-efficacy for cooking, and food preparation techniques in children and adolescents. *Journal of Nutrition Education and Behavior, 45*(4), 296–303. doi:10.1016/j.jneb.2012.11.006 PMID:23481894

World Health Organization (WHO). (2011). *Summary world report on disability*. Geneva, Switzerland: WHO.

World Health Organization (WHO). (2018). *Disability and health – key facts*. Retrieved March, 30, 2019 from https://www.who.int/en/news-room/fact-sheets/detail/disability-and-health

Wu, W. H., Wu, Y. C. J., Chen, C. Y., Kao, H. Y., Lin, C. H., & Huang, S. H. (2012). Review of trends from mobile learning studies: A meta-analysis. *Computers & Education, 59*(2), 817–827. doi:10.1016/j.compedu.2012.03.016

Wu, Y., Yao, X., & Li, S. (2018). Mobile App-Based Interventions to Support Diabetes Self-Management: A Systematic Review of Randomized Control Trials to Identify Functions Associated with Glycemic Efficacy. *JMIR mHealth and uHealth, 6*(1), e20. doi:10.2196/mhealth.8789 PMID:29334479

Xiong, S., Berkhouse, H., Schooler, M., Pu, W., Sun, A., Gong, E., & Yan, L. (2018). Effectiveness of mHealth Interventions in Improving Medication Adherence Among People with Hypertension: A Systematic Review. *Current Hypertension Reports, 20*(86). PMID:30088110

Yang, F, Wang, Y., Yang, C. Hu, H., & Xiong, Z. (2018). Mobile Health Applications in Self-Management of Patients with Chronic Obstructive Pulmonary Disease: A Systematic Review and Meta-Analysis of Their Efficacy. *BMC Pulmonary Disease, 18*(147).

Yang, X., Kuo, L. J., Ji, X., & McTigue, E. (2018). A critical examination of the relationship among research, theory, and practice: Technology and reading instruction. *Computers & Education, 125*, 62–73. doi:10.1016/j.compedu.2018.03.009

Yang, Y. T. C. (2012). Building virtual cities, inspiring intelligent citizens: Digital games for developing students' problem solving and learning motivation. *Computers & Education, 59*(2), 365–377. doi:10.1016/j.compedu.2012.01.012

Yellephant Co. (2015). Jobi's Bistro (Version 1.2) [Mobile application software]. Retrieved from http://www.yellephant.net

Yogome. (2015a). Healthy Heroes 1: Nutrition for Kids (Version 1.2.1) [Mobile application software]. Retrieved from http://yogome.com/

Yogome. (2015b). Healthy Heroes 2: My Plate for Kids (Version 1.0.3) [Mobile application software]. Retrieved from http://yogome.com/

Yosso, T. (2005). Whose culture has capital? A critical race theory discussion of community cultural wealth. *Race, Ethnicity and Education*, *8*(1), 69–91. doi:10.1080/1361332052000341006

Yousuf, I. M. (2007). Effectiveness of Mobile Learning in Distance Education. *Turkish Online Journal of Distance Education.*, *8*(4), 9.

Zeiger, S. (2015, January 12). Toca Kitchen 2 [Web log post]. Retrieved from http://www.bestappsforkids.com/2015/toca-kitchen-2

Zembylas, M., & Keet, A. (2018). *Critical human rights, citizenship, and democracy education: Entanglements and regenerations*. London: Bloomsbury Academic.

Zewig, B. (Writer), Henson, B. (Director) & Calamandrei, C. (Director). (2008). My mushy banana [Television series episode]. In B. Henson, L. Henson, H. Stanford, & B. Zweig (Executive producers), Sid the science kid. Hollywood, CA: The Jim Henson Company.

Zheng, R. (2009). *Cognitive effects of multimedia learning*. Hershey, PA: Information Science Reference. doi:10.4018/978-1-60566-158-2

Zurita, G., & Nussbaum, M. (2004). A constructivist mobile learning environment supported by a wireless handheld network. *Journal of Computer Assisted Learning*, *20*(4), 235–243. doi:10.1111/j.1365-2729.2004.00089.x

About the Contributors

Dominic Mentor (Ed.D.) directs and creates corporate training. He initiated national and international systems of self-directed, digitally infused autonomous learning systems. Dominic has systemically transformed corporations, national and international organizations. At Teachers College Columbia University he initiated and co-designed the USA's first mLearning graduate class and a social media fellowship for the NYC Mayor's Office of Adult Education. His awards include the Optimas Gold Vision 2014 award from Workforce Magazine, the IELA 2015 blended eLearning award and others. He was keynote speaker at various conferences and served as Editor and contributing author of the Handbook of Research on Mobile Learning in Contemporary Classrooms, the Computer Mediated Learning for Workforce Development, as well as chapters in Tablets in Education, and the Encyclopedia of Mobile Phone Behavior. Other publications include articles in TC Record, and Educause Journal. Presentations include UNESCO Mobile Learning Week in Paris (France), Harvard's 21st Century Academic Forum Conference (Boston, USA), and eLearning in the Workplace (New York) as well as a host of other conferences.

* * *

Tseday Alehegn has a doctorate in Health & Behavior Studies, an M.S. in Neuroscience & Education from Columbia University, Teachers College, as well as an M.A. in Education and a bachelor's degree in Human Biology from Stanford University. She currently works as Associate Dean for Academic Affairs at Marymount Manhattan College and previously worked as Associate Director at the Office of Access & Services for Individuals with Disabilities (OASID) at Teachers College. Dr. Alehegn also has prior experience working as a clinical database programmer in the biotech and medical devices industry in Silicon Valley, California.

Tara Ashmore Callen is the founding director of Aurelius for the Arts, a 501c3 that operates teaching artist run workshops within Latin America. The aim of her nonprofit is to bring ideas of art, education, and functional use of the arts to orphanages where very few resources exist. She received her doctorate degree in art education from Teachers College, Columbia University and holds a Masters degree in Education from New York University. Her current research is focused primarily upon sustainability efforts in Latin America and the role of art education in enriching the lives of children living within underserved communities.

Richard Keller obtained a B.S. degree from NYU in 1985 and a PhD in Counseling Psychology from Columbia University, Teachers College in 2004. He is an Assistant Professor of Psychology & Education and Director of the Office of Access & Services for Individuals with Disabilities at Teachers College. Dr. Keller has been conducting research on disabilities for the past 20 years, and has been involved in the disability rights movement for the past 30 years.

Jamie Krenn holds a Ph.D. in Educational Psychology: Cognitive Studies from Teachers College, Columbia University. She also holds three masters degrees in developmental and cognitive psychologies as well as a Bachelor of Science in Art Therapy. Jamie also leads the Children & Media: Analysis & Evaluation area of focus within the Educational Psychology: Cognitive Studies graduate degree program at Teachers College, Columbia University. Her research interest includes cognitive media processing, children's television and culinary cognition.

Brenda I. Lopez Ortiz is an Instructional Designer at Florida International University (FIU). She earned her doctorate in Instructional Technology and Media from Teachers College, Columbia University. Her research interests include problem-based learning in online contexts and mobile learning. Before joining FIU, she also worked as faculty, multimedia production specialist, professional developer, and English teacher.

Mara J. Luna is the director of the Upward Bound Program at the University of Puerto Rico, Río Piedras Campus. She is working towards her doctorate in Curriculum and Teaching from the Interamerican University of Puerto Rico, Metropolitan Campus. Her research interests center on mobile learning and the teaching of English to second language learners. Before joining the University of Puerto Rico, she also worked as adjunct faculty and English teacher.

Shawn McCann retired from the military after twenty years of service. He holds a doctorate in Adult Learning and Leadership from Teachers College, Columbia University in the City of New York.

Joél-Léhi Organista is the National Board President for the Education For Liberation Network. He is the elected National Vice President of Youth in the League of United Latin American (LULAC). Joél-Léhi was a high school health teacher in Salt Lake City and also was an appointed state advisor on equity issues to the Utah State Board of Education. He co-founded his non-profit, Casa Quetzalcoatl, which uses a cross-generational indigenous epistemology approach to empowering the Latinx community in Salt Lake City. He's the founder and designer of Machitia, an educator focused mobile app. Joél-Léhi has his Masters degree from Columbia University, Teachers College in Sociology and Education. His undergraduate degree was in Social Justice Pedagogy from the University of Utah. Joél-Léhi is originally from La Ciudad de México Chapultepec. He's a polyglot who speaks Spanish, English, Japanese, Portuguese, and Nahuatl.

Füsun Şahin (Ph.D.) is a researcher and psychometrician currently working at American Institutes for research, a national non-profit research institution in the United States. Dr. Şahin earned his doctorate degree in Educational Psychology and Methodology from University at Albany, State University at New York. Her professional interests rely on innovative ways to benefit from big data from large-scale assessments for learning about test-takers' cognitional and emotional states, as well as for integrating assessment into learning experiences and data collection. Dr. Şahin's work related to large-scale assessments focuses on using log data (i.e., process data), that is detailed keystrokes and associated timestamps collected in the background as examinees take digitally based assessments, for the benefit of all learners. Her efforts in this line of work serves various purposes, including but not limited to: a) learning about examinees' test-taking behaviors, such as timing strategies and problem-solving efficiency, b) identifying undesired testing behaviors, such as rapid-guessing, c) investigating examinees' emotional states that may inhibit them performing on par with their ability, such as time-pressure and cognitive load, and d) providing evidence for improving design of large-scale digitally-based assessments, such as examinees' use of calculator when an on-screen calculator is available. In this line of work, Dr. Şahin works with data from internationally known large-scale assessments Programme for International Student Assessment (PISA) and The National Assessment of Educational Progress (NAEP). She also has experiences in working with the psychometrics teams in various recognized institutions including New York City Department of Education, New York State Education Department, and American Board of Internal Medicine. Dr. Şahin has also published on innovative ways to integrate technology and evidence-based decision-making for creating meaningful learning opportunities for learners and enhance their experiences. Dr. Şahin's efforts for this purpose include contributing to Handbook of Research on

Mobile Learning in Contemporary Classrooms, Encyclopedia of Mobile Phone Behavior, and International Journal of Cyber Behavior, as well as presenting at various national conferences including Modeling and Simulation (ModSim) Conference 2017 - Modeling and Simulation in the Age of Data.

Cristina A. Sumilang graduated with a Bachelor of Science in Nursing from the University of South Florida in 2010. She worked as a Registered Nurse for seven years at the Mount Sinai Hospital in New York City, where she primarily cared for post-surgical patients who had undergone invasive thoracic and vascular procedures. She is board certified in cardiac-vascular nursing. Cristina received her Master of Science in Nursing from Columbia University in 2018, and works as a Family Nurse Practitioner while pursuing a Doctor of Nursing Practice degree at Columbia University. She is a member of Sigma Theta Tau International Honor Society of Nursing, the American Nurses Association, the American Association of Nurse Practitioners, and the Nurse Practitioner Association of New York State. She is currently doing extensive research on mobile health delivery and actively participates in advocating for nurse practitioner rights.

Jeanny Vaidya is a product manager and instructional designer who specializes in product development and building technical curriculum. Her interests include mobile learning, programming and games. She has a masters degree for Instructional Technology and Media from Teachers College, Columbia University. She enjoys delving into the various facets of educational technology including teaching web design. She has worked on strengthening programming curricula in many organizations. She also has a background in film and sociology and engages in creating change through multimodal learning communities.

Kristin Villanueva is a practicing Learning Experience Designer with a demonstrated history of working in the educational technology field. She earned a Master's degree in Cognitive Studies in Education from Teachers College, Columbia University with a concentration in Intelligent Technologies. She has a research background in cognition and human development, educational psychology, and neuroscience-based educational interventions focused around implicit theories of intelligence. She has designed a range of learning experiences that include games, e-learning modules, blended learning courses, and Instructor-Led Training (ILT). She has a particular interest in game-based learning, VR, AR, children's media, wellness, mindfulness, meditation, social change, culinary arts, and conservation science.

Index

www.ingramcontent.com/pod-product-compliance
Lightning Source LLC
LaVergne TN
LVHW080607300926
849162LV00005B/1077